JEREMY NAYDLER, PH.
theology and religious studies
tural historian and gardener who lives and works in Oxford, England. He has long been interested in the history of consciousness and sees the study of past cultures—which were more open to the world of spirit than our own, predominantly secular, culture—as relevant both to understanding our situation today and to finding pathways into the future. His longstanding concern about the impact of electronic technologies on our inner life and on our relationship to nature has found expression in numerous articles contributed to magazines such as *New View, Self and Society* and *Resurgence*.

OTHER BOOKS BY JEREMY NAYDLER:

Goethe on Science (1996)
Temple of the Cosmos: The Ancient Egyptian Experience of the Sacred (1996)
Shamanic Wisdom in the Pyramid Texts: The Mystical Tradition of Ancient Egypt (2005)
Soul Gardening (2006)
The Future of the Ancient World: Essays on the History of Consciousness (2009)
Gardening as a Sacred Art (2011)

IN THE SHADOW OF THE MACHINE

The Prehistory of the Computer and the Evolution of Consciousness

Jeremy Naydler

TEMPLE LODGE

Temple Lodge Publishing Ltd.
Hillside House, The Square
Forest Row, RH18 5ES

www.templelodge.com

Published by Temple Lodge 2018

A CIP catalogue record for this book is available from the British Library

ISBN 978 1 912230 14 3

Cover by Morgan Creative incorporating an image by Eduardo Paolozzi, *Newton After William Blake* (1997), plaster relief
Typeset by DP Photosetting, Neath, West Glamorgan
Printed and bound by 4Edge Ltd., Essex

CONTENTS

PREFACE AND ACKNOWLEDGEMENTS

Why is it that the computer has assumed a position of such predominance in our culture? What has produced the situation we now find ourselves in, with contemporary life so deeply reliant upon, and so deeply interwoven with, electronic technologies? How has this happened? What processes in human consciousness have led to this taking place? How can we make sense of it from a spiritual perspective? And what philosophical and psychological insights can be brought to illuminate the digital revolution that continues to unfold with such astonishing rapidity?

This book, many years in gestation, and many more years in the writing, is the result of having lived with these questions, which have only grown more insistent, more pressing, as each year has passed. None of these questions can be answered easily or simply. The computer was the product of a long historical development, which culminated in the scientific revolution of the seventeenth century. It was during this period that the first mechanical calculators were invented and the project to create more complex 'thinking machines' was embarked upon in earnest. But the seeds of this historical development were sown many hundreds of years earlier, deep in antiquity.

In order to understand the emergence of the computer, we need to place it in the wider context of the evolution of human consciousness over millennia. It then becomes apparent that modern consciousness has evolved in conjunction with the evolution of machines, and under their intensifying shadow. Today the computer, with its highly refined logical 'intelligence', has become more than just an indispensible aid to living in the contemporary world. It has become a kind of icon of modern consciousness, and as such presents a profound challenge to us all. It asks us ever more urgently: what does it mean to be human, and what, if anything, distinguishes us from machines?

The present study is both the product of much research and rumination pursued in solitude, and also countless conversations and exchanges with many different people who have shared with me their concerns. Of these, my conversations with Sam Betts, who also generously offered detailed comments and suggestions on every chapter of the book, have been invaluable. Without his measured and wise guidance, the book would have lacked the degree of focus and rigour that I have endeavoured to give it. For his friendship and for his full-hearted support of the creative effort involved in writing this book, I am profoundly grateful.

I am also grateful to Tom Raines, editor of *New View* magazine, for his warm encouragement and helpful feedback on the final draft of the book. His willingness to publish exploratory articles on the spiritual impact of our growing intimacy with electronic technologies, as well as his organizing the 2010 conference at Rudolf Steiner House in London on this question, provided the soil in which the seeds of the present study could be sown. As these seeds germinated into chapters, I would read them out loud, and my heartfelt thanks go to Louanne Richards for her willingness to listen to them as they emerged, and for her consistent and loving support for the whole enterprise. I have also enjoyed many interesting and fruitful (and quite often hilariously funny) conversations with Dr George Burnett-Stuart, who was kind enough to read certain chapters and to bring his far superior knowledge to bear on my attempts to gain a spiritual perspective on electricity.

I would also like to record my gratitude to Paul Ferguson who, in response to my plea for help, graciously and enthusiastically embraced the task of translating from the Latin several tracts by the sixteenth century alchemist Gerhard Dorn, who perceived the danger of the ingress of the archetype of the Binarius into European culture.

Over the years it has been my good fortune to have been invited to speak on the subject of technology to many diverse groups and organisations. To list them all would be tedious, but I would like to record my thanks to Lindsay Fulcher for inviting me to speak at the Clarendon Centre, London, both on the subject of computer technology and on the nature of electricity, and for her interest and encouragement over many years. I am also grateful to Stephen Overy, Professor Keith Critchlow and Dr Stephen Cross for their support of my forays into the question of technology at the Temenos Academy; and to Robert Chamberlain for twice risking the exploration of this subject at Freeman College in Sheffield. I have also benefitted greatly from presenting lectures and study days to Jungian groups in London and Bristol. It has been immensely helpful to have had the chance to approach the technological question from within the different contexts of the perennial philosophy, anthroposophy and depth psychology, each of which has so much to offer in furthering our understanding of this most fundamental of all questions that we face today, which is ultimately the question of our human future.

Jeremy Naydler, December 2017

Part One

THE ANCIENT WORLD

Chapter One

PARTICIPATIVE CONSCIOUSNESS
IN DEEP ANTIQUITY

The Computer and Consciousness

Some years ago, I came across a book with the curious title, *The Prehistory of Aviation*. Written by an anthropologist named Berthold Laufer in 1928, it was a survey of the many and varied accounts of human flight before the invention of the aeroplane. These included Taoist sages rising up to heaven in broad daylight having drunk the elixir of immortality, Indian magicians able to soar into the sky at will, legendary Chinese emperors flying on the backs of dragons, and a mythical ancient Babylonian king who rose up to heavenly heights on the back of an eagle. This is to mention just a few of a very large number of intriguing accounts of human flight that Laufer details, none of which involved climbing aboard an aeroplane.

In the introduction to his book, Laufer explains that it would be a serious mistake to conceive of aviation's apparently mythical and fanciful prehistory as totally distinct from the history of actual flying machines. The two things are so closely interwoven that the historical development of physical flying machines would be unintelligible without knowledge of their 'prehistory'. In the 1920s, when Laufer conducted his study, aeroplanes were a fairly recent phenomenon and still had the ability to make people marvel in a way that, probably for most of us today, they do not. The possibility of flight has become something we take almost for granted. And in our overly technologised age, we are perhaps less able to see the connection between the machines that have become so familiar a part of our external environment and the inner life of myth, fantasy, thought and desire that characterizes their prehistory. And yet this connection can hardly be denied: without the fantasies, thoughts and longings of human beings, no amount of mere technical skill could have brought forth the physical aeroplanes that are now so integral to modern life. A merely external explanation of the origin of these machines that leaves out this inner dimension cannot satisfy the deeper quest to understand why they came to be invented. Laufer puts the argument in relation to the origin of aeroplanes very succinctly as follows:

> Describing merely the gradual perfection of mechanical devices does not make a complete history of aviation. It is the spirit and the idea behind the devices that count, the idea itself means everything.[1]

Something similar must also be said of the invention of computers. Many histories of the computer have been written, but few attempt to look behind the history of actual physical devices, such as the early mechanical calculators, and consider 'the spirit and the idea' that gave rise to them. As a result, very few take hold of the question as to *why* the computer was invented. If we are properly to grasp this question, then we have to see that the origin of the computer entails something much more profound than simply a series of technical advances in engineering and design. There is also a 'prehistory' of the computer, that is inextricably linked to the emergence of certain ideas, to the sway of powerful fantasies, and above all to underlying changes in human consciousness over the last hundreds, and indeed thousands, of years. While the physical computer has entered the world as a result of modern scientific and technological skill, modern science and technology are themselves manifestations of a type of consciousness that has taken millennia to develop. As we shall see, this consciousness is quite different from the type of consciousness that prevailed in past times. Unless we take this fact into account and view the origins of the computer within the historical perspective of the evolution of consciousness, we shall fail to see the deeper significance of the computer and the specific challenges that it presents to us today.

The reader will no doubt appreciate that, in terms of *content*, the prehistory of the computer bears little comparison with the exotic theme of the prehistory of aviation. In exploring the realms of imagination and the life of thought that preceded the actual physical construction of computers, we may have the uncomfortable sense of human consciousness succumbing increasingly to the dominion of gravity, as it became more and more overshadowed by the mechanistic outlook. The prehistory of the computer involves the progressive negation and denial of the earlier spiritual viewpoint in which such ecstatic experiences as human flight could still be countenanced. One of the last believable accounts of this kind of experience is Dante's description in *The Divine Comedy* of his ascent through the heavens, written at the beginning of the fourteenth century. By the end of the same century, the rise of the philosophical movement known as 'nominalism' set the course of European history in a direction that would lead to the total collapse of the metaphysical understanding that underpinned Dante's extra-terrestrial flight. With the growing obliviousness of people towards the spiritual, it became impossible to conceive of how such an ascent could really take place; but it did become possible for them both to conceive and to manufacture the mechanical clock, and relatively soon thereafter the mechanical calculator, the seventeenth century precursor to the computer.

Unholy Twins: Logic and Electricity

Computers work with logic, and are essentially an advanced type of logic machine. For this reason, the prehistory of the computer must include a consideration of our historical relationship to logic. Furthermore, because electricity has been crucial in the development of fully operational computers, a consideration of our historical relationship to electricity is also necessary if we are to understand their emergence into the world. These are two major strands in the story of how computers came into being. On the one hand we need to understand how logic entered into human thought processes, subsequently becoming assimilated into the human mental fabric over many thousands of years, until a point was reached when it became possible to extract it from the human realm and introduce it into machines. On the other hand, we need to understand humanity's changing relationship to the powerful force to which we today give the name 'electricity', which for the greater part of human history was left to itself and, apart from its natural manifestation in the thunderstorm and the occasional occurrences of static electricity, remained in a physically dormant state, unharnessed by human ingenuity.

Typically, the advent of logic is associated with the beginnings of philosophy and abstract thought in the sixth century before the present era, in Greece. But the fact that the great civilizations of ancient Egypt and Mesopotamia were, with relatively simple technologies, capable of erecting monumental buildings and producing exquisite artefacts, both of which depended as much on the application of systematic reasoning as technical skill, demonstrates that the capacity to think logically existed long before the beginnings of philosophy. The difference between the earlier, pre-Greek logical thinking, and what began to unfold in sixth century BC Greece was that the earlier thinking was always applied to *practical* tasks. Broadly speaking, the ancient, pre-Greek cultures did not utilize logic for the attainment of wisdom or spiritual understanding, which was acquired primarily through visionary experience and direct intuition. Nor was logic studied as a discipline in its own right before the Greek period. Even the disciplines of omen interpretation and divination in ancient Mesopotamia, which were in large part empirical sciences employing a primitive form of syllogistic reasoning to interpret the will of the gods, never broke free of their dependence on the association of images, each invested with symbolic meanings. And it was these prior associations that determined the way in which divinatory observations were interpreted.[2]

At the same time, when we try to ascertain how the ancient, pre-Greek world related to electricity we find that, compared to our present electricity-dominated society, in antiquity this force was almost completely hidden from view. It would, however, be wrong to conclude that there was therefore no knowledge of it. Precisely insofar as it could be regarded as a spiritual force, rather than as simply a material process, the way in which electricity was understood and encountered in ancient cultures such as Mesopotamia and Egypt, with their profound initiatory rites, was quite different from how it is understood and encountered today, with the purely technical approach so characteristic of modern consciousness. Rather than through a technical knowledge suited to its harnessing and control, that deliberately brings it into relationship with the physical world in order to serve human needs and desires, electricity was understood as belonging to the domain of invisible powers, known primarily through *inner experience*. Such an inner experience would have been closely guarded in the mystery religions and, as we shall see, whoever wished to acquire it would have had first to show that they were both morally and spiritually mature enough to receive it.

Unlike in our own times, then, in deep antiquity people experienced logic less in the inner life of thought than in working practically with such simple technologies as the lever or the *shaduf* (which applied the principle of the lever for raising water). By contrast, electricity was experienced less as an outer force of nature than as a spiritual power encountered in the mystery rites. Just as logic was held fast within the practical sphere, and not permitted to cross over into the spiritual, so on the other side electricity was held within the spiritual sphere and not allowed to cross over into the practical. The power of logical thinking was, in deep antiquity, always kept subservient to spiritual understanding and visionary insight. At the same time, the forces of electricity were contained within a purely interior domain, and knowledge of these forces was reserved for those who had undergone a prior moral and spiritual preparation. Such considerations as these need to be taken into account if we are to understand why computers did not develop in the pre-Greek era.

In order for us to grasp how and why the computer originated, it is necessary to gain some sense of the kind of consciousness that prevailed in deep antiquity. Outwardly, people at that time may seem to have been little different from how we are today and, apart from lacking a few modern conveniences, they might appear in other respects to have been much the same as us. But it is important to see that inwardly they inhabited a very different world. Their consciousness was not the same as our modern consciousness. In this and the following chapter we shall see that the cultures of ancient Egypt and Mesopotamia, which preceded the civilization of the Greeks, were characterized by a consciousness that was not only unlike our own but also unlike that of the Greeks.

The focus of these chapters on the two earlier cultures of ancient Egypt and Mesopotamia is not intended in any way to disparage other contemporaneous cultures (for example the Minoan and Hittite in the Mediterranean region or further to the south and east the civilizations of the Indus Valley in India and the Yellow River in China), but has been chosen in order more clearly to trace the trajectory of the evolving consciousness that would eventually give birth to the computer. We may think of Mesopotamia, Egypt, Greece and Rome, the subsequent culture of medieval Europe, and then the periods of the Reformation, and the Scientific and Industrial Revolutions as belonging organically together, like a plant with roots, stems, leaves and flowers. Much more detail could of course be added to the picture than has been possible here, but that would make a very long and complicated book, and one far beyond the competence of this author. Both for the sake of the reader and of the author, this study is intended to be more suggestive than comprehensive.

Participative Consciousness in Ancient Mesopotamia and Egypt

When we contemplate the consciousness of the peoples of antiquity, we have the strong impression that for this consciousness everything in nature was experienced as being far more alive than it is for us today. In Mesopotamia, natural phenomena such as wind and rain, the sky overhead, sun, moon and stars, were all deities. In Egypt the river Nile too was a living deity, the starry sky was a goddess, the earth was a god from whose ribs the barley grew, and every plant and animal was sacred to some deity or other. The ultimate source of all life was not so much the sun as *the sun god* Ra, whose life-giving rays warmed every living creature. For the Egyptians, even the air was a god. In the animistic worldview of antiquity, which characterized practically all ancient cultures, nature was not 'objectified' as it is today. It was rather imbued with soul-qualities to which people felt they could relate themselves inwardly. Relationship to the natural world was based on an inner participation in these soul qualities, which meant that the gods were experienced as residing in the inwardness of all natural phenomena. In relating to the material world one was at the same time relating to a world that was both living and ensouled.[3]

This participative consciousness can best be caught sight of when we read the accounts of some of the ancient Greek observers of the much older civilizations that preceded their own. Plutarch, for example, describes how the Egyptian farmers sowed their seeds 'like those who bury and mourn'.[4] The reason for this attitude of lamentation was that they saw the god Osiris as being buried in the seed. The sowing of the seed was the interment of the god, whose death was the necessary precursor to his resurrection when the seed germinated. This was not an intellectual understanding, nor was it a question merely of religious belief. Rather it was a direct emotional engagement in the drama of the death of the god, experienced as the inner reality of the act of sowing the seed. Amongst the people, every stage of the farmer's labour was connected with the passion, death and resurrection of Osiris.[5] So too were the cycles of the inundation and retreat of the Nile through the seasons. When the Nile began to rise in midsummer, this was jubilantly hailed as the 'finding of Osiris', who was 'in' the life-giving, fertilizing waters.[6] The Festival of the Rising of the Nile (Akhet) was celebrated with great emotional fervour. Writing in the fifth century BC, Herodotus vividly describes the vast crowds that would attend such festivals, singing, clapping, dancing and consuming huge quantities of wine.[7] According to Heliodorus, even as late as the fourth century AD, 'no other festival [than that of Akhet] aroused in the Egyptians so much zeal' and 'the inhabitants abandoned themselves to the jubilations'.[8]

The evidence from both ancient Egyptian and later sources points to a consciousness for which deeply felt participation in nature's cycles was the norm.[9] Whether celebrated with lamentation or joy, the whole population shared in the processes of nature because these were also the sufferings and ecstasies of the gods. The same was true also of the ancient Mesopotamians, who at the height of the hot dry summer would give themselves to ritual wailings and lamentations for the lost Tammuz.[10] The mood of the people closely participated the mood of nature, which at last turned to relief and joy when the waters flowed once more throughout the land and Tammuz was 'found'. Henri Frankfort, a historian sensitive to the character of the ancient consciousness, makes the following observation concerning the prelude to the New Year festival in Mesopotamia, which marked the finding of Tammuz:

> The strength of feeling, the realization of utter bereavement, was of the essence of the celebration; without it the feast would have no virtue. For the feelings thus aroused, and intensified by being shared, are potent beyond the experiences of daily life. Therefore when the ritual

reached its turning-point, when the mood changed from desolation to rapture at the 'discovery' or the 'liberation' of the god, the very violence of feeling created the conviction that something marvellous had been achieved, the god resuscitated, salvation found.[11]

To such a consciousness, the mechanical logic that is required to construct and to operate even the so-called 'simple machines' (lever, wheel, pulley, etc.) must have presented a very particular inner challenge. Such technologies, no matter how simple, interposed a dead mechanism between the human being and the living, god-infused world of nature. Here, however, we notice a difference between the Mesopotamians and the Egyptians. In Mesopotamia, there was a greater willingness to adopt simple technologies in order to exercise more human control over the unpredictable and often harsh conditions of the Tigris and Euphrates valleys. In Mesopotamia, too, the degree of human autonomy in relation to the gods was greater as, for example, the formulation of law codes at an early date testify.[12] The prevalence of divination techniques (that were scarcely used at all in Egypt) also indicates a greater distance between the human and divine realms, for the acts of divination were a relatively intellectualized technique in the hands of the Mesopotamian scholar priests.[13]

And so in Mesopotamia, which in many respects was more 'modern' than ancient Egypt, a simple technology like that of the *shaduf* was readily accommodated. So also was the wheeled vehicle. But in Egypt, as we shall see in the next chapter, neither the *shaduf* nor the wheeled vehicle were adopted with great alacrity. In the cultural milieu of Egypt, wide-awake to the realities of the spirit world, even simple technologies had moral, and indeed cosmic implications. This was because they required treating matter as if it were no longer the dwelling place of spirit, but as if it was simply mass with little more to it than extension, density and weight. Thus the adoption of a technology like that of the *shaduf* could not have appeared to the Egyptians as morally neutral. It must have felt in some degree sacrilegious to the Egyptian sensibility.[14]

It is characteristic of the Egyptians that in the construction of buildings and artefacts, and in the crafts and arts, all of which depended in varying degrees on the ability to think logically and consistently, an element of religious constraint is never far beneath the surface. This can be noticed, for example, in the careful choice of materials in buildings, in craftwork, sculpture and the visual arts. Different kinds of stone—granite, basalt, alabaster, sandstone and so on—were employed with an acute sensitivity to their perceived spiritual qualities, as were the various metals and woods.

In all kinds of skilled craft, colour, form, substance, size, location and orientation were to a large extent determined by religious require-ments.[15] Religious considerations may also have lain behind the extra-ordinarily small range of tools, all of utmost simplicity, used by ancient Egyptian carpenters, cabinet-makers and stonemasons. The tools were restricted to the axe, saw, adze, knife, scraper, mallet and chisel.[16] This limited range of hand held tools ensured that whatever a craftsman was working on, there was a direct human relationship to it. Thereby the craftsman's sensitivity to the spiritual qualities of the materials was retained.

In a society in which every aspect of life was directed by religious considerations, and in which the temple and priesthood were the dominant social and economic forces, it is not hard to see how social and economic life was to a large extent deliberately protected from the incursions of mechanistic and utilitarian thinking. In Mesopotamia, by contrast, where palace and temple had more clearly delineated spheres of influence and where the king was never—as in ancient Egypt—regarded as the actual incarnation of a god, and where the state of war was as much the norm as was the state of peace in Egypt, mechanistic and utilitarian thinking asserted themselves more strongly. The reason for the failure of the Egyptians to embrace early technologies was not, therefore, that the Egyptians lacked the technological ability of the Mesopotamians. But nor was it simply that, in the more benevolent conditions of the Nile valley, there was not the same pressing need to develop more sophisticated technologies as in the harsher and more unpredictable lands of Meso-potamia. It is rather that the priesthood and initiate kings of ancient Egypt deliberately resisted the ingress of mechanical logic, for they knew it would only encourage the tendency to step back from the participative relationship with the divine-natural world. They understood that, if the culture became permeated with applied logical thinking, this would weaken their relationship with the gods. And should this happen, the spiritual basis of the whole culture would be undermined.

Chapter Two

THE GODS AND TECHNOLOGICAL
CONSCIOUSNESS
IN THE ANCIENT NEAR EAST

Logic and Religious Awareness in Ancient Egypt and Mesopotamia
That logical thinking was not regarded as an appropriate mode of consciousness with which to approach the gods can readily be appreciated by reading any number of texts that form the sacred literature of ancient Egypt. One example that well demonstrates the attitude of the ancient Egyptian priesthood is Chapter 17 of the *Book of the Dead*. This is an important and very ancient theological text, of which versions exist that go back to the Ninth Dynasty, in the last century of the third millennium (circa 2100 BC).[1] It consists of a series of enigmatic statements made by the sun god Ra, to which are appended explanatory glosses. The glosses, however, elucidate the text not by adding anything that comes close to a logical explanation but by piling on further layers of imagery and mythical references. Consider the following: 'To me belongs yesterday, I know tomorrow,' states Ra in the text. 'What does it mean?' asks the first gloss, and gives this response: 'As for yesterday, that is Osiris. As for tomorrow, that is Ra on that day in which the foes of the Lord of All were destroyed and his son Horus was made to rule.' A second gloss elaborates as follows: 'Otherwise said: That is the day of the "We-remain" festival, when the burial of Osiris was ordered by his father Ra.'[2]

To amplify the meaning of the text, a vignette is added that depicts two black-spotted lions (perhaps best described as leopard-lions) named Yesterday and Tomorrow. They sit back to back under the hieroglyph of the sky. Between them a sun rises over the horizon (Fig. 2.1).

In order to begin to understand the meaning of a text such as this, along with its enigmatic illustration, one has to abandon any idea that it might be possible to express it in more abstract terms, let alone as a clear philosophical statement. The more one attempts to translate it into something that would conform to modern standards of what constitutes 'clear thinking,' the further one actually moves from its original sense, for its original sense is multi-levelled and full of symbolic resonances. The modern reader of ancient Egyptian sacred texts is obliged to get used to repeatedly experiencing the defeat of the logical mind as he or she

Figure 2.1

The two leopard-lions, named Yesterday *and* Tomorrow *sit either side of the rising sun.*
Vignette to part of The Book of the Dead, *Chapter 17.*

grapples with dense imagery, thick with layer on layer of meaning. This meaning can only be penetrated through an imaginative effort in which one steeps oneself in the myths, symbolism and ritual life of the culture to which the text belongs.

Suffice it to say that the text is concerned with the rebirth of the soul and its spiritual illumination or 'solarization'.[3] It would take us too far from the theme of this book to attempt a fuller explanation of the words uttered by Ra, and the glosses appended to them. The point of citing this passage from the *Book of the Dead* is simply to highlight a difference that exists between contemporary intellectual consciousness and the consciousness that prevailed in antiquity. It was not that the Egyptians were unable to think logically, but rather that they did not see the sphere of religious reflection as one in which logic had a legitimate place.[4] This is because logic is inherently divisive: it distinguishes one thing from another, introducing separation just where the symbolic consciousness perceives apparently disparate things as belonging together, and seeks to penetrate further the links between them. In this respect, logical thought would surely have been regarded by the Egyptians as ruled over by the god Seth, who above all was the instigator of division and separation, whereas the spiritual realm is governed by the principle of unity.[5] For this reason the domain appropriate to logical thinking was seen as the material, not the spiritual, world. Logical thinking was tied to the achievement of specific material goals, and guided people only in so far as they were seeking physical solutions to practical problems. It served as the means by which people learnt to manipulate and master the material conditions of physical life, but even then such application of logical thinking was generally constrained by religious sensibilities. This is because the material world was experienced as imbued with spiritual qualities, and therefore practical activities were seen as having spiritual implications.

While in many respects the kind of consciousness that we meet in ancient Mesopotamia parallels that of the ancient Egyptians, we cannot fail to notice that in ancient Mesopotamia there was a significantly different attitude towards logic. Most striking is the place of logical, or perhaps it would be more accurate to say *quasi-logical*, thought in the practice of divination. In ancient Mesopotamia, divination was the primary means of communication between human beings and the world of spirit, in contrast to ancient Egypt where it seems scarcely to have been practised at all. We know from various ancient accounts that during the Sumerian period, in the third millennium BC, dream interpretation was regarded as one of the most reliable channels of communication between gods and humans. The interpreters of dreams at this early period tended to be women, who were seen as having superior intuitive and clairvoyant powers to the men who would come to them in their perplexity in order to seek spiritual guidance.[6] In the course of the next millennium, however, as the practice of divination through the examination of entrails became increasingly widespread, the male *barû* (literally 'inspector') priest became steadily more important. With this transition, the direct channelling of communications to and from the gods, characteristic of the priestess, or *sha-ilu* (literally 'she who asks questions of the gods'), was gradually replaced by the more indirect method of inspecting the entrails—particularly the liver—of sacrificed animals, practised by the *barû* priest.[7]

The significance of this transition is that communication between human beings and gods through divination developed into a highly complex and sophisticated science, involving precise empirical observation and logical deduction conducted within a symbolic framework of interpretation. Thus logical thought underwent a degree of liberation from the strictly practical sphere, and was exercised with the aim of gaining spiritual knowledge. During the Old Babylonian period, in the early centuries of the second millennium, the codification of divination procedures began, and this led to the amassing of an extremely large body of divination records, which would regularly be consulted to assist in the interpretation of new results.[8] To this end, rules were established on the basis of empirical observation that would guide the diviner. For example, if the umbilical fissure of a sacrificed sheep's liver, where left and right lobes meet, was observed to be long on the right side and short on the left, this should be regarded as a good omen. If it was long on the left and short on the right, it should be seen as a bad omen.[9] This was because of the symbolic significance attached to the umbilical fissure (which was associated with the palace, its administration, income, and personnel) and the symbolic meanings associated with left (usually bad) and right (usually good). Since the logical thought processes that went into the interpretation of a divination ritual, moving from observation to conclusion, always rested on the premise of a prior symbolic significance attached to what was observed, we may say that logical thought never truly broke free of its dependence on the association of images and the symbolic meanings that adhered to them. Nevertheless, both the science of divination, and also the closely related science of omen reading which emerged during the same period of the early second millennium BC, may be understood as entailing the emancipation of logic from the arena of applied thought and its incursion into a sphere hitherto regarded as the preserve of intuitive and mediumistic states of consciousness.[10]

Logic and Technology in Ancient Mesopotamia and Egypt

At the same time, in the cultural milieu of ancient Mesopotamia we may detect a greater willingness than in ancient Egypt to relate to the natural world through mechanical technologies. The *shaduf*, for example, which was a mechanical apparatus for raising water from a river, canal, or some other water source, was employed in Mesopotamia from the Akkadian period at the end of the third millennium (circa 2400–2200 BC).[11] It consisted of a long wooden pole that rocked on a pivot, and which had at one end a receptacle and at the other a large stone or lump of clay that acted as a counterweight. When the receptacle was lowered into the water, the counterweight swung up into the air. Once the receptacle was filled, lifting it was made much easier because the counterweight at the other end of the pole balanced out the weight of the water in the receptacle. In Figure 2.2, the receptacle is on the left nearest to the human figure and the counterweight is on the right.

Based on a slightly off-centre fulcrum, the *shaduf* applied the principles of leverage to increase the physical force exerted on the load to be raised. Thus a so-called 'mechanical advantage' was gained through the fulcrum establishing a relationship between the two ends of the pole balanced on it, such that the weight on one end facilitated raising the weight at the other. But the leverage principle itself depends on another more elementary principle, which is that either end of a rigid pole is raised only when the other end is lowered. While this may seem too obvious to warrant our consideration, the arrangement actually embodies in a

Figure 2.2
An early Mesopotamian shaduf.
From an Akkadian cylinder seal.
Late third millennium BC.

physical apparatus what is known in logic as the *exclusive disjunction*, expressed verbally as 'one or the other but not both!'[12] Clearly, either one end of the pole is up or the other end is up, but both ends cannot be up at the same time.

It is significant that the *shaduf* was only adopted in ancient Egypt towards the end of the eighteenth dynasty (circa 1300 BC) during the reign of the pharaoh Akhenaten, after it had already been in use in Mesopotamia for roughly a thousand years.[13] Even if we accept that contact between the two cultures during this period was minimal, thus reducing the possibilities of Mesopotamian cultural influence on Egypt, this does not really explain why the Egyptians were so slow to adopt the *shaduf*. The principles of leverage on which the *shaduf* was based were almost definitely known and applied in the construction of pyramids during Old Kingdom times (circa 2550 BC).[14] To extend these principles to irrigation by means of an apparatus such as the *shaduf* would not have been particularly remarkable. The Egyptians were an intelligent, imaginative and highly practical people: it was not lack of any of these qualities that held them back from introducing the *shaduf*. It seems far more likely that they felt that fetching and carrying water by hand was, despite being more onerous, nevertheless a more appropriate, perhaps a more respectful, relationship to the life-giving sources of water.

The scene depicted in Figure 2.3 was painted about 100 years before the introduction of the *shaduf*, and beautifully conveys the intimate, participative relationship of the Egyptian peasants to the pool from which they collect water, one joyfully immersing himself in the pool in order to fill his water-pot. In a country upon which very little rain fell, the source of the water in such pools was the River Nile, which was regarded less as merely a 'water resource' (as we might describe it today) and more as a living being, a deity, whom the Egyptians knew as the god Hapy. To Hapy songs were sung, offerings were made and rituals performed.[15] The people lived with Hapy as with a mighty, and for the most part gentle and

Figure 2.3
Drawing water in pots from a lily pond.
New Kingdom tomb decoration. Thebes,
Egypt. Circa 1450 BC.

benevolent, giant in their midst. Portrayed as a male figure with breasts, Hapy is shown in Figure 2.4 bearing gifts of flowers and vases of water.

In order to explain the late adoption of the *shaduf*, this context of an intensely felt relationship of the people to Hapy, along with the whole

Figure 2.4
Hermaphrodite figure of Hapy, the god of the
River Nile, bearing vases of water and flowers.
Temple of Horus, Edfu.

tenor of the sacred culture, so closely guarded by the priesthood, must be taken into account. The fact that the *shaduf* embodied a mechanical logic that was not itself constrained to serve a religious or spiritual end, but a purely practical one, and the fact that its introduction corresponded to a period in Egyptian history when, during the religious upheavals of the reign of the pharaoh Akhenaten, the worship of the traditional gods was suppressed, are both acutely relevant. With the power of the priesthood and temples severely weakened by Akhenaten, the traditional religious constraints were also weakened. Under these circumstances there would have been less opposition to, and therefore a greater opportunity for, the introduction of a mechanical apparatus such as the *shaduf*. Central to Akhenaten's revolution was his placing himself as sole mediator between the Egyptian people and the divine Aten, or sun disc, thereby displacing the ancient relationship to the gods. With the traditional relationship to the gods spiritually dislocated, the way was opened for the mechanical apparatus of the *shaduf* to be interposed between the people and their beloved river god, Hapy.

Figure 2.5, one of the earliest depictions of the *shaduf* in Egypt—from an Eighteenth Dynasty tomb, painted less than ten years after Akhenaten's death—begs us to ask what has happened to the peasant's relationship to the life-giving waters with which he used to fill his water-pot. Plant-life is

Figure 2.5
The shaduf *depicted in an Eighteenth Dynasty tomb painting. Tomb of Neferhotep, circa* *1325* BC.

as abundant as ever in this picture, and yet the peasant has taken a step back from the older animistic covenant with nature that he used to enjoy. Instead of immersing himself in the water in order to fill his water-pot, the peasant now stands on the canal bank working the machine. Here we see that a more detached relationship to the natural world, mediated by this new technological contraption, has surely occurred.

The delayed introduction of the *shaduf* in ancient Egypt bears comparison with the late introduction of the wheel, which was used for the purposes of transportation from as early as the mid fourth millennium in Mesopotamia.[16] It is interesting that in Mesopotamia, the earliest actual wheeled vehicles that survive are hearses for the transportation of the dead, which were buried with draught animals and human attendants in the royal tombs at Kish, Susa and Ur.[17] In other words, they served a spiritual purpose. It seems that wheeled vehicles were associated with both royalty and with death. In time, however, the use of wheeled vehicles broadened out to include the transportation of the living, and they were seen as especially advantageous in military campaigns (Fig. 2.6).

Figure 2.6
Four-wheeled war chariots. Royal Standard of Ur. Circa 2700 BC.

While early wheeled vehicles used 'solid' wheels, by the beginning of the second millennium spoked wheels were invented, which made possible both more speed and manoeuvrability. What is of particular interest is that although the Egyptians knew of the wheel from circa 2750 BC, they seem deliberately not to have developed it for transportation until circa 1600 BC (more than 1,500 years later than in Mesopotamia).[18] Its adoption in transportation was forced on the Egyptians at this time by their wars with the Hyksos, whose military superiority depended largely on chariots. As this was long after the Egyptians were aware of the principle of the wheel, we may surmise that it was precisely *the principle* of the wheel that was the reason for their reluctance to adopt it. The wheel may best be understood as a one-armed lever capable of rotating through

360° around a fulcrum that is the axle (Fig. 2.7). In other words the wheel and the *shaduf* belong to the same family: both are essentially levers.[19] Both adopt the same mechanical logic that is directed to the achievement of practical ends that are not necessarily contained or guided by religious considerations.

Figure 2.7
New Kingdom wheelwrights.

For the ancient consciousness, the ongoing task of harmoniously aligning the human to the spiritual world was an unquestioned priority. Both in ancient Mesopotamia and in Egypt, the state was organized precisely in such a way as to guarantee harmony with the realm of the gods, and this responsibility devolved largely on the king.[20] Logic, along with abstract, imageless thinking in which ideas can be associated in accordance with practical, utilitarian considerations, carried the danger of disrupting the inward alignment of the human being with the divine world. This is why it was felt that practical undertakings needed to be dedicated to the gods. By setting them into a greater sacred context, the logical element in thinking was subsumed under the imaginative, for in those days it was ever the image that mediated the divine. This was the case in all the ancient Middle Eastern cultures, with the sole exception of the Israelites, who waged war on sacred images and steered themselves away from the kind of imaginative and associative thinking that elsewhere characterized relationship to the world of spirit.[21]

Storm Gods and Initiation Rites

We have already seen that the ancient approach to electricity was not (as it is today) based simply on the observation of outer phenomena, but also on inner experience, and that a deeper knowledge of it was restricted to the limited circle of those who underwent initiation in the Mysteries. While within nature the experience of electricity was available to all people in the tumultuous phenomenon of the thunderstorm, for the ancient consciousness this phenomenon was never experienced as wholly *outward*, but regarded as a manifestation of the divine powers of gods. In Mesopotamia, these gods included the Sumerian Ishkur (forerunner of the Akkadian god Adad), and his sister Inanna, the warlike Ninurta, the Babylonian warrior-god Marduk, and so on. Such gods as these were not so much 'storm gods' as gods whose character was, or could be, violent and destructive. It was the soul-forces of violence, anger and destructiveness that human beings would initially have to face in the Mystery rites connected with these gods. But beyond this encounter with soul-forces was a deeper reckoning with powerful life-negating energies that emanated from the depths of the Underworld.

In Mesopotamia, gods who manifested thunder and lightning were typically at the centre of Mystery rites involving descent into these dark Underworld depths, where the gods themselves would succumb to inertia, barrenness and death. This was the case with all the deities mentioned above: Ishkur, Ninurta, Inanna and Marduk.[22] They all underwent a descent into the Underworld, where they had both to meet and to overcome those hostile powers which would otherwise destroy them. They had to find within themselves a spiritual counterforce strong enough to enable them to master inwardly these destructive powers. On their resurrection from the Underworld, they emerged not only with their creative potency renewed and restored, but also with a new ability to wreak death and destruction. Thus when the Sumerian goddess Inanna emerged from her Underworld ordeal, for which a remarkably full description has survived, she was utterly terrifying, for she had now acquired 'the eye of death' and anyone on whom she turned this eye was doomed.[23] Likewise, in an important ceremony following the 'liberation' of Marduk from his three day incarceration in the Underworld, which was celebrated annually in the Babylonian New Year Festival, the god was formally bestowed with the power to 'command annihilation' as well as to give life.[24] We know that both Ninurta and Inanna, like Marduk, had command over the lion-bird Imdugud, the personification of thunder and lightning. Imdugud was one of the main symbols of their violent, destructive, and warlike prowess.[25] Thus Inanna is portrayed in Figure 2.8 with lightning bolts in her hands riding on the back of Imdugud, who spits fire from his mouth. In the chariot is Inanna's brother Ishkur.

The ordeals that these deities underwent in their annual rites of death and regeneration, so as to re-emerge with formidable and indeed dreadful

Figure 2.8
Inanna, clutching lightning bolts, stands on the back of the winged thunder-lion Imdugud,
while her brother Ishkur rides on the storm chariot. Mesopotamian cylinder seal, circa 2500 BC.

powers, were precisely the experiences undergone by those human initiates who assumed their roles in the Mysteries. This is why moral preparation was considered essential for those who underwent initiation.[26] If we are correctly to ascertain what the ancient relationship to electricity was, it is therefore in the direction of the Mystery rites that we

Figure 2.9
The storm-god Ninurta or Marduk grasps thunderbolts as he does battle with a monster,
possibly Tiamat or Anzu (Imdugud). Temple of Ninurta. Nimrud, Iraq.

must look. The lightning flash or thunderbolt was regarded as one of the mightiest weapons of the gods.[27] It was a force of annihilation whose terrible power issued from realms that were beyond, or strictly speaking *beneath*, the physical world, along with all else inimical to life that was kept securely confined in the Underworld. And it was this force the initiate inwardly encountered.

While the Mesopotamian storm gods were certainly terrifying, and the power they wielded was violent and destructive, they were nevertheless also connected with the energies of life and fertility. They wielded both the power of destruction in their thunderbolts and the power of regeneration through the fertilizing rain that accompanied the storm. But in Egypt, where the fertility of the land was primarily dependent not on the fall of rain but on the annual inundation of the River Nile, there was a different perception, which more sharply distinguished the provenance of atmospheric electricity from that of the forces of life. The electricity experienced in the thunderstorm unquestionably belonged to the sphere of Seth, the god of destruction, illness, strife and confusion, to whom we have already referred in relation to the divisive nature of logic.[28]

For the Egyptians the energies of life, which infuse all creatures with strength of form, health and vitality, streamed into the natural world from the otherworldly domain of Osiris, which was the spiritual source of the Nile flood, and within nature they were under the guardianship of the god Horus.[29] By contrast, atmospheric electricity was perceived as belonging to the forces of chaos and disorder, under the command of Seth, a desert god who was *the opponent* of life, fertility and health (Fig. 2.10), and who 'thunders in the horizon'.[30]

Because of his association with atmospheric electricity, it may be appreciated that knowledge of electricity, in so far as it was connected with knowledge of Seth, would have belonged within the context of the Osiris Mysteries. Seth was the murderer of Osiris, and in this respect could be understood as the demonic initiator of all those who underwent the Osiris initiation. To experience the death-dealing power of Seth was an essential aspect of the initiatory experience. By contrast, the god Horus, the son of Osiris, was the guardian of life, vitality and the flourishing of living forms in nature. He was the inveterate opponent of, and victor over, Seth who was associated with the desert wastelands, and it was with the god Horus that the spiritually reborn initiate would identify.[31] Those who underwent the Osiris initiation, and experienced rebirth as Horus, came to know and to take command of the energies of Seth, in a way that was not so dissimilar from the Mesopotamian experience. Thus while the force of electricity, in so far as it was identified as Sethian, was assigned a

Figure 2.10
Seth.

position within the ancient Egyptian mythological worldview that was antagonistic, and indeed violently opposed to, the fertilizing and life-giving energies mediated by such benevolent divine figures as Osiris and Horus, in the person of the initiate-king and others who underwent the initiatory rites both these powers were known and mastered.

Chapter Three

POETS, VISIONARIES AND THE RISE OF THE CLEVER MAN

Prophecy and Poetry in Ancient Greece

When we turn in our imagination to consider the mental world of the Greeks at around the time of the birth of philosophy in the sixth century BC, we need to remember that the civilizations of Mesopotamia and Egypt were already as old to the Greeks who lived at that time as the Greeks are to us today. Greek civilization was a newcomer on the scene. It had nothing like the historical and cultural background that the earlier civilizations of the Middle East had. The earliest written inscriptions in Greek, for example, date back only to the eighth century BC, whereas the earliest Sumerian and Egyptian inscriptions were made during the last centuries of the fourth millennium, roughly 2,500 years earlier.

Much, however, was shared in the religious outlooks of the Greeks and the Mesopotamians. We find remarkably similar practices of divination and omen reading to find out the will of the gods, a similar view of the fate of the soul after death in a dark and dismal Underworld region, and a similar pantheon of gods and goddesses—with Ishtar, for example, corresponding to Aphrodite and Marduk to Zeus. The consciousness of the Greeks in the Archaic period (late seventh to early fifth centuries BC), when philosophy first began to emerge, was in certain important respects *not so different* from the consciousness that prevailed throughout the ancient world at that time. The revolution in consciousness that brought about the birth of philosophy occurred through a crisis in perception and thinking that was accompanied by considerable anguish and conflict.

As the birth of philosophy in ancient Greece is one of the major milestones in the emergence of the computer, it is essential to look further into what it entailed. Before the eighth century BC, Greece was an entirely oral culture, and it remained a predominantly oral culture even during the period when the great literary works of Homer, Hesiod, Pindar and Aeschylus appeared. Right into the Classical period (from the early fifth to the early fourth centuries), the proportion of the population that was literate was tiny.[1] The literature that came into existence at this time was meant to be read out loud or recited, rather than read privately: the written word always took second place to the spoken. At the time of the advent of philosophy, poetry, music and dance were the main cultural media, and all of these were essentially performing arts.[2]

Poetry, then, was not so much for reading as for hearing, and for learning by heart and reciting. The recitals would have had a musical accompaniment and often would have been sung. Greece was a 'song culture', in which not only individual singing but also choral singing and dancing played a central part in civic life.[3] Poetry, music, song and dance were the inseparable ingredients that fed the souls of the Greeks. Music and song were a normal part of social life, whenever and wherever people came together. Every educated Greek was expected to know a large amount of poetry by heart, and to be able to sing and play the lyre (Fig. 3.1) Furthermore, no festival in honour of the gods was complete without singing, which was often performed in contests between rival choirs. These choirs were big—at Athens ten choirs, each consisting of 50 boys and 50 men, competed at the dithyrambic contests held in honour of the god Dionysus.[4]

Figure 3.1
Greek pottery painting depicting a music class. Fifth c. BC.

We can still experience today that when poetry is lodged in the memory, to evoke it is to evoke far more of its creative power than when it is simply read from a book. One has to bring the words forth from out of oneself, and in bringing them into being the poem is, in a sense, brought to birth anew. The reading of words from the printed page requires nothing like the same degree of engagement and participation in the creative energy that lies at the heart of a poem and gives it its life. In ancient Greece, really good performers (known as 'rhapsodes') would totally immerse themselves in the poems they were reciting so that they too touched the creative fount from which the poem sprang. Rhapsodes would perform before vast audiences, sometimes as many as 20,000 people in Plato's lifetime, throwing themselves into the parts of the characters and stirring the emotions of the crowds.[5] As much as ancient Greece was a 'song culture' it was also a *feeling* culture, even at the time when philosophy and the habit of abstract thought were making inroads into the general consciousness. When people attended poetry recitals, or dramatic performances, they did not sit back as spectators but were stirred up emotionally, becoming active and enthusiastic participants in what was unfolded before them.[6]

This emotional engagement was also an imaginative engagement, for poets like Homer and Hesiod wrote in images; they painted pictures with words. Whether these pictures portrayed the succession of gods in the creation of the world, as in Hesiod's *Theogony*, or the deeds of the warrior-heroes of old, as in Homer's *Iliad*, we meet in these poems composed in the eighth century BC nothing approaching abstract thought and reflection. We only begin to touch on it in the figure of Odysseus, to whom we shall soon return, but ancient Greek consciousness did not begin to be reflective in a philosophical sense until the sixth century BC. Homer's Odysseus was not a philosopher: as we shall see, the love of wisdom was less a motivating factor in his psychological constitution than his love of the cunning stratagem.

Reason and the Visionary Imagination

Before the sixth century, wisdom was not seen as the province of philosophers, or as something to be won through disciplined intellectual thought, for before the sixth century no philosophers existed. Rather, wisdom was transmitted in the images and verses of the poets, who occupied the place that during the sixth century BC began to be claimed by philosophers. That is to say, it was the *poets* who were traditionally regarded as 'wise men' or *sophoi*.[7] The reason why they were seen as wise (*sophos*) is that their verses were divinely inspired by the Muses, and it was this that gave to their poems their spiritual authority. For through such inspired consciousness, they were able to become aware of an order of being that transcends time and the limitations of material existence. Poets, therefore, did not just write from their own skill (in Greek, *technê*), but uttered revealed truth.[8]

In this respect, poets had a natural affinity with seers, who likewise had the ability to see into the supernatural, and to draw wisdom from the world of gods and spirits. It was thus as poet-seers that the first philosophers made their appearance. Pythagoras, for instance, whose life spanned the sixth century, belonged to the small class of those who were known as 'divinely inspired' or *theios*. But where Pythagoras differed from the poet-seer was that he taught that the element within the human being that gains knowledge of the world of spirit is the *thinking* soul.[9] It was therefore a kind of knowledge in which clarity of thought was paramount. It was not enough to know through image and feeling; for Pythagoras, one had also to be able to cast one's knowing into the form of pure concepts. Instead of referring to the gods, Pythagoras referred to cosmic principles, such as the Monad (the originating source of the universe) and the Dyad (the principle of contraction and separation through which the universe came into manifestation).[10]

Another early philosopher was Parmenides, who lived in the second half of the sixth century. Still writing out of the poetic tradition, he presented his philosophy in a mystical verse poem, and yet he too emphasized the importance of thinking in a way that none of the traditional poet-seers did. For Parmenides, the element within us that is able to make contact with the realm of spirit is none other than the capacity to think. Through thinking we lay hold of the source of Being, which transcends the world of changing forms. Paradoxically, this insight was not arrived at by discursive thinking or rational argument but was mystically revealed to Parmenides by the goddess who dwells beyond the gates of Day and Night. But what she revealed to him was a teaching specifically concerned with the nature of thinking itself. She tells him that intrinsic to thinking is its capacity to grasp Being as such, 'for the same thing exists for thinking and for Being'.[11] In other words, Parmenides is led to see that there is a more profound reality than the realm of the gods, which is *Being as such*. But this greater reality can only be grasped by thinking.

The writing of Parmenides was nevertheless inspired through his visionary encounter with a goddess, and its authority derives from the force of what was revealed to him as *inspiration* rather than from the logical reasoning that he employed to support it. Other early philosophers, such as Pherecydes of Syros, were still to a large extent thinking pictorially, with the gods as central players in their cosmologies. There was at this period an ambivalence towards the embrace of purely abstract thinking, because it threatened attunement with the world of spirit conceived as the community of gods. For the older consciousness, it was *the image* that mediated the divine, and the cultivation of imageless thinking in preference to image-saturated thinking placed the traditional pious relationship with the gods in jeopardy. It must have seemed to this more traditional consciousness that what was being advocated was an approach to spirit with a type of thinking appropriate only to the crafts. For just as in the older cultures of ancient Egypt and Mesopotamia, the primary use of logical and rational thought in pre-philosophical Greece was in making and constructing things.[12] This type of practical thinking was denoted by the word *techně*, which meant 'craft' or 'skill'. It is, however, significant that *techně* also meant 'cunning'.[13]

The Figure of Odysseus

In Homer's *Odyssey*, we see how Homer, who so vividly articulated the traditional pious relationship to the gods, also gave to the Greeks a new kind of hero in the 'crafty' man. While Odysseus was courageous, strong, noble and daring like the warrior heroes of the *Iliad*, these were not the qualities that Homer most celebrated in his hero. What was special about Odysseus was the fact that he was *polymêchanos*: 'a man of many devices'.[14] Odysseus survived on his wits, and his exercise of cunning was for the most part not god-inspired, but due entirely to his own human 'craftiness'—his ability to work out how to get out of many a tight situation.[15]

One of the stories told in the *Odyssey* concerns the fateful encounter of Odysseus and his companions with the Cyclops—a giant with just one eye in the centre of his forehead.[16] The Cyclops, whose name was Polyphemus, dwelt on an island with others like him, living separately in caves, herding sheep and goats, and feeding on wild plants, milk and cheese. Polyphemus was himself the son of the god Poseidon, and the text makes it clear that the race of Cyclopes felt themselves to be equal to the race of divine beings. The crudeness of their dwellings and the simplicity of their lifestyle, so close to nature, reminds us of our own paleolithic ancestors, as does the eye in the centre of their foreheads, for it evokes the ancient clairvoyance of pre-technological humanity (Fig. 3.2).

When Odysseus and his companions arrived at the island, they stumbled upon the cave of Polyphemus, and marvelled at the crates laden with cheeses, the milk pails and the lambs and kids penned in a corner. While they were still inside the cave, the giant returned with his flocks and a big bundle of firewood, and rolled a mighty boulder across the cave entrance, to keep his animals from wandering away. When Polyphemus noticed the presence of Odysseus and his friends cowering in the recesses of the cave, he saw them simply as food to be eaten, and grabbed two of them, ate them for his dinner, and then fell asleep. When morning came he ate two more men for his breakfast, rolled aside the huge boulder, and went forth with his flocks, making sure to put the boulder once again back in place.

Figure 3.2
Roman mosaic of a Cyclops. Villa Romana del Casale, Sicily. Fourth c. AD.

Reflecting on the difficulty he was in, Odysseus realized that it would be impossible to get out of the cave with the Cyclops alive, because he would physically prevent any escape; but he also realized that should he manage to kill Polyphemus while the giant was sleeping, he and his men did not have the strength to roll back the huge boulder, and so would be trapped inside the cave forever. Odysseus therefore devised a cunning plan. When the Cyclops returned in the evening, he once again grabbed two men and ate them for his dinner. But now Odysseus bravely came forward and offered the Cyclops wine that he had brought from his ship. Polyphemus gladly drank the wine and found it pleasing. He then asked Odysseus his name. 'My name is *Oudeis*' replied Odysseus, deliberately mispronouncing his name so it meant 'nobody'. Polyphemus replied that as a special favour he would eat *Oudeis* last. Under the influence of the wine, Polyphemus very soon fell into a deep slumber, and while sleeping Odysseus and his men managed to drive a stake into his single eye, blinding him. In terrible pain, the giant rose up and yelled so loudly that neighbouring Cyclopes soon came hurrying to the cave to ask what was the matter: 'Are your flocks being stolen, or are you being attacked by guile or by force?' With achingly significant words, from within the cave Polyphemus replied:

It is Nobody (*Oudeis*) that is slaying me by guile and not by force.[17]

Alas for Polyphemus! His words only ensnared him more deeply in Odysseus's trap, for his neighbours then went away, thinking that if nobody was trying to kill Polyphemus or doing violence to him, then he had no need of their help.

Polyphemus was defeated by a guile that he was utterly helpless to resist. As the wounded giant spoke the fateful words that sent his neighbours back to their caves, Odysseus tells us: 'my heart laughed within me that my name and my flawless scheme had so beguiled'.[18] Odysseus then concealed his men underneath the sheep of the Cyclops. The blinded and enraged Polyphemus, having rolled back the boulder and placed himself at the entrance with the intention of catching the Greeks as they tried to escape, was then once again tricked. For as the sheep passed out of the cave, he stroked the back of each one with his hands but did not discover Odysseus and his companions clinging to their undersides. And so they made their escape.

The blinding of Polyphemus should be regarded as a key incident in the mythical prehistory of the computer.[19] It marked the emancipation of 'craftiness' from the crafts, that is to say from *the making of things*. It thus cleared the way for cunning to take its place as an independent faculty of cognition within the human psyche, to be applied for the achievement of human aims, whether or not these were put at the service of the higher will of the gods. Although Polyphemus was rough and gruff, and although he lived in a cave and had no qualms about eating human beings, he lived both closer to nature and closer to the gods than his Greek visitors. His lumbering and intellectually unrefined state of consciousness—similar in many ways to that of the Olympian gods—was, however, no match for the agile intelligence of the Greek hero. And so the giant succumbed to the clever hero, his clairvoyant eye was put out, and as is the fate of the conquered, he was vilified by the victorious consciousness that now began increasingly to assert itself in human evolution.[20]

To judge by the amount of times that the blinding of Polyphemus was depicted in Greek art, the deeper significance of this event was well understood by the Greeks (Fig. 3.3). In Homer's story, the blinding of Polyphemus had enormous consequences. Odysseus was in a sense every Greek—he was the new representative of humanity, not just in so far as he was 'a man of many devices', but also because his surfeit of cleverness had the tendency to spill over into hubris. And in so doing it then alienated him from the divine powers. This tendency was ever recognized by the Greeks, and yet they—like Odysseus—could often not help themselves from giving into it.

As Odysseus sailed away from the island of the Cyclopes, he could not resist the temptation to hurl insults at Polyphemus from the relative safety

Figure 3.3
The blinding of Polyphemus.

of his ship, cruelly taunting the wounded giant. He also revealed to him his true name: not *Oudeis* but Odysseus. Thereupon, the enraged Polyphemus stretched both his hands to the starry heavens and invoked the wrath of his divine father, the sea-god Poseidon, that it fall upon Odysseus and cause him endless trouble, suffering and distress. It was a kind of baptism for Odysseus, for the meaning of the name *odysseus* that until this point he had kept secret from Polyphemus is in fact 'man of pains'.[21] Against this curse Odysseus had no power, and as a result he was condemned to wander the seas constantly opposed by Poseidon, the spirit of the ocean depths, and it was another ten years before he finally arrived at his homeland.

Chapter Four

HARNESSING LOGIC
TO THE PURSUIT OF TRUTH

Philosophy and Technical Thinking
The story of Odysseus related in the last chapter was extremely popular
during the sixth and fifth centuries BC, when it was learnt by children,
recited by rhapsodes and its many episodes depicted on vases. It is no
coincidence that this was also the time when philosophy first began to
emerge in Greece. The *Odyssey* extolled living on one's wits, which
during this period was just what the Greeks were coming to value more
and more. There was an increasing admiration for cleverness, or *technê*, a
word which, as we have seen, can be translated both as 'art' or 'skill'
applied to the making of things, and equally as 'craft' or 'cunning'.[1] Our
word 'technology' derives from this Greek word *technê*, and from its
etymology we can see that it lies in the nature of technology to embody
not only physical skill but also mental cleverness or cunning.

Another word close in meaning to *technê* was *mêchos*, 'an expedient' or
'a means'. *Mêchos* is the root of *mêchanê*, a 'contrivance' or 'device.' We
have already noted that Homer described Odysseus as *polymêchanos*, 'a
man of many devices'.[2] When it was later translated into Latin, *mêchanê*
became *machina*, and so it is the origin of the modern 'machine'. Like
technê, the range of meanings denoted by *mêchos*, and its derivatives,
covered not just something objective that one could point to, but also a
mode of consciousness. *Mêchos* was the root of several verbs, adjectives
and adverbs, which all carried the connotation of being crafty, clever or
cunning.[3] This duality of meaning was carried over into the Latin word
machina, which meant both 'a contrivance' or 'a device' and also 'a trick'
or 'a stratagem'. The Latin brought out the moral ambivalence that was
now seen to characterize this mode of consciousness. The Latin verb
machinor, for example, meant both 'to devise' and also 'to plot' with
malicious intent.[4] This moral equivocation continued in the English
word 'machine', which until the seventeenth century was both a noun
denoting an engine or apparatus and also a verb meaning 'to plot', 'to
scheme'.[5]

All of this points to the fact that the advances in technology made by
the Greeks (to which we shall shortly come) depended on the devel-

opment of *a type of thinking* that it was their destiny to cultivate both to a greater degree than, and in a different way from, the cultures that pre-ceded them. This type of thinking was characterized by a specific kind of intellectual focus that was originally confined to the making of things (craftsmanship) in the pre-Greek cultures, but was liberated from this confinement by the Greeks. Once released into the sphere of the daily conduct of life, it came to undermine the older ethos of pious living, so well summed up by Hesiod when he wrote:

> That person is truly blessed and fortunate
> who works blamelessly before the gods,
> observing birds and not overstepping taboos.[6]

The observation of birds was of course nothing to do with 'bird watching' in the modern sense, but was rather a divinatory practice for ascertaining the divine will by carefully taking note of their direction of flight, their movements and cries.[7] The more the exercise of human thinking and planning came to be valued in its own right, the more the need to check the disposition of the gods and align oneself with it came to be seen as superfluous. Indeed, the very notion that the flight, movement and cries of birds could indicate the will of the gods, came increasingly to be seen as questionable.[8]

To understand the profundity of this shift in consciousness, we must realize that thinking in abstract concepts, devoid of any pictorial, symbolic or religious content, and driven with increasing momentum by the logical association of one thought with another, had the effect of creating an ever widening gulf between human beings and the gods. Once the mythical and religious element was expunged from the thought process, and the association of one thought with another was no longer infused with the piety of old, a certain hardness insinuated itself into the human soul. Inevitably, the religious outlook would soon itself become the target of the new thinking.

The sarcastic remark of the sixth century BC philosopher Xenophanes, who said that if cattle or horses had gods, they would imagine them in the form of cattle and horses, is an early example of how undermining of traditional religious piety the new breed of thinkers could be.[9] In the hands of some of these early philosophers, the burgeoning capacity to think critically became like an acid that was thrown mercilessly at the old religious consciousness. Later philosophers, like Critias and Prodicus, who lived during the following century, sharpened the attack on the foun-dations of the traditional religious worldview. Critias argued that the gods were simply a human invention, created to deter people from doing wrong in secret.[10] Prodicus attempted a rational explanation of the origins of the gods as the mere personifications of what human beings found useful—the god Dionysus personified wine, Hephaistos personified fire, the goddess Demeter personified bread, and so on.[11] In each case, the attack on the gods was at the same time an assault on the imaginative and feeling consciousness, which was pierced by the barbs of an ever more fiercely acerbic intellect.

This intellect had no time for religious sensitivities. It needed to cut through the older, more spiritually attuned awareness that, being tied to images, was more dreamlike, and less intellectually acute. But it had a devastating effect on the life of feeling at the heart of religious piety.[12] For this reason, it provoked a strong reaction to philosophy from the religious establishment, which saw the emphasis that philosophers placed on the individual's use of reason as a kind of impiety. In the heightened atmosphere of late fifth century BC Athens, with philosophers and priests sometimes in open contention, logical reasoning—especially when bent towards giving rationalistic explanations of the gods—was regarded by traditionalists as neither trustworthy nor reliable. It characterized a type of consciousness that, in claiming a new autonomy from the world of spirit, thereby courted terrible danger for the human community.

It was in fact primarily religious hostility that stood behind the late fifth century prosecutions of such philosophers as Anaxagoras, Protagoras, Diagoras and Socrates, leading to their exile or death.[13] Behind the hostility to the new thinkers was the perception that there was something of the trickster about logical reasoning, something mercurial, which would ride roughshod over the honest life of religious devotion, and it was this that led to a widespread distrust of it.

This distrust of logical reasoning was aggravated by the activities of the Sophists, who taught the art of logical argument and successful disputation for a fee. Protagoras was one of the more famous Sophists, well known for his boast that he could always make the worse argument prevail. His attitude was typical of Sophism, which viewed logical reasoning as an art or skill (i.e. a *technê*) that one could use to gain a desired result, whether or not this was true.[14] In the hands of clever Sophists, paradoxical or false propositions were made to seem true by means of spurious logical arguments, which were like traps into which skilled logicians led unsuspecting enquirers. In his dialogue, *Euthydemus*, Plato gives a lively portrait of two Sophists who run rings around one such befuddled seeker after truth, who finds himself having to assent to such propositions as 'no one can tell a lie' and that 'the father of this man is a dog'.[15] The Sophists were less concerned about using logic to serve truth than they were about the art or *technê* of winning debates. This is why they were regarded as teaching people how to lie and deceive, rather than as guiding them towards goodness and truth.

It was for this reason that Plato distinguished mere technical knowledge from knowledge of the Good. He pointed out that it is the nature of *technê* to apply to a particular field of knowledge or expertise, rather than to the moral development of the practitioner, who is at liberty to misuse his knowledge. For Plato, the skills of the Sophist in argument were no different in kind from the skills of artisans and builders, who by virtue of their knowledge acquire the ability to intentionally make things badly just as much as to make them well. Thus, according to Plato, the skill or *technê* of the doctor or the teacher, like that of the Sophist, can be used for good or ill: in itself, *technê* neither assumes, nor does it necessarily embrace, morality. It is without conscience. By contrast, when we really and truly know the Good, which for Plato is a spiritual principle or *archê*, it possesses us as much as we posses it. True knowledge means integration of what is known, and so those who know the Good must find it impossible to act badly or immorally. Knowledge of the Good is therefore not a *technê* like the knowledge of building, carpentry or medicine. It is a different *kind* of knowledge, and indeed a different *level* of knowledge, higher in the scale of value than *technê*, for without it no *technê* is of any genuine worth.[16]

A similar view is to be found in Aristotle. 'All *technê*', he says, 'is a type of making under the guidance of reason.'[17] Craftsmen know the rational steps necessary to arrive at what they set out to produce, and this is why they can teach others. But *technê* does not lead to knowledge of ultimate principles or to any real wisdom, for the study of Being as such lies beyond the scope of any particular science or skill. It is beyond the reach of *technê*.[18] We have already seen that Parmenides, who lived a century before Plato and Aristotle, and wrote at just the time when the 'crafty' thinking of Homer's Odysseus was being so widely lauded, also insisted that beyond craftiness there is a further capacity intrinsic to thinking: its capacity to grasp Being as such, 'for the same thing exists for thinking and for Being'.[19] Thus in sixth and fifth century BC Greece, alongside the emergence of the new technological attitude with its accentuation of cleverness and instrumental thinking, there also emerged a deeper metaphysical attitude that saw the capacity to think as leading us to an apprehension of the ground of Being itself.

Harnessing Logic to the Pursuit of Truth

For philosophy to establish itself as genuinely directed towards the attainment of wisdom, it needed to clarify its relationship with disputational reasoning. On the one hand the art of disputation was quite distinct from the essential enterprise of the pursuit of wisdom, which was understood to be the true meaning of philosophy. On the other hand, it was indispensable to it, for philosophy was a path of knowledge based on the rigorous application of thought rather than on the activation of the visionary imagination. In so far as the philosophical impulse expressed a new desire to orientate to life through the faculty of reasoning, rather than through the more ancient means of divination, dream interpretation, prophetic awareness and the reading of omens, logical thinking was critically important. But in so far as philosophers saw the kernel of the philosophical life as the intensification of the activity of thinking to the point at which it becomes spiritual insight, logical reasoning was seen as a limiting factor to be transcended. This is why logic was never regarded as part of philosophy, but rather was viewed as a capacity of the mind that could be developed, or equally as a skill (or *techné*) to be learnt.[20]

It was only through the efforts of Socrates, Plato and above all Aristotle that the burgeoning ability to persuade by reasoning ceased to be a loose cannon firing in all directions, and was constrained by explicitly formulated laws of thought to serve the truth. The three main laws of thought formulated by Aristotle are referred to as the Law of Identity, the Law of Contradiction and the Law of the Excluded Middle. The Law of Identity simply states that a thing is identical with itself, and is not another thing, as expressed in the statement 'A is A'. The Law of Contradiction goes a step further by affirming that a thing cannot both be and not be the same thing at the same time and in the same respect. Thus, according to the Law of Contradiction, the two statements 'A is B' and 'A is not B' are mutually exclusive. The Law of the Excluded Middle goes an incremental step further by affirming that it is not possible that there should be anything *between* the two parts of a contradiction. In other words, there can be no *third* option in a logic built up on the basis of the Law of Contradiction. A thing either is or is not the case. Either A is B or A is not B: there is no third alternative.[21]

In Chapter Two, we saw how the principle of 'exclusive disjunction' was applied to the sphere of technology in simple machines such as the lever and the *shaduf*. In the three laws of thought formulated by Aristotle, we see it applied to the sphere of thinking and the quest for truth. We would do well to pause to consider the impact that such laws of thought would have had on the old consciousness for which the gods were still able to manifest in many different guises, and be experienced in many different phenomena. For the ancient devotional consciousness the Law of Identity was constantly contravened, for nothing was ever securely just one thing, but could very well be a manifestation of something else. The presence of the gods in the midst of the ancient world was similarly dependent on there being no Law of Contradiction or Law of the Excluded Middle to curtail awareness of the mysterious ambiguity of all things. If someone saw a horse beside a river, they might well suppose it was not simply a horse but the god of the river in the form of a horse. The rainstorm was not presumed to be merely a physical phenomenon of rain, but was regarded as potentially a divine manifestation. And so it was for all natural phenomena. Things could both be and not be what they were. There was an inherent fluidity in the perception of phenomena that allowed them to continually suggest 'a third alternative'—a *tertium quod non datur.*[22]

The three laws of thought constituted one pillar of the new discipline of logical thinking. The other pillar was constituted by the principles of syllogistic reasoning, in accordance with which one thought could legitimately be associated with another. This question of how one thought could legitimately be associated with another thought was crucial to Socrates, Plato and Aristotle. Unlike the Sophists, they saw the need to bend the art of disputation to the service of truth rather than to that of personal advancement. Truth had to be made the criterion for winning an argument. Thus in Aristotle's *Organon* (his collected treatises on logic) the process of associating one thought with another was brought under completely impersonal rules that either granted or withheld validity to any given thought sequence, being concerned with its logical form rather than its actual content. And so people were shown how to conform their thinking to the objective figures and moods of syllogistic reasoning.[23]

Consider, for example, the following sequence of thoughts concerning horses (now without any reference to possible divine associations): 'All horses have four legs; this animal has four legs; therefore this animal is a horse.' While, on first hearing, this argument may seem to have an air of plausibility about it, we (hopefully) sense that something is wrong with it. For while the first and second statements may both be true, it is not necessarily the case that the third statement is true: simply having four legs is not sufficient to make an animal a horse. Were we, however, to put the statements in a different order, the sequence of thoughts could be made valid as follows: 'All horses have four legs; this animal is a horse; therefore this animal has four legs.' The key principle of syllogistic reasoning is that whatever is asserted (or denied) about any whole is asserted (or denied) about any part of that whole.[24] In our example 'four-leggedness' characterizes a much larger group than simply horses, including lions, dogs, sheep and many other animals. So while we may say that all horses have four legs, because horses are one member of the whole group of 'four-legged animals', it would be illegitimate to conclude that every four-legged animal is a horse. When, however, we place before our mind the two propositions 'All horses have four legs' and 'This animal is a horse', we know with complete certainty that (unless the animal has been in an accident or was born deformed) *because* this animal is a horse it *must* have four legs. Something lights up in our minds when we think it through, and this light is the light of compelling truth to which we, as rational beings, are obliged to submit. That is to say, we have an inner experience of necessity that pertains to the process of thought.

It was Aristotle who formulated the rules of syllogistic reasoning, by which we may legitimately move from a major premise to a minor premise and then to a conclusion. He saw exactly what it was that made one sequence of thoughts valid and another invalid, and he devoted several treatises to explaining these fundamental rules of logical thinking. And yet, if we stop to reflect on what an ancient Egyptian would have made of the *Prior Analytics*, Aristotle's major treatise on syllogistic reasoning, we would undoubtedly find an enormous resistance to the introduction of an entirely impersonal and non-divine factor into the manner in which thoughts are associated with each other. The rules of syllogistic reasoning served to exclude the gods from the processes of consciousness. While this may have been experienced as a liberation by the early philosophers, because the innate authority of reason seemed to them greater than that of any of the gods, the question of the ultimate identity of this new factor to which the human mind was now obliged to bow was of course crucial. For it was as if the human mind came under the influence of something that claimed greater authority than the gods, and asserted its hegemony over conscious thought-processes. Through it, a new certainty in thinking became available to us, enabling greater clarity and independence of thought to become the norm for the inner life. And yet, in the depths of the soul, where knowledge is experienced *feelingly*, a new coldness crept in.

Logical thinking, which had in earlier epochs been regarded as pro-fane, was perceived by the Greek mind as having the potential to serve a spiritual purpose, namely to guide human beings to an experience of the truth. And this inner experience is what Aristotle feasted upon in his elaboration of the rules of syllogistic reasoning. Through the differ-ent figures and moods of the syllogism that Aristotle was able to articu-late, a certain demand was placed on the intellectual acuity that so typified the consciousness of the Greeks at this time. The demand was that—at least in the process of reasoning—this intellectual consciousness should be obliged to submit itself to criteria that transcended not only the archetypal influence of the gods within the psyche, but also the personal desires and preferences of the now much stronger individual ego.

Nevertheless, the introduction of formal logic constituted an important step towards the establishment of something alien and mechanical in the midst of the realm of human thought. Aristotle's three laws of Identity, Contradiction and the Excluded Middle imported into the inner life of the human soul criteria of thinking that apply to the relationships of physical objects considered merely as mass occupying space, stripped bare of meaning. In the same way, syllogistic reasoning imported into the inner life an essentially mechanical activity, albeit driven by the human will to truth. To the extent that logic was used as a means to attain truth, the mechanical coldness of syllogistic reasoning was requisitioned by the genius of Aristotle to serve a spiritual end. But when eventually logic was released from this spiritual constraint, then the way would be open to build a 'thinking machine' that would utilize the power of logic to serve a thousand other ends.

The Light of Contemplation

Both Plato and Aristotle held that the ability to think is what makes human beings spiritual beings, but they did not see this ability as confined merely to logical thinking. It is important to bear this in mind when contemplating the claims that are often made today that computers replicate the process of human thought. Logical reasoning was regarded neither by Plato nor by Aristotle as the full extent of the human capacity to think. Both saw logic as needing to be shepherded by a more profound intuitive experience of the nature of thinking as such, in which an intimation of our spiritual essence is given to us. This experience cannot be captured by a mechanism, for it transcends the mechanical just as surely as logic veers towards it. Plato and Aristotle regarded *contemplative thinking* as our spiritual ground, through which we come to ourselves—to that part of ourselves which is at the centre of our being, referred to as the *nous* or deeper intelligence. For Plato, only those decisions and actions that spring from this deeper intelligence are truly free.[25] Likewise for Aristotle, our freedom lies in carrying back the origin of our actions to the 'dominant part' of the soul, the *nous*, for 'it is this part that chooses'.[26] To live from this deeper intelligence, rather than simply to acquire skill in logic, is the real key to both human happiness and to inner knowledge of our spiritual essence. Consider this statement of Plato:

If people who have been earnest in the love of knowledge and of true wisdom, and have exercised these aspects of themselves above all [by engaging in contemplative thinking], then their thoughts cannot fail to be immortal and divine, should the truth have come within their grasp. And in so far as human nature is capable of partaking in immortality, they cannot fail to achieve this; and since they are constantly caring for the divine power, and have the divinity within themselves in perfect order, they will be singularly happy.'[27]

This statement could not have been made by an ancient Egyptian or Mesopotamian, but neither could it have been made by a Sophist. It was only possible for someone who had had an experience of the nature of thinking as an activity beyond mere craftiness or cleverness (however well-disciplined). It involved the realization that the activity of thinking is a matter not just of successful disputation, but also of the inner experience of a transpersonal power: the divine or daimonic within us.[28] This is why Plato distinguished between disputational reasoning that proceeds through logical deduction or inference, which he termed *dianoia*, and intuitive understanding or insight, which he termed *noêsis*.[29] *Dianoia* was for Plato a skill or *technê* that is a means to attaining a given result. By contrast, *noêsis* is an end in itself, achieved through an intensification of the thinking process to the point at which it becomes a contemplative 'dwelling within' its object, a transcendent Idea or Form.

For Aristotle, too, mere logical argument (*dialektikê*) which is forever caught in the manifold nature of reality, has to be distinguished from the metaphysical thinking of the contemplative mind (*nous*), which is directed toward Being as such.[30] Like Plato, Aristotle was certain that the exercise of contemplative thinking (*theoretikê*) was the key to human happiness, precisely because it is a divine power within us:

It is the activity of contemplation that constitutes complete human happiness ... Such a life as this however will be higher than the human level: not in virtue of their humanity will people achieve it, but in virtue of something within them that is divine.[31]

Both philosophers understood that while logical reasoning forced people to wake up and be alert, it had its limitations and certainly did not constitute the whole of what it means to be rational. Unless it was guided by a type of thinking that transcended the thought-processes of discursive reasoning, and brought the mind to a participatory indwelling in its object, logical thinking could only open the door onto, but never actually occupy, the domain of spirit.

The domain of spirit accessed by contemplative thinking was conceived as a realm of light. In the *Republic*, Plato argued that just as the sun's light is essential if we are to perceive objects in the world around us, so the Idea of the Good illumines the inner world in which the mind (*nous*) forms true conceptions and comes to know the essence of things. The Idea of the Good shines with a spiritual light, as an inner sun, which enables us to comprehend the depths of existence and reality.[32] This is not just a simile. Plato explicitly states that the Idea of the Good gives birth to the light we experience in the visible world. It is the spiritual source of the light that illumines the world.[33] Therefore in the light that permeates the visible world we have an intimation of an omnipresent world of spirit, which we may also encounter inwardly in our life of thought. Aristotle does not go so far as Plato in identifying the Idea of the Good as the origin of light, but he does refer to the activity of thinking as being like light, which he asserts is not in its essence corporeal.[34] In so far as thinking enables understanding to take place, it has a power of illumination similar to that of light. Thinking is, in other words, a luminous medium with affinities to the light that illumines the outer world, for it illumines the thoughts that we conceive in our minds. And should we become fully concentrated within the activity of thinking itself, we would experience something that lives within us that is entirely spiritual and therefore immortal.[35]

Both Plato and Aristotle, then, recognize that thinking is much more than the ability to reason logically, for at its heart it holds the possibility of becoming a contemplative act, through which we step into a purely spiritual dimension, and through which we gain our freedom. But the fact that they also associate this contemplative level of thinking with the experience of light has a special significance, for it encouraged a profound reverence for the light as a manifestation of the divine. If, for Plato, the sun is the offspring of the Idea of the Good, this also implies that the sunlight that illumines the visible world has an intrinsically moral quality. Other ancient thinkers, such as Plotinus, also held that light is essentially spiritual in its nature, and is the 'first activity' of the One, which Plotinus equates with the Good. This 'first activity' is a 'generative radiance' which he identified with the formative principles or archetypes that lie behind and within all material forms.[36] For Plotinus, the spiritual light is also the medium of contemplative thought, by which we come to know; so human knowing arises through a participation in this formative, creative and intrinsically moral principle of light.[37]

In subsequent chapters we shall see that this reverence for light continued into the Middle Ages, thanks to Christian theologians and mystics like St Augustine and Dionysius the Areopagite, who well understood the connection between light and contemplative thinking, and transmitted their understanding to the medieval world. Only with the Reformation of the sixteenth, and the Scientific Revolution of the seventeenth centuries did this understanding begin to be lost. It was during this period that a new emphasis was given to the activity of discursive reasoning and analysis, to the detriment of the faculty of contemplative thinking, which became increasingly marginalized. At the same time, as we shall see, the emerging scientific consciousness turned its attention towards the study of electricity, so vital for the development of the computer. It is therefore worth noting how profoundly the modern conviction in the fundamental role of electricity at the very heart of material existence differs from the contemplative traditions of antiquity, which elaborated a worldview in which light was assigned a pre-eminent place within the order of nature, conceived as both material and spiritual.

Chapter Five

TECHNOLOGY IN
THE GRECO-ROMAN AGE

Logic: the Driving Force of Technology in the Greco-Roman World

Once Aristotle had successfully formulated the rules of correct logical thinking, the breach between the new intellectual consciousness and the old religious consciousness based on myth, imagination and feeling could not again be closed. At one and the same time, Aristotle co-opted logic into the service of truth, and also hugely strengthened the logical component in human thinking. Logic now became the driving force that powered not only philosophical but also scientific and technological investigations. What Euclid was able to achieve with such succinctness and translucent clarity in his *Elements*, and what Archimedes was able to set forth in his treatise *On the Equilibrium of Planes* were only made possible through the prior work of Plato and Aristotle in the spheres of mathematical thought and logic respectively. If Plato made mathematics the servant of metaphysics, Aristotle enabled logic to be deployed in the service of the exact sciences.

In his treatise *On the Equilibrium of Planes*, written around the middle of the third century BC, Archimedes set out a series of postulates concerned with the determination of the centre of gravity in various geometrical figures. He thereby explained how contraptions like the *shaduf* worked, but he did so in an entirely theoretical and abstract manner. Consider the quality of thought that informs the following statements:

> Equal weights at equal distances are in equilibrium, and equal weights at unequal distances are not in equilibrium but incline towards the weight which is at the greater distance. If, when weights at certain distances are in equilibrium, something be added to one of the weights, they are not in equilibrium, but incline towards that weight to which the addition was made.[1]

In these sentences we feel ourselves to be in another world entirely from that inhabited by the ancient Egyptians. The goal of the Greek scientist (Archimedes was not the first) was to formulate in clear concepts the principles underlying phenomena such as leverage, without any concern for the realm of gods and spirits. But neither, in this treatise, was Archimedes concerned about the practical uses to which the lever might be put: his was a purely theoretical approach.

Archimedes was, however, by no means just a theoretician for as is well known he was also a superb engineer. But he did not bequeath to us any written treatises on engineering. This was because he regarded it—along with all other utilitarian pursuits—as 'sordid and ignoble' compared to theoretical science.[2] This did not stop him from creating some ingenious machines, many of which were dedicated to the military defence of his native city Syracuse. But his attitude was typical of the Greek scientist during this period. Mechanical problems were studied for their own sake as essentially theoretical questions, and it was largely this attitude—that logical thought be directed towards gaining pure knowledge or *epistême*, rather than towards useful and practical applications—that pre-dominated.[3]

Nevertheless, in Archimedes and in such figures as Ctesibius of Alexandria and Philo of Byzantium, who were both his contemporaries living during the third century BC, logical thought irrepressibly sought out and allied itself to mechanics. The result was an unprecedented flowering of inventiveness in the sphere of mechanical engineering. During the third century there was an outpouring of water clocks, pumps, valves and siphons, cogwheels, cams, watermills, screw-pumps, springs and compound pulleys. At Syracuse, Archimedes created military engines so terrifying that the Romans who were laying siege to the city believed they were fighting against gods.[4] Thus was the daemonic power of mechanical logic experienced as in some sense unleashing inhuman agencies.

Watermill and Camshaft

It has often been remarked by historians how little these inventions were put to widespread practical use. As well as certain military contraptions, however, there is a good deal of evidence to suggest that the watermill, with its cogs and wheels, was in fact put to use (but probably not in great numbers) throughout the Roman Empire from around the first century BC onwards. Vitruvius gives the earliest description of one towards the end of the first century BC, while the first archaeological evidence is from the early first century AD.[5] The mill described by Vitruvius is shown in Fig. 5.1.

Just how many watermills there were in Roman times is almost impossible to say with certainty, owing to the relative paucity of literary references and archaeological remains. To date only 56 sites have been

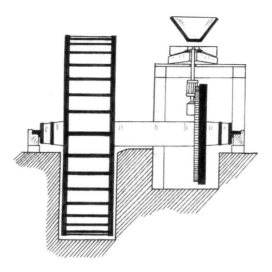

Figure 5.1

In the watermill described by Vitruvius the wheel (left) sits in a wheel-pit through which water rushes, causing the horizontal wheel-shaft to turn a second cogged wheel (right). This engages with a lantern pinion above it that turns a vertical iron spindle, which drives the millstone round and round. Above the millstone is the hopper (top right) containing the grain to be milled.

authenticated, but the fact that they were used for both the grinding of corn and for industrial purposes (for example, crushing ore) is indisputable.[6] This is important for two reasons. First of all, we see how within just a few centuries of the formalization of the rules of logic, the application of logical thinking to the construction of complex machines that substitute purely mechanical actions for human actions becomes a reality. Secondly, we may also notice that the functioning of the industrial watermill depended upon a primitive form of programming, and should therefore be regarded as a distant precursor of the computer. This needs a little explanation.

In order for the watermill to become an effective tool in industrial processes, the rotary motion suitable for grinding grain had to be converted into linear motion. For this to happen the cam, which had already been invented during the third century BC, had to be employed. The cam is a lobe, or projection, on a rotating shaft that has the effect of translating rotary motion into a linear (up and down or back and forth) motion. Through incorporating cams, the watermill could be used to operate triphammers and thereby carry out pounding actions (Fig. 5.2).

In the pounding of cloth in the fulling industry and in the crushing of ore in mines, up and down motion was required. In the sawing of wood

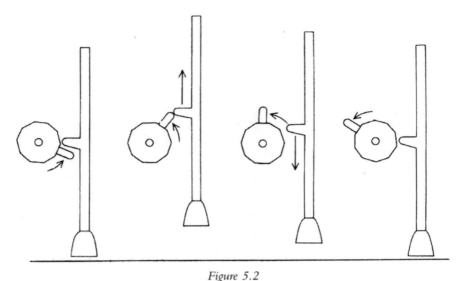

Figure 5.2

The operation of the cam turns rotary motion into linear motion. The rotating cam engages a matching cam projecting from a hammer, first raising and then dropping it.

in sawmills, back and forth motion was needed. All three utilized water power for cam-shaft mediated mechanized action.[7] Cams do not necessarily have to come in one shape and size. Both the number and the shape of the cams determines the movements the trip-hammer makes, and the cams may be so designed that the movement of the trip-hammer to which they are in relationship is effectively programmed by them. If, for example, the hammer meets two cams per cycle, this will result in two hammer blows; if it meets three cams, three hammer blows. Furthermore, depending on the shape of each cam, the movement of the hammer up and down will be more or less punctuated. Depending on the speed of the rotation of the shaft on which the cams are mounted, the sequence of hammer blows will be faster or slower. As one commentator has observed:

> The camshaft provides a natural potential *to program movement*. It involves a kind of mechanical memory for cyclical automated processes (the cyclicity being provided by the rotation of the shaft).[8]

Already then, in ancient Roman times, through the deployment of the cam, this potential to program movement was being actualized in the industrial watermill. While the evidence is undoubtedly thin both on the ground and in the literary sources, there is nevertheless sufficient evidence from both sources to make it highly probable that watermills were used in

Roman times at least to a limited extent in the fulling, mining and timber industries.[9] A significant breakthrough into automation had been made, even if it was not widely exploited. We shall return to consider further the significance of the cam in Chapter 8.

Why Was There No Industrial Revolution In Roman Times?

By the first century AD, the range of mechanical contraptions that had been invented is quite extraordinary. Hero of Alexandria notably designed a number of automata, employing cog-wheels and gears, cranks and cam-shafts. He even went so far as to invent a primitive steam engine, but it seems not to have been put to any practical use.[10] One has the feeling that all this engineering brilliance (of Hero and others before him) was to the inventors more a kind of play than a serious attempt to address industrial needs. Certainly the watermill stands out as an application of technological thinking to increase industrial efficiency, but one is struck by how many of the inventions were, as historian Henry Hodges observed, merely 'interesting gimmicks', for use in temples or theatres to mystify or amaze those present.[11] 'At no time,' writes Hodges, 'does it seem to have crossed the minds of these ingenious men that their inventions could have been used to provide new sources of power or to make industry more efficient, and yet they came within a hair's breadth of a real industrial revolution.'[12]

In this 'hair's breadth' lies all the difference in consciousness between the ancients and ourselves. Why was all this technological inventiveness not put to greater practical use? Clearly the Romans appreciated the practical value of the watermill, but the evidence suggests that it was not exploited nearly to the extent that it was in the Middle Ages. Why not? And why, after its discovery by Hero of Alexandria, did the Romans not go further and develop steam-power? One reason is that the ancient experience of nature was permeated by an awareness of—or at least a residual belief in—gods and spirits. Because of this, there must have been considerable resistance to the introduction of machines in place of human labour and the unmediated human relationship to the natural world.[13] One of the early examples of a literary reference to a watermill is in a poem by Antipater of Thessalonica (written between 20 BC and AD 10), which refers to the water nymphs that drive it, under the command of the goddess Demeter.[14] Despite the fact that the poet is describing a mechanism, the gods and spirits are to him the main forces involved in its functioning. Thereby a limit to the effectiveness of machines and a merely mechanical approach to nature was presupposed.

Another reason is that even for those thinkers who were thoroughly at home in mechanical thinking, the design and production of machines was not seen as a noble or worthwhile end to be pursued for its own sake. We have already noted that Archimedes believed that the only worthy goal of intellectual knowledge was theoretical understanding, not practical application. And this was typical of the ancient authors, for whom contemplative thinking was considered far superior to technical thinking. The practical uses to which mechanical devices could be put was less important to scientists like Ctesibius, Philo and even Hero of Alexandria (who lived much later than them during the first century AD), than the development of pure theory. Typically, what interested Hero was *why* mechanics works. He was looking to understand the motive force behind mechanics, which he perceived to be so closely bound up with the nature of logical reasoning that he referred to it as no different from reasoning itself or *logismos*.[15]

For the early mechanical engineers, true knowledge involved rendering in conceptual form the principles underlying the realm in which mechanics operated, which brought them into the sphere of geometry rather than technology. Archimedes' analysis of the laws of leverage in his *On the Equilibrium of Planes* is a case in point. At the same time they could not fail to see that the logical thought which they employed to reach such theoretical understanding was equally applicable to the practical engineering challenges of building physical machines. And yet they regarded this as a debasement of true knowledge because what was true on the level of pure theory only worked practically when certain adjustments were made to take account, for example, of friction and air resistance.[16] For Archimedes and the others, *ideas* had more reality, and pure theory had more value, than the concrete things in which the ideas became embodied. Thus mechanical thinking was itself held spellbound by a mentality that despised practical applications. For in that direction lay the labour of slaves, not free men.[17]

Ancient Astronomical Computing Devices

Along with the watermill, the other invention of significance for the development of the computer was the first astronomical computing device, capable of predicting the movements of the Sun, Moon and five planets. Several of these were produced in the third century BC. Archimedes constructed one of them, which survived for at least 150 years, for it was shown to Cicero in the first century BC.[18] It was the precursor of the so-called Antikythera mechanism, which was retrieved from the sea in 1902, and which was constructed circa 100 BC, more than 100 years after Archimedes' death. Thus the Antikythera mechanism was the product of a roughly 150 year tradition of complex mechanical technologies going back to the third century BC.[19] The mechanism (Fig. 5.3) demonstrates a very high degree of technological proficiency, for in its compact design it had over 30 gears, three dials and a crank, by means of which it was able to calculate the position of the Sun and Moon, and hence solar and lunar eclipses, and probably also the location of the visible planets.[20]

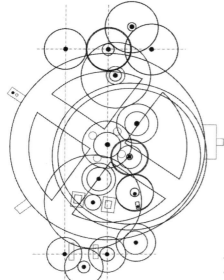

Figure 5.3
Schematic diagram of the Antikythera
mechanism. Circa 100 BC

Apart from the intricate craftsmanship that went into its construction, the purpose that the Antikythera mechanism was designed to serve was extremely ambitious. By materializing the contemporary mathematical models of the movements of the celestial bodies, it sought to capture the motions of the Sun, Moon and planets in its mechanism, thereby enabling predictions of eclipses, and so on, to be made. This, of course, was something that astronomers could already do through mathematical calculation. But now the mechanism, through its crank, gears and dial, freed them from the mental labour of having to make their calculations. In this respect, the mechanism should be regarded as a sophisticated mechanical calculator. As with the other inventions already discussed, like the cog-wheel, cam-shaft and steam engine, which brought the Greco-Roman world to within a 'hair's breadth' of an industrial revolution, we stand amazed before the Antikythera mechanism. We feel compelled to ask: Why were there so few of them? Why did they not develop further?

Why did we have to wait until 1642 until Pascal would begin again with a much cruder mechanical calculator?

One answer is that the intellectual atmosphere in 1642 was totally different from that of the early centuries BC. The mechanistic philosophy had triumphed and most educated people viewed both the natural world and the whole cosmos as a mechanism set in motion by God and then left to carry on by itself. By contrast, for the major philosophies of the Greco-Roman period (Platonism, Aristotelianism, Stoicism and Neoplatonism), the cosmos was a living organism endowed with soul, and the habitation of gods and spirits. Right through to the end of the Roman Empire, there was a vestigial, but nonetheless powerful, sense of the sacredness of nature and the cosmos. Even amongst the Atomists, there was no thought of the cosmos as a mechanism. As historian R. J. Forbes observed concerning the ancients:

> None of them would have accepted the Newtonian world-machine created by God and left to run its course, the concept so dear to the eighteenth-century scientist. Even Democritus, Epicurus, or Lucretius [all ancient Atomists] would have shuddered at such blasphemies.[21]

The Antikythera mechanism does not therefore provide evidence for an early mechanistic conception of the cosmos. The mathematical models used by astronomers at the time were regarded simply as aids for calculating the positions of the Sun, Moon and planets, and were not meant to describe their actual motion. The mathematical and the physical approaches to the cosmos were quite distinct from each other in the Hellenic period. Subsequently, in the second century AD, this distinction was more or less set in stone for more than 1,000 years by Ptolemy.[22] On the one hand, then, the models used in mathematical astronomy were meant simply to 'save the appearances' of planetary motion without any claim being made that these models gave a physical explanation of planetary motion. On the other hand the mechanistic worldview, which in the seventeenth and eighteenth centuries was embraced by the modern consciousness and which provided the matrix within which technical thinking was given free rein, was utterly alien to the Greco-Roman mentality. The spirit of the times was not yet ripe for such a development. Here we see that a history of physical computing devices that ignores the history of consciousness cannot give us any real insight into the origins of the computer. If we were looking for just one example to show why this is the case, then we would not need to look any further than the Antikythera mechanism.

Chapter Six

THE ECLIPSE OF THE MYSTERY
KNOWLEDGE OF ELECTRICITY

The Sub-Earthly Source of Electricity

Despite the enormous differences between the Greco-Roman era and the preceding era of the Mesopotamian and Egyptian civilizations, neither the Greeks nor the Romans were ready to fully exploit the new technologies that they discovered. Similarly, they were not ready to lay hold of electricity and exploit it for human benefit. It is interesting to compare the advances made in the development of logical thinking (and mechanical technologies) with the changes that occurred in the understanding of electricity in the Greek and Roman period. While in the relationship to electricity there was a similar liberation of human thinking from the earlier mythical and religious consciousness, at the same time the Greeks and Romans remained very far from modern attitudes. It would have been impossible for an ancient Greek or Roman to imagine that electricity could be generated by human beings and its energy harnessed for human benefit.

We have seen in Chapter Two that in ancient Mesopotamia and Egypt atmospheric electricity—especially as experienced in the vivid and terrifying manifestations of lightning and thunder—was understood as a destructive power wielded by the deities connected with violent storms. In Mesopotamia, these deities all had a relationship to the Underworld, the realm of death. In Egypt the god Seth, with whom the power of electricity was associated, was a desert god implacably opposed to the forces of life. In neither culture was electricity regarded as a simply 'natural' manifestation: it had a provenance that both lay beyond, and was inimical to, the form-giving forces of nature. This is why in Mesopotamia and Egypt knowledge of this power was restricted within the sacred confines of the Mysteries.

A similar attitude can also be detected in ancient Greece, at least up until the end of the seventh century BC, after which time it became gradually more difficult to sustain. Rudolf Steiner observed that 'the Greeks held all that was connected with electricity in secret within the Mysteries', and this would seem to be borne out by the evidence.[1] We tend to think of Zeus today as the mighty sky god of the Greeks, who in the modern imagination is more or less indistinguishable from the Roman god Jupiter. It may come as a surprise to learn that just as the Mesopotamian storm gods underwent a death and descent into the Underworld, so too did Zeus, at least in the earlier period of his cult, traces of which remained long afterwards. Zeus was not just a sky god but also had a chthonic aspect. In his chthonic aspect he was worshipped as Zeus Chthonios, the subterranean counterpart to the sky father Zeus and a god of the dead.[2] The initiates of his early Mystery cult, like those of the Mesopotamian and Egyptian Mystery rites, had to come face to face with the destructive forces of thunder and lightning as inner spiritual realities. We know, for example, that Pythagoras had to spend 27 days (three times nine) in an underground cavern in Crete before being permitted to look upon the throne of Zeus and witness the dazzling destructive power of the thunderbolt.[3] The Cretan thunder-rites involved three stages, beginning with purification, probably in silence and darkness, followed by hearing at close quarters the terrifying sound of thunder (simulated by various instruments, notably the bull-roarer or *rhombos*), and possibly climaxing in the blinding flash of the thunderbolt itself.[4] Both the long initial purification and the dramatic way in which the second and third stages of the rites were conducted ensured that a powerful inner experience of these Underworld forces was achieved.

In Hesiod's *Theogony*, written some time towards the end of the eighth century BC, the story is told of Zeus attacking and defeating both the primeval Titans and the monster Typhoeus (or Typhon) using thunder and lightning as his main weapons. These battles were not disimilar to those fought in Mesopotamia by the god Marduk against Tiamat, which formed part of the Babylonian Mystery rites. Hesiod's account has much in common with the earlier Babylonian creation epic.[5] The battle of Zeus and Typhoeus that he describes is depicted in Figure 6.1 below, from a mid-sixth century BC vase painting. The thunderbolt-wielding Zeus and the serpentine yet winged Typhoeus clearly resembles the depiction of the battle of Marduk and Tiamat (Fig. 2.8), and suggests a commonality of imaginative experience that the myths shared.

According to Hesiod, the Underworld forces were personified in three primeval Cyclopes who represented the threefold being of electricity as experienced in sound (thunder), light (the flash of lightning) and the annihilating power of the thunderbolt.[6] Along with the Titans and the hundred-handed giants, these Cyclopes were the offspring of the first generation of gods, Uranus and Gaia, and were confined by their jealous father Uranus within the depths of the earth. There they remained until Zeus freed them when, in his struggle against his father Chronos and the

Figure 6.1
Zeus hurls a thunderbolt at Typhoeus. From a black-figure vase. Circa 550 BC.

Titans, he sought their allegiance. In gratitude for being set free, the Cyclopes bestowed their powers—the thunder, the lightning bolt and the eerie flash of light—upon Zeus, who was consequently able to achieve victory.[7]

That the provenance of the three primeval Cyclopes was a region deep within the interior of the earth confirms the sub-earthly or sub-natural source of electricity: it was precisely because it belonged to the Underworld that to know it required a descent into the interior of the earth. According to Hesiod, the region in which the Cyclopes dwelt was as far beneath the earth's surface as the heavens are above it. This region was named Tartarus, and was located 'in the depths of the ground (*chthôn*)'.[8] Although located within the earth, in using the word *chthôn* (which prefigures the modern conception of 'matter') rather than *Gaia* (the earth goddess), Hesiod implies that Tartarus does not belong to the living organism of the goddess Gaia. For even the roots of the earth (Gaia) do not extend as far as Tartarus: it is located so deep within the interior of the earth that it is not *of* the earth (Gaia).[9] In the *Theogony*, Tartarus is a kind of anti-world, a mirror image of the world above, for the nine-stage journey to the heavenly realm above is exactly mirrored in a nine-stage journey down to the gates of Tartarus beneath the earth.[10] This presumably is why Pythagoras was required to spend 27 (three times nine) days in purification rituals before being exposed to the full force of this power.

Papyrus fragments discussed by Carl Kerényi in his study of the

Eleusinian Mysteries provide further evidence that knowledge of electricity belonged to the Mysteries and was gained only through an initiatory descent into the interior of the Earth.[11] According to Kerényi, an enormous gong called an *echeion*, which had a nerve-shattering effect on those who heard it, was employed in the Eleusinian rites to simulate the thunder that emanated from the Underworld regions. In Sophocles' play, *Oedipus at Colonus*, the playwright treads carefully around these rites so as not to betray the Mysteries, yet conveys sufficient for us to understand that the provenance of the mighty force of electricity is the Underworld. Terrifying peals of thunder resound as the Underworld opens to receive Oedipus at the end of his life, then lightning and more thunder, drawing closer as he prepares to begin his descent into the realm of death.[12]

Magnetism also featured in Mystery rites. It featured both in the Mysteries of the Kabeiroi (Cabiri) celebrated at Lemnos and Thebes, and in those of the unnamed gods of Samothrace (an island in the northeast Aegean sea). These unnamed gods are often identified with the Kabeiroi, who were the sons or grandsons of Hephaistos, and their rites were connected with the metallurgical crafts.[13] We know most about the rites at Samothrace, which were presided over by the 'mistress of Samothrace' whose name was Elektra, and whose initiates would wear iron rings.[14] At a certain point in the Samothracian Mysteries, it is said that a magnet was exhibited, implying that knowledge of magnetism was regarded as intrinsically sacred and therefore it too was in need of protection from profane understanding.[15] The fact that the mistress of Samothrace was named Elektra suggests the possibility that within the Samothracian Mysteries, and no doubt elsewhere, there was knowledge that the atmospheric phenomena of thunder and lightning were connected with the static electricity generated by rubbing amber, which in Greek is *elektron*. The myth of Phaethon, the son of Helios who drove his father's sun-chariot so wildly that the earth was alternately frozen and scorched, supports the view that such knowledge existed, for the story ends with Zeus hurling a thunderbolt at Phaethon so that he falls dead into the River Po. Significantly, his two grieving sisters are then changed into poplar-trees on its banks, and these trees weep amber tears.[16] In this way, three key factors—the disordered powers of the sun, destructive atmospheric electricity and amber—were brought into relationship with each other by the Greek mythical consciousness.

The Naturalistic Approach to Atmospheric Electricity

From the seventh century onwards, the ancient taboo that saw knowledge of electricity as the preserve of the Mysteries was weakened. One early indication of this is the decline of the old Zeus Mystery-rites, in which his worshippers underwent a ritual death and rebirth. Essential elements of his cult were transferred to the god Dionysus, including the ritual death and rebirth and the eating of the sacred animal.[17] Although his connection to the Underworld powers was preserved in his chthonic aspect as Zeus Chthonios, Zeus lost the important characteristic, which Dionysus retained, of being a dying and resurrecting god. This loss suggests that, by the beginning of the Archaic period (late seventh century BC), his worshippers no longer underwent the initiatory descent into the Underworld and the accompanying encounter with the spiritual forces that underlie the manifestation of electricity in the thunderstorm.

The growth of the new intellectual consciousness during the following century brought about a momentous shift in the way in which leading thinkers came to understand what constituted an appropriate relationship to such primal forces. In the writings of Anaximander, who was born in the last decade of the seventh century, the figure of Zeus himself became drained of life in the abstractly conceived principle of an indefinite and unlimited divine power, which he named the 'Boundless' (apeiron).[18] The mighty god, who used to be a living presence in the darkly gathering storm clouds, was now not only deprived of his chthonic aspect but also distanced to a much more remote, intellectually defined mode of existence. Anaximander felt the need to conceive of the godhead independently of any natural manifestation. But as much as this had the effect of detaching the god from the human being's experience of natural phenomena, so it also had the effect of detaching the experience of the natural phenomena from the god.

And so at the same time as he redefined Zeus as the 'Boundless', we find Anaximander putting forward the first naturalistic explanation of atmospheric electricity—thunder, lightning, and thunderbolts—without any reference either to the god or to the Underworld. Anaximander proposed that:

> all these things occur as a result of wind; for whenever it is shut up in a thick cloud and then bursts out forcibly, through its fineness and lightness, then the bursting makes the noise, while the rift (diastolê) against the blackness of the cloud makes the flash.[19]

Zeus was no longer to be regarded as a cause of the thunderstorm. It was the purely natural interaction of wind and cloud that caused the sound of the thunder and the flash of light. For Anaximander, knowledge of atmospheric electricity could both be sought and found without the seeker of knowledge being obliged to participate in Mystery rites.

Anaximander's approach was typical of the new, more intellectual mode of consciousness that gradually made inroads into the culture.[20] The attempt to understand electricity outside the sacred context of the Mysteries was one of the most important deeds of the pre-Socratic philosophers. Today few historians concern themselves with it. But what it meant was that the older spiritual understanding that electricity originated in the complex interactions of the supernatural or heavenly Zeus and the chthonic forces of the Underworld, and that it was within the Underworld realm that the powers of electricity were ignited, was almost entirely lost sight of. Only Aristotle retained some of the old understanding of the sub-earthly provenance of electricity. His discussion of electricity, in the *Meteorologica*, is full of subtle and penetrating insights into the ways in which electricity should be understood.[21] But his immediate successor, Theophrastus, failed to build on the insights outlined in the *Meteorologica*, and they were effectively lost from view.

Theophrastus' work, of the same title, we only know from later Syriac and Arabic versions. Anything that remains there of Aristotle's potent guiding thoughts is subsumed by a proliferation of different theories which Theophrastus borrows from earlier pre-Socratic thinkers. These emphasize the idea that the cause of atmospheric electricity is frictional (for example, the collision or rubbing of clouds together, the rotation of wind within hollow clouds, the rending of clouds when wind enters or leaves them and so on).[22] Thereafter, the most common explanation of atmospheric electricity, which the Romans subsequently embraced, was that thunder and lightning are simply physical phenomena and nothing more. They are caused either by the friction of colliding clouds or else by the violent interaction of wind and cloud.[23] For Cicero, thunder and lightning were due to both causes:

> When the winds enter a cloud, they begin to break up and scatter its thinnest portions; if they do this very rapidly and with great violence, thunder and lightning are thereby produced. Again, when clouds collide their heat is forcibly driven out and the thunderbolt is the result.[24]

And so the spiritual perspective was lost.

Amber and Lodestone

Along with atmospheric electricity, the pre-Socratic philosophers sought to de-mystify static electricity generated by rubbing amber (*elektron*), as well as the phenomenon of magnetism. We have already noted that a connection may have been made in the Mysteries between atmospheric electricity and the peculiar propensity of amber to generate static electricity. We also noted that there may have been some understanding in certain Mystery centres concerning the relationship between electricity and the forces of magnetism. During the sixth century such perspectives were lost sight of as the electrostatic propensity of amber came to be regarded as belonging to it uniquely, as one of its properties. It was seen as similar to the intrinsic property of lodestone to attract iron.

Thales, who lived during the first part of the sixth century BC, was one of the first to consider these two phenomena of electricity and magnetism outside the context of the Mysteries. He was credited with the view that, just as the Magnesian stone (or lodestone) attracts iron, so amber becomes 'magnetic' when rubbed.[25] Thales assumed that the magnet's power to attract iron was due to an inherent virtue existing in the magnet itself: it was a manifestation of its 'soul' (*psuchê*).[26] So, likewise, amber's ability to attract light objects when rubbed was due to the inherent nature of the amber. Just as with the early naturalistic accounts of atmospheric electricity, this way of thinking is no longer able to conceive how the sources of these phenomena could emanate from the subterranean Underworld regions. The spiritual understanding of electricity and magnetism through the Mysteries was gradually displaced by a mode of understanding that could only see these forces as physically confined within, and belonging to, certain natural substances. Thales, along with Anaximander and Anaximenes, was at the forefront of a movement whose aim was to free human knowledge from the older spiritual context of the Mysteries, which demanded that the human being participate inwardly in the arising of the phenomena.[27] Now the phenomena were to be encountered wholly outwardly.

Over the coming centuries, this mode of thinking became well established in educated circles. Static electricity came to be regarded as but a curious effect produced by the physical action of rubbing the specific physical substance, amber. The approach to both magnetism and electricity was to tie these phenomena to the substances with which they were associated, so that rather than attention going towards the magnetic or electrical effects themselves, it went instead to the object that produced them: to the lodestone that attracts the iron, and to the amber that attracts a range of different things to it. And so for Theophrastus, writing in the last years of the fourth century BC, amber has an *inherent* power of attraction, like the lodestone, but different from it in so far as 'it is said to attract not only straws and small pieces of sticks, but even copper and iron, if they are beaten into thin pieces'.[28] Thereby, those who sought knowledge of the forces of electricity and magnetism were relieved from the previous obligation to subject themselves to the unsettling experience of the Underworld initiation. But the new objectifying consciousness (with the exception of Aristotle, although as we have seen few were able to follow up the indications he gave) lacked the capacity to grasp the underlying principles at work in the generation of electrical phenomena. As a result, there was little interest in them. Static electricity came to be seen as just a property of amber, in the same way as atmospheric electricity was just a product of the friction of clouds and winds.[29] The genie of electricity was safely confined to its bottle, and there it would remain for hundreds and hundreds of years.

At the beginning of the chapter, I suggested that there was a parallel between the unfolding development of logical thinking and technology on the one hand, and the unfolding development of the human relationship to electricity on the other. Both are symptoms of an underlying evolution of human consciousness away from an earlier participative awareness of a world saturated with the felt presence of the gods, towards a more objectifying and detached awareness in which the realm of the gods could no longer be experienced as interweaving the phenomena of nature. As the rational function gained in strength within the human soul, and was increasingly granted priority over the feeling, imaginative and intuitive functions, so the human being achieved independence from the realm of the gods. But so also did human awareness suffer a drastic reduction in its scope, for the spiritual dimension—once so tangibly present to human beings—slowly but surely faded from view. It is interesting that a number of thinkers during the Roman period felt inwardly moved to write treatises on the gods, a favoured title being *On the Nature of the Gods*. The gods, whose presence everyone used spontaneously to acknowledge and revere, had now become so far removed from human sensibility that they could become subjects of learned treatises.[30] But if the old dispensation of consciousness could not be sustained, more than 1,000 years would have to pass before human beings would be able to begin to work systematically towards the development of a thinking machine. In the following chapters, we shall look at how the gathering impetus towards this development was held in check during the medieval period, and how it was only with the dissolution of the medieval outlook that the way was then open for the idea of a thinking machine to be realized.

Part Two

THE MIDDLE AGES

Chapter Seven

GRAMMAR AND LOGIC IN THE MIDDLE AGES

Intellectus and Ratio

Despite the tumultuous events that led to the eclipse of the Roman Empire and classical civilization, the intellectual consciousness that had been developed by the Greeks and sustained by the Romans endured into the Middle Ages. The medieval period is often referred to as an 'age of faith', but it was also an age in which the intellect was cultivated to an extraordinarily high degree, especially following the growth of the universities during the thirteenth century. And yet the intellectualism of the Middle Ages differed from modern intellectualism in one very important respect. In Chapter Four, we saw that both Plato and Aristotle distinguished two different aspects of thinking: one was the process of reasoning things out using logic, the other was the act of understanding through which insight is gained into the nature of reality. These two aspects usually work together, but by making a distinction between them Plato and Aristotle were able to point to the fact that logical reasoning does not constitute the whole of the human capacity to think. Neither does it constitute the essence of thinking, for this resides rather in the act of understanding and the possibility of gaining insight into that towards which our thinking is directed.

This distinction was transmitted to the Middle Ages through such writers as St Augustine and Boethius. Plato's *dianoia* (the reasoning that proceeds through logical deduction or inference) was translated into Latin as *ratio*, while *nous* (the faculty that becomes activated as intuitive insight or *noêsis*) was translated as *intellectus*. The distinction was helpful in identifying two different stages to the acquisition of knowledge: first we reason things out and then we gain understanding and insight. But it was much more than merely a description of stages in the process of reasoning. It also indicated two different, but complementary, approaches to the pursuit of knowledge: one was through argument and disputation, the other was through contemplative thinking. These two approaches really corresponded to two different *levels* of thinking, each aiming at a different level of reality, the former transitory and the latter eternal.[1]

Generally speaking, in the medieval period, it was understood that if we function solely with logical, analytical and adversarial reasoning, which thrives on argument, we may indeed have the ability to arrive at a certain degree of knowledge or 'science' (*scientia*). We may be able to solve practical problems and settle legal disputes, but we will not arrive at wisdom (*sapientia*). Wisdom requires a stilling of our mental activity to a point at which we become receptive to a deeper level of meaning and truth. And that kind of inwardly active, yet also receptive, mental activity is what characterises the *intellectus*, which was widely understood as an interior 'seeing' by means of an incorporeal or spiritual light.[2] The distinction between science and wisdom was based on an experience of the reality of contemplation as an intensification of the activity of thinking. In this activity human beings experience their freedom, for the exercise of free will requires both reasoning things out and the intuitive intelligence of the *intellectus*.[3] In the Middle Ages, there was widespread acceptance of St Augustine's view that the two kinds of thinking are not equal to each other. The *ratio* should submit itself to the higher guidance of the *intellectus*, as this gives access to a deeper level of truth. For example, meditation upon sacred Scripture and the writings of the Church Fathers leads the seeker to a wisdom whose transcendent source can only adequately be grasped by the *intellectus*. A merely logical, analytical approach cannot give the same degree of insight, and so the *ratio* should follow where the *intellectus* leads.

It will be seen, then, that the word *intellectus* did not have the modern connotations of somewhat dry, abstract thought that the modern term 'intellectualism' has, but was much closer to being a faculty of living spiritual intuition, born of contemplation. For the medieval mind, it was rather the *ratio* which was dry and abstract, as its domain was the syllogism and syllogistic reasoning.[4] The general acceptance in the Middle Ages of St Augustine's view that the *ratio* should submit itself to the higher guidance of the *intellectus* meant that the discipline of logic, although it was developed to a high degree of sophistication, was nevertheless held to account by an overarching moral demand that it serve the interior life through its being dedicated to the pursuit of truth. As we shall see in later chapters, it was only when this moral demand on logic was withdrawn that it would become possible for logic to be requisitioned in the service of information technology. This did not happen until after the medieval period, and yet during the Middle Ages certain developments occurred that helped to pave the way for it to happen, by loosening the bonds that tied logic into the specifically human realm. We shall discuss these developments later in the chapter. First we need better to understand how the recognition in medieval times of the priority of the contemplative *intellectus* over the discursive *ratio* impacted on the whole tenor of the culture.

During the early medieval period, there were no universities. Intellectual enquiry was pursued mainly in the monasteries and cathedral schools, the latter receiving an important impetus during the reign of Charlemagne. For the most part the monasteries were the dominant centres of learning until the twelfth century, and although the growth of urban populations in the twelfth century meant that the cathedral schools grew in prestige, the monastic influence was the determinative one, suffusing the whole culture.[5] It is, perhaps, difficult for us to re-imagine just how all-pervasive was the influence of the monasteries throughout the medieval period. In England alone at the time of the Norman conquest, there were more than 60 monasteries, compared to just over 100 towns, the vast majority of which we would today regard as little more than villages. While the number of towns doubled during the next century, the number of monasteries increased tenfold to nearly 600, far outnumbering urban settlements.[6] Wherever one lived, one was never far from a monastery. The monasteries held the society together: they looked after the poor, they gave shelter to travellers, they nursed the sick and at the same time they were the custodians of early medieval culture. It was in the monasteries that books were produced, that libraries were slowly built up, that standards of scholarship were upheld, and learning cultivated.

Monastic learning was always pursued in the religious context of the desire of the soul to unite with God. Within the standpoint of this religious approach to knowledge, a third level of thinking was also recognized: a higher illumination of the mind (*mens*), which awakens to awareness of the immanence of God within us. This higher illumination occurs when the thinking of the *intellectus*, intensified to prayerful contemplation, opens itself to the divine. It is a special kind of attentiveness in which the mind is not only completely stilled but, as it were, slows down to the point at which it dwells in God, or equally becomes aware of God dwelling within it. The mind or *mens* was traditionally located in the heart rather than the head, and so contemplation was understood to involve an opening of the heart, or the so-called 'eye of the heart' (*oculus cordis*), which gives us insight into transcendent meaning, truth and value.[7]

In the Middle Ages the quest for wisdom, based on the prayerful contemplation pursued in the monasteries, had religious experience at the very root of it. The underlying attitude was that true knowledge unites us with God, and for this reason philosophy and theology could not realistically be separated. This attitude, so characteristic of the monasteries, also predominated in the cathedral schools and it profoundly impacted on the way in which the role of logic was conceived. As the emphasis on the study of logic grew stronger in the schools, and became stronger still when the first universities emerged towards the end of the twelfth and the beginning of the thirteenth centuries, logic nevertheless continued to be regarded as subservient to the pursuit of religious goals. This is because the new universities were essentially religious institutions. According to St Thomas Aquinas, writing in the thirteenth century, the function of logic is that it serves the quest for wisdom, providing the mind with the necessary tools to think clearly.[8] Such clarity of thought is necessary for the discursive *ratio* to arrive at correct conclusions, but for St Thomas the clear thinking of the *ratio* is itself dependent on the more comprehensive understanding of the *intellectus*, which is concerned with theology or 'divine science'. In the hierarchy of subjects studied in the university, theology was at the summit. For St Thomas, theology or divine science provides the starting points for every other science.[9]

The Centrality of Grammar in Early Medieval Culture
Throughout the Middle Ages, the basic curriculum in both the monastic schools (for children destined to become monks) and cathedral schools (for the education of the secular clergy) was the seven Liberal Arts. The seven were divided into two groups. First were the three foundational studies, known as the *Trivium*, consisting of Grammar (the study of language), Dialectic (the art of logical reasoning), and Rhetoric (the art of eloquent expression). The second group was the *Quadrivium*, consisting in Arithmetic, Music, Astronomy and Geometry, taken up only after the *Trivium* had been mastered. In both the monastic schools and the cathedral schools in the early period, Grammar was the core subject. It was considered more important than either Dialectic or Rhetoric, and both Dialectic and Rhetoric were to a certain extent beholden to it. Rhetoric was more or less assimilated to Grammar in the Middle Ages, since it was concerned less with the art of public speaking than with the art of composing different kinds of written text, each of which had their own rules and styles.[10] But Dialectic always retained its independence and would in due course become the rival of Grammar.

One reason for this pre-eminence of Grammar in the early Middle Ages is that it was regarded as essential that all monks and secular clergy acquired the ability to read and write, for otherwise it would have been impossible for them to study the Bible. But as well as the Bible, two other kinds of non-scriptural text were also studied: firstly, texts written by the Church Fathers, like Cassian and St Basil, whose meditations on the spiritual life were vital sources of sustenance for those living under monastic vows; and secondly, the works of pre-Christian classical authors, who were regarded as models of good writing style. Poets like Virgil, Lucan, Statius and Ovid, historians such as Livy and Sallust, and philosophers such as Seneca and Cicero were all studied in both the cathedral and the monastic schools as an integral part of learning Grammar.[11] Through studying Grammar, then, many monks and clergy came to love classical poetry, history and philosophy.

But the main reason for the pre-eminence of Grammar both in the monastic and the cathedral schools lay in an attitude towards language that most of us have lost today. Language was seen as ultimately sacred, for it was an articulation of the divine creative Word or Logos. Medieval authors often surprise us by their veneration for Grammar. But this was due to its role as the custodian of language, considered as something intrinsically holy. Grammar not only entailed the study of correct forms of linguistic expression, but precisely in doing so it orientated the human being towards fundamental aspects of Creation. In the outlook of the Middle Ages, the natural world was endowed with profound spiritual meanings, for ultimately it mirrored the divine Logos that is its creative source. For Hugh of St Victor (an abbey in Paris, which became renowned in the twelfth century as a centre of learning) 'the entire sense-perceptible world is like a sort of book written by the finger of God'.[12] Hugh and others believed that human language had the capacity to draw forth and re-express the divine language already embodied in nature. Grammar, when deeply entered into, could thus lead to a sense that the human being is inserted into nature in such a way that language is as much *given to us* by the world that we dwell within as it is *produced by* us. That is to say, language is not simply a human invention arbitrarily imposed upon nature, but rather re-expresses the God-given way things are. For John of Salisbury (also writing in the twelfth century) at the very root of language, elementary vowel sounds rise up within us as from the very depths of nature herself.[13] John saw Grammar as the guardian and guarantor of the spiritual integrity of language, and referred to it as language's 'sanctuary' or 'holy place' (*sacrum*). It was the function of Grammar, in other words, to guard the holiness of language.[14] In Figure 7.1, we see a tenth century depiction of Grammar as a beautiful young lady seated before the holy place of language, which she guards with her rod in one hand while extending her other hand in an open gesture, offering her teaching to a class of students.

This quasi-mystical view of Grammar meant that Dialectic could hardly hope to compete with it as the core subject in the curriculum of

Figure 7.1
Grammar personified as a beautiful young lady both guards the sanctuary of language and offers her teaching to a class of students.

the Liberal Arts in the early Middle Ages. Grammar was regarded as the foundation of, and the key to, all the other Liberal Arts, and was considered to be more foundational than Dialectic for the study of philosophy.[15] This is not to deny that Dialectic had its role in education in both the monastic schools and the cathedral schools, for as the tool of discursive reasoning (*ratio*), the specific function of logic was to ensure that the activity of reasoning was correctly conducted. Only then could human thinking arrive with certainty at knowledge of what is true and what is false.[16] But logic was conducted within the matrix of ordinary language, which was the province of Grammar. Thus in Aristotle's *Categories*, an important source for the study of logic in the early Middle Ages, various logical categories such as Substance, Quality, Quantity, Action and so on were closely related to their respective grammatical forms. It was understood that, for example, the subject of predication (Substance) was also a noun, and the predicates applied to it were usually adjectives (falling under the logical categories of Quality or Quantity) or verbs (falling under the categories of Action or Passivity).[17] This ensured that a relationship was maintained between Dialectic and the actual world in which people lived, for each part of speech is bound to an aspect of nature. By means of nouns we name things and living creatures; by means of adjectives and adverbs we characterize their specific qualities and attributes; by means of verbs we describe how they are acting or being acted upon, and so on. For Boethius, whose translation of, and commentary on, Aristotle's *Categories* was hugely influential during this period, the two disciplines of Dialectic and Grammar were deeply intertwined. Dialectic, although a distinct and independent subject in its own right, could not be separated from the study of Grammar.[18] And precisely because it was embedded in human language, Dialectic was constantly drawn into the service of those disciplines that study *the nature of the world*, and whose forms of thought require linguistic articulation.[19]

Whereas in later centuries, the fundamental gesture of logic became one of enmity towards ordinary language because ordinary language was seen as embroiling the rational mind in ambiguity and imprecision, this was not the gesture of logic in the Middle Ages when the sciences were pursued within a religious and symbolic rather than within a mathematical framework. The distinctive function of logic was to serve as an instrument of the mind in the pursuit of wisdom and knowledge. This remained the overriding view throughout the Middle Ages. It was, for example, the view of St Thomas Aquinas (writing in the thirteenth century) that logic should be concerned with the right operations of the reasoning faculty, ultimately directed towards knowledge of what is real. His understanding of the role of logic is well summed up in the statement, 'Logic is ordered towards obtaining knowledge of things' (*Logica ordinatur ad cognitionem de rebus sumendam*).[20]

In post-medieval times it was the destiny of logic to be severed from the real world of things and it was precisely this severance that enabled it eventually to be deployed in the service of information technology. But thereby it also broke free of the human realm, to function autonomously within machines. In the Middle Ages, this could not happen because Dialectic was held fast within a human embrace, and constrained to serve human ends, notably ascertaining the truth about real being.[21]

The Conflict of Grammar and Dialectic

One of the differences between the educational curriculum of the cathedral schools and that of the monasteries was that the cathedral schools often went beyond what was required for the training of the clergy, giving more space to the subjects of the Quadrivium, and including as advanced studies both Law and Medicine along with Theology. In both the monastic schools and the cathedral schools Grammar held a pre-eminent position in the curriculum, but in the cathedral schools Dialectic was given considerably more emphasis than it was in the monasteries. The growing emphasis given to the study of logic during the twelfth century became the cause of increasing tensions between the monasteries and the cathedral schools. Already in the eleventh century St Anselm had applied logical analysis to theological problems, but this tendency developed further in the following century, most famously in the hands of Peter Abelard. The monastic attitude was that theological mysteries are not susceptible to logical analysis and it was regarded as irreverent to make them the subject of rational disputation. Monks were anyway simply not interested in a merely intellectual or speculative approach to theology: their focus was on knowledge born of inner experience.[22]

In the later decades of the twelfth century, the growing demand for higher studies in the curriculum offered by the cathedral schools led to the formation of the first European university at Bologna, and subsequently the universities of Paris and Oxford. The first universities emerged out of already existing schools, which now organized themselves as institutions independent of the cathedral, but not of the Church: at both Paris and Oxford not only the teachers but also the students were still clergy and were governed by canon law. Learning was very much regarded as a sacred vocation. However, the period of the emergence of the universities corresponded exactly to the period during which works of Aristotle, hitherto unavailable—including several more of his treatises on logic—became available in Latin translation. The impact that this had on learning was far-reaching. During the early thirteenth century, Dialectic rapidly displaced Grammar as the foundation of the Liberal Arts studied at the universities. The classical authors—the poets, historians and sages of Roman times like Virgil, Statius, Livy, Cicero and Seneca—so prominent in both the monastic and the cathedral schools of previous centuries, were all but abandoned at Paris in the early decades of the thirteenth century, as Dialectic swept all before it.[23]

An important change of sensibility thus took place at this time. The ancient authors, or *auctores* as they were known, were held in great reverence in the early Middle Ages, and were regarded not just as 'authors' but also as 'authorities'. Of all the *auctores*, it was the poets in particular that had tended most to be revered. Poetry was the highest reach of Grammar, for through poetry we experience the Word or *Logos* as creative power.[24] The domains of poetry and philosophy were felt to border each other. For example, John of Salisbury regarded it as axiomatic that 'Poetry is the cradle of philosophy'.[25] And many early medieval philosophers wrote in a style that mingled both prose and verse (so-called *prosimetra*). This was partly in emulation of influential writers of late antiquity, like Boethius and Martianus Capella, in whose writings verse and prose interweave, but it particularly suited the contemplative and imaginative (rather than argumentative) tenor of twelfth century thinkers.[26] This acceptance, and indeed celebration, of the proximity of poetry and philosophy could be sustained so long as the contemplative *intellectus* was able to retain its position of supremacy over the discursive *ratio* as the main organ of knowledge.

As the new educational establishments of the thirteenth century championed logical technique over Grammar, displacing the Classical authors from the curriculum, people in the milieu of the universities less and less saw the value of literature or poetry.[27] What this meant was that they could only appreciate less the contribution of contemplative thinking to the quest for knowledge. The mainstay of the monastic outlook had no real place in the universities, because the scholastic method of disputation, in which theological questions were debated using rigorous logical analysis and argument, excluded the poetic sensibility of the contemplative. An antagonism between Dialectic and Grammar grew up that can at times be discerned not only between the universities and the monasteries, but also between those universities (such as Orleans) that held on to the traditional curriculum emphasizing the study of the classical authors, and those universities (such as Paris) which so strongly embraced Dialectic, developing the art of disputation as a central educational skill.

The tensions between Grammar and Dialectic can be seen in an illustration from a late twelfth century manuscript (Fig. 7.2), in which the personifications of Grammar and Dialectic are positioned either side of Philosophy. Dialectic, to the left, holds a great serpent with knots in its body and with a dragon's head, while Grammar, to the right, holds a sheet of parchment. In the central figure of Philosophy, whose head is turned towards Dialectic but whose heart is turned towards Grammar, we feel the predicament of Philosophy in the late twelfth century, having to hold the balance between Dialectic on the one side and Grammar on the other.

Figure 7.2
Grammar and Dialectic, either side of Philosophy. From a late twelfth century manuscript.

Such depictions were not unusual in the Middle Ages. The serpent-dragon motif imaginatively represents something that was perceived as belonging to the nature of the faculty of logical thinking, hidden from view yet experienced nevertheless as an essential part of its character. We see a similar depiction of a serpent-dragon in Figure 7.3, from a mid-twelfth century manuscript version of a commentary by Boethius on a logical treatise of Porphyry. Both these manuscripts would have been produced in monastic *scriptoria*, where monks worked tirelessly to make copies of ancient manuscripts, many of which they also illustrated. In Figure 7.3, the knotted dragon-snake is now held in the left hand of a woman who represents Dialectic, while in her right she holds a floral sceptre in which the hierarchical order of the world is summarized as a sequence of genera and species.[28] So although in one hand she holds the threatening reptile, in the other it is as if she offers the viewer the promise of enlightenment on the fundamental structure of the world through mastering the categories of thought. Dialectic is thus portrayed as highly ambivalent.

A comparable image of Dialectic is shown in Figure 7.4, which is a detail of a twelfth century illumination that appears in a manuscript of two treatises on Grammar by Priscian. Here the snake is held in the right hand of the figure representing Dialectic, while in her left she holds a bundle of

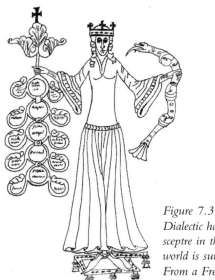

Figure 7.3
Dialectic holds a dragon-snake in one hand and a floral
sceptre in the other in which the hierarchical order of the
world is summarized as a sequence of genera and species.
From a French manuscript of Boethius' Commentary on
Porphyry's Isagoge. AD *1140.*

Figure 7.4
Dialectic, with a snake in one hand and a bundle of five discs,
representing various logical texts, suspended by threads in the
other. From a twelfth century manuscript of two treatises on
grammar by Priscian, made in the Benedictine monastery of Saint
Sulplice in Bourges.

threads from which hang five discs, with the names of various logical
treatises written on them.[29]

These images of Dialectic belong to an iconographical tradition that
has its origins in the detailed descriptions of each of the personified Liberal
Arts given by Martianus Capella, in his influential book *The Marriage of
Philology and Mercury*, written in the fifth century AD.[30] Capella describes

Dialectic as a pale-faced lady whose flashing eyes are in constant movement, who holds in her left hand a snake twined in immense coils,
concealed under the sleeve of her cloak. In her right hand she holds wax
tablets inscribed with formulae or patterns made beautiful by their contrasting colours. Behind these patterns, however, is a hook. While she
offers the attractive patterns to all whom she meets, anyone who takes
them is soon caught on the hook and dragged towards the hidden snake,
which first bites and then entangles its victim within its many coils.[31] For
the monks copying dry logical treatises, Capella's unflattering description
of Dialectic must have been warmly appreciated, for it resonated with the
dim view of logic that prevailed in the monastic tradition. The fact that
the serpent-dragon was in medieval iconography also associated with the
serpent in the Garden of Eden, who with subtle arguments persuaded Eve
to eat of the fruit of the forbidden tree, must have struck the monks as
particularly apt.[32] The widespread perception of logic as potentially
divisive and morally duplicitous made the snake with its forked tongue a
perfect symbol of Dialectic, and lay behind the deep suspicion that many
felt towards it.[33] There was something inherent in logic that was
apprehended as working against humanity, and separating us from God.
Some depictions of Dialectic with her draconian beast could easily be
mistaken for Eve. So, too, depictions of Eve could readily be taken as
depictions of Dialectic.[34]

Despite these negative associations and the general suspicion with
which logic was regarded outside of university circles, during the first part
of the thirteenth century Dialectic was elevated to primacy within most
universities, while Grammar was gradually reduced to having a status no
greater than any of the other remaining Liberal Arts.[35] The struggle
between Dialectic and Grammar during this period comes vividly to light
in a poem written by Henri d'Andeli, called *The Battle of the Seven Arts*
(1250), misnamed because the central protagonists in the battle are in fact
'Dame Logique' on the one side and 'Dame Grammaire' on the other. In
this rather extraordinary poem, Dame Grammaire hurls verses and rhymes
at the standard bearers of Dame Logique, while the latter defend themselves with sophisms and logical paradoxes. The champions of Dame
Grammaire include the grammarians Donatus and Priscian, whose text-
books were for centuries staple in the teaching of grammar in both the
monastic and cathedral schools. These heroes bravely take on famous
logicians such as Socrates, Plato and Aristotle, but they don't stand a
chance. The battle ends with Dame Logique victorious, and the poet
commenting prophetically: 'Sirs, the times are given to emptiness; soon
they will go entirely to naught.'[36]

For the poet, Henri d'Andeli, the triumph of Dame Logique was the triumph of intellectualism emptied of real meaning. The battle, which he saw being played out in European intellectual life at the time, represents another milestone in the prehistory of the computer. It was in reality a battle between two different mentalities, one based on the *intellectus* and the other based on the *ratio*, and its outcome had profound implications for the future. While no one doubted that skill in logic was essential to be able to think clearly and to discriminate between truth and falsehood, it was also understood (especially in monastic circles) that logic does not of itself provide thinking with any real content. Unlike poetic inspiration or the insights born of contemplation, logic is not fecund. As scholastic thought became increasingly removed from the inner life grounded in actual experience, a certain barrenness of thought began to characterize university learning. To the extent that thinking became more logically sophisticated, it also became from a spiritual point of view ever more sterile. A growing sense of disconnection from the world of spirit ensued, which was subsequently expressed in the rise of nominalism in the fourteenth century, a subject to which we shall return in Chapter Ten. First of all, we must explore further the implicit tendency of logic towards the mechanical, as this unfolded in the Middle Ages.

Chapter Eight

THE LOGIC MACHINE AND THE CAM

Logical Devices

The fact that logic concerns itself with the *forms* of thought rather than their content means that the rules by which sequences of thought can be judged as either logically valid or invalid can quite easily be represented in diagrams. We know that in antiquity various logical diagrams were employed, which represented for example the relationships between different types of statement.[1] And this diagrammatic tradition continued into the Middle Ages. The reason why this tradition is significant for the prehistory of the computer is that, while these diagrams were not in themselves actual physical machines, they were nevertheless aids to thinking mechanically. And for this reason in the medieval period they were referred to as machines or devices (*machinae*).[2] They provided a means of conforming the mind to the rigid rules of thought, which apply no matter what subject one is considering or how one feels about it. The logic diagrams were mental maps that exposed logically invalid arguments and indicated the pathways by which the completely impersonal reasoning process would reach logically valid conclusions.

One of the most common diagrams is known as 'The Square of Opposition', which shows the relationships between four different kinds of statement. The four kinds of statement are the universal affirmative, the universal negative, the particular affirmative and the particular negative. In Figure 8.1 we see a Square of Opposition (actually an oblong rather than a square so as to be able to fit in examples of each of the types of statement) from a ninth century manuscript of the Carolingian era. The

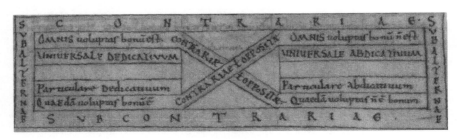

Figure 8.1
The Square of Opposition from a ninth century manuscript.

universal affirmative (upper left) is exemplified by the phrase 'All delight in the good.' Its contrary, the universal negative (upper right), reads: 'None delight in the good.' Below left, the particular affirmative reads: 'Someone delights in the good', and the diagonal between it and the universal negative at the top right shows that it contradicts it: for if someone delights in the good then the statement 'None delight in the good' is clearly contradicted. Below right the particular negative reads: 'Someone does not delight in the good' and the diagonal leading from it to the universal affirmative at the top left shows that it contradicts the statement: 'All delight in the good.'

The logical relationships in the diagram are perhaps more easily seen in Figure 8.2, which gives them in English, and uses the standard convention of representing the four types of statement as A, E, I and O. There we see that the relationship between the particular affirmative (I) and the particular negative (O) at the bottom is referred to as 'sub-contrary' because while they may both be true together they may not both be false together ('someone does delight in the good' and 'someone does not delight in the good'). We also see that the relationship between the universal affirmative (A) and the particular affirmative (I) beneath it, and the relationship between the universal negative (E) and the particular negative (O) is called 'subaltern', which means that the universal implies the particular, but the particular does not imply the universal.

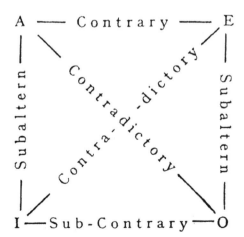

Figure 8.2
The Square of Opposition.

To those for whom logic does not come easily, such a diagrammatic aid is of value, for it helps the mind to shift its focus from the meaning of the words to their underlying logical form. It doesn't matter what one is arguing about: it could be wisdom, it could be rats or roses. The logical form will not change. Through such a diagram, one learns to put one's own feelings aside—one's warmth of interest in, or equally personal dislike of, the content of what is being stated—in order to consider in a detached way the purely logical structure of the statements that are being made. In doing so, we can then have a double experience: on the one hand, we may experience the utter rigidity of the rules of logic, which operate in an entirely external way, with absolutely no concern for the content of what is being thought about, but only for the correctness or incorrectness of the logical forms that thinking is forced to assume. But on the other hand we may also experience something light up in our thinking, as the truth of a certain line of reasoning and the falsity of another line of reasoning becomes clear to us. The experience then has a moral component to it, for by enabling us to see what is true and what is false the process of logical reasoning stimulates in us moral feelings.

With such a diagram committed to memory, the student of Dialectic had as a resource a useful 'device' or *machina* that could be mentally consulted when seeking to construct or validate certain arguments.[3] It is not hard to imagine that what the diagram depicts are logical pathways, through which a current of thought can either flow or not flow. Despite the pathways being mechanical, the living energy of thought that is conveyed by them is fired by the human desire for truth. The *machina* serves the truth, and herein lies the difference between the logic *machina* of the middle ages and the logic machines of later times that prefigured the modern computer. Because the latter were based on a mathematicised logic, they were no longer concerned with the discovery of the truth as a moral value, but only with making correct calculations. It was a recurrent preoccupation of early medieval thinkers to ensure that logic was applied to the truth about real being. There was an acute awareness that the tendency of logic is to lead the mind away from relatedness to the actual world and towards fruitless disputes, in which the skill of marshalling arguments claims greater attention than the search for wisdom based on experience. While it is always hazardous to generalize, it would seem that through the greater part of the medieval period, the logic *machina* was primarily co-opted into the service of training thinking for the genuine search for wisdom, for there was widespread recognition that once torn out of the context of genuine moral, metaphysical or theological striving, logic is at best sterile, at worst destructive.[4]

The Evolution of Camshaft Technology
We have already observed how in the ancient world, once the discipline of formal logic had become established, logical thought tended to seek

out and ally itself to mechanics. And so from the third century BC onwards, there was a surge in the invention of complex machines.[5] Most of these machines remained a rarity in the ancient world, but in the field of military weaponry (ballistas, catapults, siege engines and so on) the machines were widely put to use. Likewise, the spread of watermills— although slow at first—became more rapid towards the end of the Roman period (in the fourth century AD). Whether rare or common-place, the appearance of complex machines was due primarily to the habituation of the psyche to logical thought-patterns, for the thought-patterns of formal logic are intrinsically mechanical. For this reason, machines should be viewed as essentially logical thought-patterns exteriorized in physical form. In the medieval period, when logic was cultivated to such a strong degree, it is only to be expected that there should have been an equally strong tendency towards an effusion of machines. This was certainly the case, but the tendency was to some extent held in check by the spiritual constraints that we have already noted in the previous chapter.

One of the most prevalent of machines in the medieval period was the watermill, its use being widespread from the ninth century onwards. In 1086, the Domesday Book recorded 5,624 watermills in England. The number was probably considerably greater than this since much of northern England was excluded from the Domesday survey.[6] As the population grew, so too did the number of watermills, so that by the year 1300 there were between 10,000 and 15,000 watermills in England alone. Other north European countries would have been similarly well serviced by watermills. A vivid impression of their cultural impact can be gained by considering the industrial scale of their deployment in Paris in the early 1300s, when there were 68 watermills in the upstream section of the main branch of the Seine alone—a distance of under a mile.[7] In Figure 8.3, floating watermills under the arches of a bridge over the Seine in Paris are depicted in a manuscript of this period. They enable us to glimpse how the medieval world was by no means innocent of machines, the product of applied logical thought and technical skill.

While in the early Middle Ages watermills were used almost exclu-sively for grinding grain, towards the end of the eleventh century the development of the cam enabled watermills to power a variety of industrial processes. We have already seen that the cam was known in antiquity and was utilized (albeit relatively rarely in Roman times) for example in crushing ore (Chapter 5, pp. 45–47 with Fig. 5.2). What is particularly significant about the cam is that it represents a primitive form of mechanical memory. In cyclical automated processes (such as those

Figure 8.3

Floating watermills under the arches of a bridge over the Seine in Paris. From an early fourteenth century manuscript.

generated by watermills) which set a shaft rotating, the cam enables the cyclical movement of the rotating shaft to be translated into a linear movement (vertical or horizontal) of the so-called 'follower'—i.e. whatever it is the cam connects with, for example a trip-hammer. The number, shape and size of the cams along with the speed of rotation of the shaft on which they are mounted enables the movement of the follower effectively to be programmed. It is a very simple form of programming but it is nevertheless a means of 'instructing' the machine to perform its simple actions in a particular sequence. If, for example, the watermill sets a camshaft in motion and this camshaft connects with four trip-hammers placed side by side, the sequence of hammer blows can be adjusted in a variety of ways. One hammer might be pounding up and down four times per revolution, another only twice, and the hammer blows can be synchronized through the exact placement of the cams on the camshaft (Fig. 8.4).

From the end of the eleventh century, water-driven cam-operated hammers were used for the treatment of hemp for making rope, for fulling cloth, tanning leather, pulping paper and crushing ore.[8] If we

A—Mortar. B—Upright posts. C—Cross-beams. D—Stamps. E—Their heads.
F—Axle (cam-shaft). G—Tooth of the stamp (tappet). H—Teeth of axle (cams).

Figure 8.4

A watermill rotates a camshaft that operates four trip-hammers (referred to as stamps) used to crush ore. Although the illustration is from the mid-sixteenth century, water-driven hammers like this were operating as early as the eleventh century.

briefly consider the fulling of cloth, we find that by the early fourteenth century there were 130 fulling mills in England.[9] The fulling of cloth, by which it was scoured, cleansed and thickened, was originally carried out by human feet trampling the cloth in a trough of water. The trip-hammers replaced this rhythmic human action with the mechanical action of wooden hammers being raised and dropped. In this way, the human being was displaced by the machine. But it was tiring work and we may imagine that most people would have welcomed the machine taking over their labour in this way. In the Middle Ages, the domain of the machine was limited: it was not at this time threatening. Its first incursion was into those productive activities like fulling which required repetitive human physical action, and which the machine could simulate. But in time the machine would come closer and closer to the consciousness which gave it birth. It would seek its source, and then it became a threat to the human being. But in the Middle Ages this had not yet happened.

A typical camshaft in a mill would contain a relatively small number of cams. But the key characteristic of the cam is that it is a type of information storage. A certain 'intelligence' is instilled into the machine, but at this stage it is very primitive. The more cams there are on a camshaft, the richer the information content of the shaft, but in practice this remains very straightforward. It is not complicated. This may be why the cam was so readily accommodated within the medieval sensibility whereas, as we shall see in the next chapter, the introduction of the mechanical clock conspired to undermine it. Even when in the fourteenth century cam technology became more sophisticated, it did not unsettle the medieval sensibility.

The more sophisticated cam technology of the fourteenth century made its appearance in the first programmed musical instruments using pegged cylinders. These had already been developed in the ninth century in Baghdad, but now they appeared in Europe and they were really the first automata to be witnessed by Europeans.[10] One such device was an organ that 'played itself' by means of a series of cams on a rotating drum that operated levers on organ pipes. The cams were of different shapes and sizes, and this enabled different tunes and harmonies to be played (Fig. 8.5). The same principle could also be applied to a self-playing flute.[11]

The transformation of the camshaft into a rotating drum or cylinder with wooden pegs inserted into holes on the cylinder surface brought out

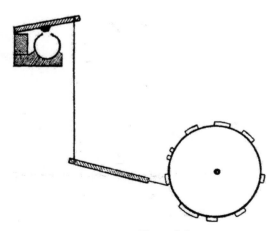

Figure 8.5

Cross-section of the mechanism of the self-playing organ and flute invented in the ninth century in Baghdad. It depends on a rotating cylinder (bottom right) with cams of different shapes and sizes operating levers that close or open the holes of an organ pipe or flute (top left). A series of such cams and levers would enable tunes and harmonies to be played.

an unsuspected potential of the camshaft. Once there are pegs that fit into holes drilled into the surface of the cylinder, the cylinder can be re-pegged with ease by simply lifting the pegs from the pre-arranged holes and placing them in different holes. In this way the instrument can be 'reprogrammed'. In the case of a musical instrument, each hole corresponds to a sound, so the placement of the pegs determines the melody played. As the cylinder rotates, the lever is either engaged or not engaged, and a note is either played or not played. Here is a machine governed by the binary logic of the lever which, in eschewing any third or intermediate position, embodies 'the law of the excluded middle'. Through the pegged cylinder this binary logic begins to sing. But notice that, just as in the Hellenistic world, the application of this clever new development in cam technology is limited to the sphere of court entertainment. It is not regarded as anything more than a plaything.

We find pegged cylinder technology applied to the mechanical performance of carillon bells early in the fourteenth century in Spain.[12] In time metal pins would replace wooden pegs. If one were to project the cylinder surface onto a plane, then the rows of holes would represent the presence or absence of bell sounds and the different sections of the cylinder would become columns representing points in time (determined by its rotation speed). The projection would look extremely similar to the punched card that was used to program the earliest computers (Fig.8.6).[13]

Despite the fact that in the late Middle Ages pegged cylinder technology was applied to things apart from music, it was not taken up as the foundation of a new 'information technology'. No punched card was developed out of it. No attempt was made to apply it to industrial processes. It was directed rather towards creating spectacles that would fascinate and charm those who saw them. For example the earliest astronomical clocks, which were positioned on church towers, would sometimes have figurines controlled by this technology—the so-called Jacks or Jacquemarts—that would strike the bells of the clock. Displays could become quite elaborate. In the 1320s an automated procession of monks was introduced to accompany the striking of the carillon at Norwich cathedral, while at St Jacques in Paris a procession of the Three Magi appeared.[14]

Why was the principle of the adjustable cam not developed any further? Why did the medieval mind fail to use it as the basis for devising increasingly sophisticated ways of programming machines? Why, indeed, did it not make further strides towards inventing the computer? The answer surely lies in the fact that the medieval mentality was unwilling, perhaps even constitutionally unable, to devote the necessary energy to

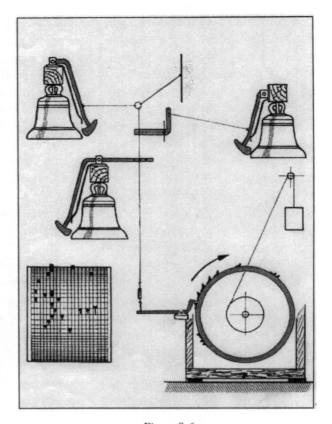

Figure 8.6

The melody of the carillon bells was 'programmed' by inserting metal pins into a large iron cylinder, shown bottom left and right. As the cylinder rotates, the pins cause levers to rise and fall. Attached to the levers are wires connected to hammers, which then strike the bells. The profile of the cylinder (on the left) is an early precursor of the punched card.

such merely technological projects. The salvation of the soul concerned it to a far greater degree than the development of new technologies. In the Middle Ages, there was no concept of 'information', which so preoccupies us today. The medieval mind was more occupied than we can imagine with orienting towards and living by great religious truths. Its focus was not on accumulating empirical facts, and the concept of 'data' was totally alien to the medieval sensibility. The medieval mind had other priorities and lived by quite different values from the modern mind, being so richly nourished by religious imagery and being so firmly held within the embrace of an unquestioned spiritual outlook.

Nevertheless, as we shall see in the next chapter, the invention of the mechanical clock had a psychological and spiritual impact much more far-reaching than the cam, and was a major contributor to the dismantling of the medieval religious consciousness. It was the clock, not the cam, that paved the way towards the computer because, unlike the cam, the clock affected human consciousness far more profoundly. It is to the mechanical clock that we must therefore now turn.

Chapter Nine

THE MECHANICAL CLOCK AND HUMAN CONSCIOUSNESS

The Mechanization of the World Picture

In the preceding chapters, we have seen that during the Middle Ages the religious attitude to life had the effect of restraining the tendency towards an overvaluation of one-sided logical thinking and mechanical technology. It thereby inhibited the unfolding of the conditions that would later prove to be so essential to the development of the computer. While the ability to think logically was greatly encouraged in the universities, for clarity of thinking was seen as crucial both in the quest for truth and in assuming responsibility for our actions as autonomous moral agents, the strong orientation to the world of spirit that pervaded medieval culture tended to favour prayerful, contemplative thinking over mere logic. As we have seen, this was an area of considerable tension within medieval culture. On the one hand it was recognized that the ability to master logic was foundational in the education of the whole human being, but on the other hand logic was regarded as a potentially false friend, whose deeper allegiance might be to the serpent in the Garden of Eden than to the angels in heaven. The traditional religious outlook worked constantly to temper the tendency of those enamoured with logic to veer away from reverence for the spiritual. But the strong impulse within logic towards the mechanical proved to be much less susceptible to religious constraint. And so we find that towards the end of the Middle Ages, with the gradual disintegration of the traditional religious outlook, new intellectual foundations were laid that would eventually lead to the first mechanical calculators.

One symptom of the disintegration of the medieval outlook was that the regard for the higher contemplative faculty of the *intellectus*, which ensured that philosophy and religion were directed towards the same spiritual truths, began to wane, allowing greater license to the *ratio* to develop independently of the contemplative mind. This led to the gradual separation of the spheres of reason and religion. It also led to a growing focus on logic as an independent discipline to be studied as an end in itself, rather than as a means to securing our knowledge of the real world. The growing emphasis on the *ratio* to the detriment of the contemplative faculty of the *intellectus* meant that there was a decline in awareness of the inward aspect of the natural world, which was less and less experienced as alive, ensouled and imbued with meaning. The cultivation of the life of contemplation based on the *intellectus*, which had, for the greater part of the medieval period, conspired to hold back the impetus towards the mechanization of the worldview, gradually became weaker during the Late Middle Ages from the fourteenth century onwards.

As we have seen, watermills were an important feature of medieval life from the ninth century, but they were readily encompassed within the overall organic and spiritual cosmology of the time, and did not present a threat to it. The same cannot be said of the mechanical clock, introduced during the first half of the fourteenth century. Unlike the watermill, the advent of the mechanical clock marked the beginning of a new relationship not only to the passage of time, but also to the conception of the cosmos itself. In both respects, the medieval consciousness was challenged. The new relationship to time inaugurated by the clock was based upon the quantification of an aspect of daily human experience, to which people had previously related in a qualitative way. Through the mechanical clock, time was re-conceived in terms of measurable units, each identical to, but separated from, each other.[1] But since many of the early mechanical clocks, which began to appear in the 1330s, were astronomical clocks actually showing the movement of the planets, they presented a new image of the cosmos that encouraged people to view the universe as a great machine. The mechanical clock demonstrated that it was possible to represent the functioning of the cosmos as so many parts external to other parts, all mechanically interacting with each other. A famous example is Giovanni de' Dondi's *astrarium*—a seven-sided astronomical 'clock', the upper part of which had seven large dials, one for each of the five planets plus the Sun and the Moon (Fig. 9.1). On the lower framework were dials indicating the fixed and movable festivals of the Christian festivals as well as the 24 hours of the day and night.[2]

Already in the thirteenth century, certain thinkers were experimenting with the idea that the cosmos could be understood as a great clockwork mechanism, but it did not really catch on. Mechanical clocks were still at the developmental stage and had not yet become physical presences in town squares and palaces. The notion that the universe could be conceived in mechanistic terms was therefore a strange one: everyone knew that the revolution of the celestial spheres was caused by spiritual Intelligences (i.e. angelic beings). In the next century, such certainties were no longer quite as certain as they had been. By the mid-fourteenth century,

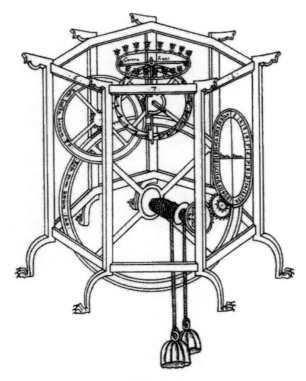

Figure 9.1
Part of the mechanism of Giovanni de' Dondi's weight-driven astronomical clock, or
'astrarium', completed in 1364, which showed the movements of the Sun, Moon and the
planets.

when Jean Buridan put forward a theory of impetus that eliminated the
need of spiritual Intelligences to explain the motions of the heavenly
spheres, the intellectual and cultural atmosphere was more receptive.[4] His
contemporary Thomas Bradwardine once more took up the idea of the
universe as a *machina mundi*, now conceived explicitly in clockwork terms.
With the first mechanical clocks appearing in the royal courts and on
public buildings of the cities of Europe, this notion was much more
favourably received.[5] This was because the new mechanical clocks artfully
simulated the movement of the planets, and in doing so they practically
demonstrated the viability of such a view. Towards the end of the
fourteenth century, Nicole Oresme would press ahead with the
mechanistic analogy, comparing God to a divine clockmaker, who
created the world as a giant clockwork and left it to run itself.[6] Between
Oresme's divine clockmaker with his clockwork universe and the earlier
conception of Dante, writing at the beginning of the fourteenth century,

for whom God was 'the eternal gardener' and the whole of creation his garden, an unbridgeable gulf can be seen to have opened up.[7]

Such radical ideas could not, however, really take root until human beings felt themselves to be sufficiently detached from their experience of the world to be able to apply mechanistic concepts to it. This would take many centuries, for the whole fabric of the medieval worldview served to protect the soul from the experience of alienation out of which the mechanistic conception arose. Only as this protective cultural fabric unravelled did the spread of the mechanistic worldview become possible. At the same time, the clock functioned as a kind of talisman of the mechanistic worldview. It was not neutral, but had a profound impact not only on the way people thought, but also on the way they experienced both the passage of time and the nature of the world. As the clock became ever more pervasive, so did the intuition of the organic wholeness of the cosmos begin to seem less credible. The human being's instinctive sense of belonging within this greater organic whole began to founder. A new malaise began to creep into the soul, an indefinable sense of being an outsider to, and an alien observer of, the natural world.

How Clocks Worked

What exactly was a mechanical clock? How did it work? Such questions need to be addressed because only then can we appreciate why the clock had such a devastating impact on the human being's inner life. The mechanism of the mechanical clock was not so different from the technology already used in watermills. But it developed it in a particular direction that was of great consequence. There were four main components of the mechanism of the medieval mechanical clock. The first was an axle to which was attached a weight that as it unwound made the axle turn; so, instead of the axle being turned by the flow of water, as in the watermill, in the mechanical clock it was turned by the tendency of the heavy lead weight to fall towards the earth. The second component was a regulatory mechanism or 'escapement' that slowed and evened out the rate of fall of the weight. It had to be introduced to prevent the weight from simply plummeting to the ground. The third component was a gear train that transmitted the rotary motion of the weight-laden axle through to the fourth component: a pointer or striking device that would indicate the hours. These components can be picked out by looking carefully at Figure 9.4.

Of these four components, the most important was the escapement. We may visualise it by considering Figure 9.2. On the left is a so-called 'crown wheel' attached to an axle that is set in motion by the falling

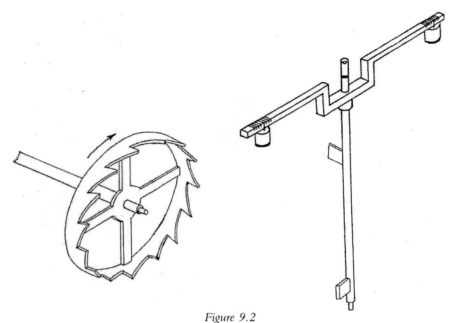

Figure 9.2

The two parts of the verge escapement. To the left, the crown wheel, with an unequal number of teeth; to the right, the verge surmounted by the weighted beam or 'foliot'. On the verge two palettes are fixed at right angles to each other.

weight (indirectly, because of the intervening gear train). The crown wheel, which is effectively an unbroken series of cams of unequal number, is one part of the solution of the problem of how to regulate the spinning of the axle. The other part of the solution is the vertical shaft or 'verge', on the right, which has two palettes projecting from it at right angles to each other. On top of the verge is a beam, weighted at either end (to determine the rate of swing), which is called a 'foliot'.

The escapement works as follows. When the crown wheel and the verge are brought together, the motion of the wheel is blocked first of all by the upper palette, which is made to swing round (by the impetus of the wheel) until the lower palette blocks the wheel again. The lower palette in turn is swung back until the upper palette engages the next tooth of the wheel, and so on. The verge is thus forced into an oscillating movement, aided by the foliot at the top of the verge. By means of this regular oscillation the motion of the wheel is checked and steadied (Fig. 9.3).

In Figure 9.4, an early tower warden clock features the escapement mechanism on the left-hand side, with the foliot oscillating above it. On the right-hand side the striking mechanism includes a second verge

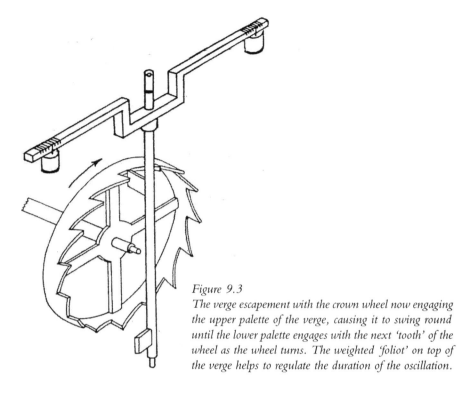

Figure 9.3
The verge escapement with the crown wheel now engaging the upper palette of the verge, causing it to swing round until the lower palette engages with the next 'tooth' of the wheel as the wheel turns. The weighted 'foliot' on top of the verge helps to regulate the duration of the oscillation.

now fitted with a hammer lever. By studying this diagram, we can see very clearly how each part of the clockwork plays its particular role. Here is a mechanism that is able to monitor and regulate itself, through the simple application of various binary technologies. The crown wheel (which as we saw is really an unbroken series of cams arranged in a circle), the two palettes of the verge, and the foliot with its adjustable weights at either end of its two arms, are all components of twofold composition. The motive force is provided by the two lead weights, one driving the pointer on the left and the other the bell-hammer on the right.

In so far as the lead weights give the clock an independence from any external source of power, we see in them the liberating potential of what we today call the force of gravity. It is as if, through the clock mechanism, human beings formed a new compact with gravity. In Aristotelian physics, the tendency of certain things to rise (referred to as buoyancy or levity) and the tendency of other things to fall due to their heaviness were understood as two complementary principles acting within the universe. Levity was assumed to be more closely associated with the spiritual aspect of existence, and heaviness with the more material aspect: the weight of

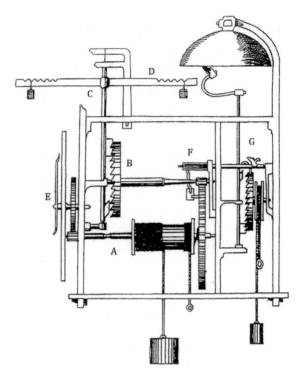

Figure 9.4

An early tower warden clock with an alarm bell. The axle (A) is rotated by the unwinding cable attached to the falling lead weight at the centre of the picture. The rate of fall of the weight is regulated by the 'escapement', consisting of the crown wheel (B), which engages a vertical shaft or 'verge' (C), surmounted by a delicately balanced swinging horizontal bar or 'foliot' (D). The clock face, with a pointer (E) to mark out the hours is to the left of the diagram. The gear train (F) linking the axle to both the clock face and the striking mechanism can be seen to the right of the crown wheel. The striking mechanism (G) with a second verge fitted with a hammer lever, and below it a second, smaller lead weight, is to the right of the diagram

material bodies (Fig 9.5).[8] Whereas in the high Middle Ages the heavenly world, conceived as a realm of light, was the focus of the soul's longing, now in the late medieval period the concept of weight begins to come into focus for Western consciousness as something that can be utilized in order to introduce more precision and control into our relationship with time. The advent of the weight-driven mechanical clock represents an important symbolic moment in the Western European psyche's awakening to what we now term the force of gravity. Indeed, it marks the commencement of a single-minded fascination with gravity, to the detriment and eventual denial of levity in the experiments of Galileo and the Accademmia del Cimento. We shall return to further discuss the sig-

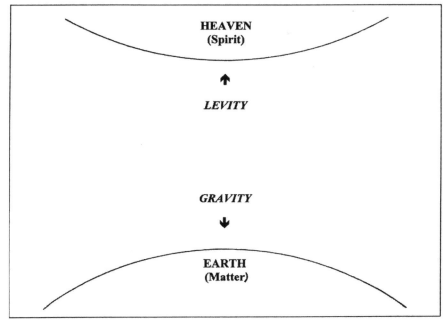

Figure 9.5
Levity and Gravity: two complementary principles acknowledged within the medieval
worldview.

nificance of this symbolic moment in Chapter Eleven ('The Renunciation of the Upper Border').

The Clock and Human Consciousness

The invention of the mechanical clock had enormous consequences for
human consciousness. As we have seen, the verge escapement was a
crucial technical intervention that enabled the mechanism of the weight-
driven gear train inside the clock to be exactly regulated. What is less
obvious is that the purely technical thinking that was responsible for this
invention had a profound impact in a quite different sphere, namely on
the general awareness of time. That is to say, the solution of the technical
problem of how to regulate the fall of a lead weight through introducing
the binary action of the verge with its oscillating foliot, had implications
that reached far beyond the original extremely limited context within
which the problem was addressed. Perhaps not for the first time, but
nevertheless now with an unprecedented impact, the effects of applied
technical thinking were to have far-reaching repercussions on how
human beings thought, felt and conducted themselves. For the

mechanical clock was destined to completely supplant not just previous time-keeping methods but *the very consciousness of time* that prevailed up until its introduction.

In the fourteenth century, when the mechanical clock was introduced, most people thought only rather loosely about the 24 hour cycle of the day and night. Since the first hour of the day was counted from sunrise and the first of hour of the night from sunset, in summer the hours were longer in the daytime and shorter at night, while in winter the daytime hours were shorter and the night hours were longer. Thus the twelve hours of the day and the twelve hours of the night expanded and contracted with the seasons. There was no sense that the hours should be of equal length throughout the year: this did not belong to the concept of an hour. An hour was less a quantity of time than a *quality* of time. And the quality that an hour had was explicitly religious.

In practice, few people reckoned time by all the twelve hours of day and night. In the Middle Ages, the seven canonical hours that signified times of prayer during the day were the major regulating influence on the rhythm both of the monastic day and the daily routine of all who lived and worked within earshot of monastery or church bells and participated in Christian life, which was the vast majority of the population. The seven times of prayer during the day were Lauds (first light, before sunrise), Prime (the first hour, beginning at sunrise), Terce (the third hour, mid-morning), Sext (the sixth hour, midday), None (the ninth hour, mid-afternoon), Vespers (late afternoon, before sunset) and Compline (after sunset).[9] The full cycle of prayer included an eighth hour of Vigils, in the middle of the night, which divided the night into two periods of sleep. Just as the seasons of the year have their moods, so too did the seven canonical hours of the day, along with the nightly Vigils. Each had their particular qualities and moods, and these were related to through the liturgical rhythm, and thereby they were brought within the religious domain.

In the medieval period, the hours were each related to key archetypal events both in the life of the Virgin Mary and in the passion of Christ, to which one oriented one's inner life. For example, the dawn hour of Lauds was associated with the visit of the pregnant Virgin Mary to her relative Elizabeth; the mid-afternoon hour of None was associated with the presentation of Jesus in the temple; and Compline was the hour of the coronation of the Virgin.[10] While similar connections were made between the hours and Christ's passion, it was the life of the Virgin which especially drew the devotion of ordinary people. The many Books of Hours that were produced for the laity from the thirteenth century on testify to the extent to which devotion to the Virgin was the emotional focal point of people's daily life. The passage of time was marked by the emotionally and spiritually charged recall of the soul to key images at the heart of Chrisitan spirituality.[11]

The awareness of the qualitative aspect of time did not suddenly disappear with the invention of the mechanical clock. Indeed, the period of the greatest flowering of the devotional Books of Hours was the fourteenth and fifteenth centuries at the time when the mechanical clock was introduced and gradually established itself in European society. For many generations, the religious hours continued to be observed even as the new purely quantitative hours of the clock also came to be observed. As the mechanics of the clock became more sophisticated and, through the process of miniaturization, became for increasing numbers of people the main reference point in the structuring of the day, so consciousness of the qualitative aspect of time was gradually eroded. The momentousness of the shift in time-consciousness that began in the fourteenth century can be glimpsed if we consider the fact that the medieval way of referring to the passage of a few minutes was in terms of how many prayers it would have taken a person to recite during that time interval. Whereas we might say, 'It took five minutes' the medieval would have said, 'It took fifteen Paternosters' or 'It took two Misereres'.[12] Once clock-time established itself, the experience of the passage of time no longer incorporated these spiritual associations. The earlier reference of the temporal to the sacred order was displaced by an entirely abstract time-reckoning.

The spread of the new mechanical clock (at the beginning just large and expensive public clocks) really took off in the second part of the fourteenth century. By the 1350s all the major cities of northern Italy had their public clocks, and by 1370 there was a public clock on one of the towers of the Royal Palace in Paris. The mechanical clock striking the equal hours was seen as a prestigious new acquisition for a city or commune, and helped to usher in an increasing level of regulation in the collective organization of time.[13] The very notion of an hour underwent radical change. The striking of the hour was no longer a call to prayer. It did not remind people of their relationship to God, but of their labour contract or a business appointment that they should not be late for. As one commentator has argued, if time used to belong to the Church now it belonged to the merchant.[14]

What is intriguing is that the Church seems to have willingly co-operated in this transition, making both buildings and bells available for the new clocks that regularly struck the equal hours.[15] The lack of resistance from the Church is all the more remarkable for the fact that the process was not politically or even economically driven, but was largely a socially driven process in which local communities provided the main impetus.[16] Clocks were prestige objects. They exercised a fascination over people. Soon models were designed for domestic use, to be fastened to a wall inside the houses of the wealthy, their weights hanging from winding barrels at the bottom of the clock. In the 1430s, roughly 100 years after the first weight-driven mechanical clocks were introduced, the first spring-driven clock—using a coiled spring instead of a weight to provide the motive force—became viable (though far from perfect).[17] The spring coil mechanism was perfected in subsequent decades. This meant that the clock could become increasingly portable. By the mid-sixteenth century clocks could be worn on a person, on a ribbon or chain around the neck.[18] By the end of the sixteenth century the hinged lid started to be used, and so during the seventeenth century the pocket watch gradually developed. We can see, therefore, a process of technological miniaturization that resulted in human beings forming an increasingly intimate relationship with their mechanical time-keepers.

But what was this new sense of time in reality? It was initially brought about by a mechanism that was based on the control of the force of gravity on the one hand and the alternate engagement and disengagement of the verge palettes with the teeth of the crown wheel inside the clock mechanism on the other. Thus the secularization of the human experience of time was the direct consequence of the functioning of a machine and the particular technical mentality that invented it, which then cast its influence more and more deeply into human consciousness and into the regulation of society.[19] This machine—through its simple yet clever mechanism—was able profoundly to affect the inner life of human beings. We have to ask: how was it that the technical mentality that worked so strongly and persistently into human evolution was able at this critical juncture at last to break through the protective sheaths of religious, philosophical and educational tradition so as to exert a direct influence on human consciousness—on the way people thought, felt and acted? For it had produced a tool that was no longer *simply* a tool that served human needs: it was an instrument that, by inveigling those who used it into adopting the abstract, quantifying mentality that it embodied, now made new demands on them. As Lewis Mumford observed, with the invention of the mechanical clock, 'Time-keeping passes into time-serving and time-accounting and time-rationing.'[20]

Thus the clock did not just affect human relationship to time, it interposed itself between human beings and their relationship both to nature and to God. It did this most obviously in so far as it became the icon of the mechanistic view of the cosmos. It also did it in so far as it directly and persistently influenced human beings in the conduct of their daily lives. In this respect it can be seen as foreshadowing the computer which, like the clock, substantializes in its binary mechanism a certain mentality, which then affects all who use it. But the most significant effect of the clock in relation to the prehistory of the computer was that it initiated a new 'reign of quantity', in which the qualitative aspect of nature slowly but surely came to be regarded as secondary to its quantitative aspect. It is this process that we must now turn to consider.

Chapter Ten

THE QUANTIFICATION OF THE WORLD

The Denial of Ideas in Nature

At the end of the last chapter, I referred to the clock as the icon of the mechanistic view of the cosmos. This remained the case for roughly 300 years—from the introduction of the early mechanical clocks in the mid-fourteenth century to Pascal's invention of the mechanical calculator in 1642. During this period, the clock was the representative machine. It provided for the mechanistic imagination a readily available image, which acquired a certain numinosity in so far as it conveyed the idea of a well-ordered cosmos, made up of myriad components, designed and set in motion by the 'divine clockmaker'.

One of the principal characteristics of machines is that they are made up of discrete parts, which are brought together in order to function as a greater whole. But the wholeness of a machine differs from the wholeness of a living organism in so far as a machine can be taken apart and put together again without permanently destroying it, for each of the parts has an existence independent of the whole. Were one to take apart a living organism piece by piece, one would surely kill it in the process. Each part of the living organism is dependent on the whole for its survival. And once dismembered into parts, the wholeness is gone, and powerless to rekindle life and unity from its fragments. There is a 'belonging together' of the parts of an organism that points to the existence of an *inherent* organizing principle, intrinsic to every part. It is this inherence of the organizing principle that a machine lacks. The organizing principle of a machine is always *imposed upon it* from the outside, by its maker.[1]

In the Middle Ages, it was widely accepted that the organizing principle in living creatures was an *idea* inherent within the organism, that actively 'informed' the organism with its specific qualities and attributes. It was regarded as much more important than a thing's material constituents in determining its nature for it has formative power. If we should consider that a horse is made up of flesh, bones, blood and sinews, none of these material constituents is what makes it a horse. The medieval mind sought out the inner idea as formative power (the 'formal cause') that endows the material constituents of the horse with the nature of 'horseness', as distinct from the nature of a dog or a rat. This idea-as-organizing-principle that lives in the horse can readily be grasped by the human mind, for it is intrinsically intelligible to human thought.

Since any number of creatures of the same species share the same essential nature, this essential nature could also be understood as a pre-existent spiritual archetype which each individual of a given species makes manifest. This pre-existent mode of being of the idea as archetype was conceived by the religious imagination to be in the 'mind of God'. The twelfth century thinker William of Conches could declare:

> God has the ideas of all things in his mind. Whence, Plato, in the *Timaeus*, calls the divine mind the archetypal world.[2]

The medieval consensus, then, was that ideas are much more than simply thoughts that human beings produce in their minds. They also exist in nature as the inherent organizing principles of creatures, and in the 'mind of God' as the spiritual archetypes of which all of creation is a manifestation.[3]

It can readily be seen that a mechanistic understanding of the world must militate against such a view, for the simple reason that it does not recognize the existence of *inherent* ideas in things. For the mechanist, there is no inner or spiritual dimension actively determining a thing's nature: its nature is entirely external and physical. For the mechanist, it does not make any sense to think of ideas as organizing principles with real formative power, active in nature independent of us. Because the mechanist can only experience ideas as products of the human mind, the perception of nature as alive and animated is lost, and nature, once experienced as a living organism, begins inevitably to die.[4] The crystallization of nature into dead mechanism belongs together with the conviction that ideas are nothing more than the product of human thinking. This conviction, which came to be known as 'nominalism', began to seem attractive to people in the fourteenth century. Before the fourteenth century, nominalsim was often discussed but it was not widely accepted, for in the earlier medieval period the world was woven through with an intrinsic coherence, meaning and spiritual significance, which the human mind could grasp in contemplation. In the fourteenth century, however, as thought was more and more claimed as the offspring of the human mind alone, so the inherent intelligibility of nature became questionable, and its coherence, meaning and spiritual significance unravelled.

William of Ockham: Herald of the New Age
One of the pioneering figures of the new consciousness was William of Ockham, who lived during the first part of the fourteenth century, and

was a contemporary of Jean Buridan and Thomas Bradwardine, two early proponents of the mechanistic view of the universe, whose ideas we touched on in the previous chapter. For Ockham, an idea (in Latin *universale* or 'universal') has no reality apart from the act of its being conceived. It is real only in so far as it is thought by the human mind. In a typically terse formula, Ockham states of ideas: 'their being consists in their being thought' (*eorum esse est eorum cognosci*).[5] There are therefore no ideas outside the human mind. What Ockham expressed corresponded to the experience of a growing number of people. It was becoming difficult to experience ideas as anything other than belonging to their own inner life of thought.

Ockham pressed upon his contemporaries the presumption that only singular entities exist outside the mind, each of which has an existence separate from every other. If ideas were to exist outside the mind, according to Ockham they too would have to be singular entities. 'Horseness', if it really existed, would then have to be a singular entity. But if it were so, how could it possibly exist in a multiplicity of creatures? Ockham's argument is simple enough: if only individuals or single things have real existence, then the very notion of an idea as a formative power in things cannot have any real existence outside the human mind that thinks it. If it did exist as real, it would have to be another individual, but then it could not possibly be actively present in a multiplicity of individuals. Thereby Ockham's world is excised of soul and intelligibility. The interior light by which medieval philosophers had hitherto come to know the intelligible order of nature is extinguished in Ockham's philosophy along with the ideas which it had previously illumined. Ideas are now just abstractions, little more than names that for our own convenience we bestow on single things that we judge to be similar to each other. For Ockham the conceptual is irremediably severed from the real.[6]

The argument reveals the pathos at the heart of the new consciousness of which Ockham was the forerunner. For this new consciousness, what previously had been experienced as pervading the world can now no longer be found there. The world presents itself to Ockham as no longer harbouring a thought-element. It is now bereft of spiritual content. For previous generations of thinkers, it was never in doubt that when we recognize something, and come to understand it, we do so because a spiritual content within the thing has been lit up in our thinking. The thoughts that we formulate in our mind were not felt to be in a different sphere altogether from the meaningful organization that we inwardly perceive in nature: therefore knowledge was experienced as occurring through a participation of the knowing mind in the object known.[7] For

Ockham, such a participative view of knowledge is no longer valid. It does not correspond to his experience, which is that the mind is *outside* what it knows, for the inner world of nature has withdrawn from him, and presents to him only its exterior countenance.

A fissure then appears in the heart of knowledge. On the one hand, we directly know only individual things, and this kind of knowledge involves an intuitive grasp of their reality. On the other hand, as soon as we reason about and try to understand the things we know intuitively, we are dealing not with the things themselves but with *propositions* about them, which may be true or false. If 'intuitive' knowledge is limited to our direct awareness of whatever it is we perceive in nature, then speculative or 'abstractive' knowledge is doomed to be one step removed from the natural world. And so Ockham must focus on what makes human thought legitimate, and this leads him to dwell intensively on the subject of logic, about which he writes a series of commentaries, expositions and treatises, notably the *Summa Totius Logicae* ('Summary of All of Logic').

Ockham concerns himself especially with the logic of terms (or so-called 'terminist' logic), in which the meaning of propositions is seen to rely entirely on how terms are used within a given linguistic context. In Chapter Seven, we saw that traditionally human language was regarded as grounded in the nature of things. Now Ockham comes to regard language as established by human convention and as having no intrinsic relationship to nature. For previous generations nature had been understood as the product of the intermingling of form (or archetypal idea) and matter, and so it was penetrable by human thought. But Ockham has stripped the formal (or ideal) element from nature. He therefore conceives that the role of logic should be restricted to the correct use of terms in a sphere that is separated from the actual knowledge of nature and really existing things: it is a 'science of reason' rather than a 'science of reality'.[8]

This divorce of the study of human reasoning (i.e. logic) from actual knowledge of reality is crucially significant, for logic now begins to break free of its traditional obligation to serve truth and to 'obtain knowledge of things' (as Thomas Aquinas put it).[9] It begins to function in a self-enclosed world disconnected from reality, a process that will lead eventually to the autonomous logic machine. This detachment could only occur within a nominalist sensibility, for only a nominalist sensibility could regard it as appropriate to free logic entirely from its relationship to the world and the quest for knowledge of things. With nature stripped of its metaphysical content, the emphasis thereby shifts towards the purely *logical* matrix within which human thought operates.[10] And so we see how the rise of the mechanistic worldview and the accompanying ascent of nominalism, both of which were manifestations of a new experience of divorce from the inwardness of nature, lie behind the origins of computer technology.

There is a further aspect of this process that is equally relevant to the prehistory of the computer. When Ockham looks beyond human thought and the linguistic signs by which it is expressed, he is confronted by a natural world that has lost every trace of inwardness, for it appears to him as purely material. In traditional medieval metaphysics, matter as such, or *materia prima*, had no positive attributes for it was a pure potentiality that was raised to actual existence through being infused by form. For Ockham, by contrast, matter exists in its own right. It is in itself (i.e. as *materia prima*) no longer a potentiality, but has specific characteristics. He states:

> It is impossible to have first matter (*materia prima*) without extension, for matter cannot exist without having parts distant from part. But this amounts to asserting that matter is extended, quantitative and has dimensions.[11]

What this means is that the form, or archetypal idea, is no longer required for things to come into manifestation. In this statement, Ockham arrives at the foundation of the later scientific worldview, which will eventually be taken up in the seventeenth century. But he arrives there several centuries prior to the scientific revolution actually taking place. We see then how, towards the end of the Middle Ages, the turn towards the mechanistic and quantitative worldview in which 'parts are distant from part' begins to take place.[12] It arises out of a new sense of divorce between the inner life of human thought and the inner life of nature, which is accompanied by the draining of spirit from nature. At the same time it involves a new accent on quantity rather than quality as a fundamental characteristic of the natural world. Thus, as early as the fourteenth century, the quantitative view of the world begins decisively to usurp the older tradition.[13]

Setting Out on the *Via Moderna*
Ockham was by no means the only nominalist of the fourteenth century. Many other late medieval thinkers were also inclined to deny the real existence of ideas outside the human mind. Nicholas of Autrecourt, Jean Buridan, and Nicole Oresme were some of the more famous of his nominalist contemporaries. All of them were attempting to express philosophically a new way of seeing things, central to which was the

emerging consciousness that only the particular is real. This meant that the natural world was becoming a conglomeration of essentially isolated phenomena for which the uniting principle had been lost. Applied to individual creatures, this manner of seeing necessarily tended to reduce them to parts external to parts, in other words to machines.[14] Once one perceives the world in this way, then the reign of quantity begins, for as Aristotle so well understood:

> Quantity means that which is divisible into constituent parts, each or every one of which is by nature some one individual thing.[15]

While it is to Ockham that we owe this critical step towards the over-throw of the traditional conception of a world infused by spirit and its replacement by a world made up of matter, characterized by physical extension and quantity, other nominalists went further than Ockham in this direction. Nicholas of Autrecourt revived the ancient Greek theory of atomism, regarding the movement of atoms as accounting more effec-tively for the qualitative characteristics of things than any appeal to forms.[16] Jean Buridan also pushed the conception of quantity almost to the point of making it the origin of quality. In his theory of impetus, the two categories of quality and quantity hitherto kept distinct collapse into each other.[17] This line of thought, in which emphasis was laid on the importance of expressing qualities in terms of measurements, was taken up by Nicole Oresme in the later part of the fourteenth century. Oresme sought to treat degrees of qualitative intensity (*intensio*) such as of heat or velocity, in terms of quantitative changes related to extension (*extensio*) of matter or time regarded as a measurable continuum. The quantitative measurement of time was, of course, made much more precise by the invention of the mechanical clock. Oresme's aim (achieved with more or less success) was to translate the 'quantity of quality' into geometrical figures.[18]

In all of these developments, carried forward not just by the above thinkers but by many others as well, we see how the nominalist viewpoint opened up the possibility of a new quantitative approach to under-standing the world. Those thinkers and early experimenters who iden-tified themselves with this new approach were characterized in the early fifteenth century as following the 'modern way' (*via moderna*) as opposed to the 'old way' (*via antiqua*) that prevailed during antiquity and for most of the Middle Ages. It is interesting that when Ockham used the word *modernus* in the first half of the fourteenth century, it meant simply 'contemporary', but as the fourteenth century moved towards its close, the meaning of the word broadened to include all those thinkers of the new nominalist sensibility, whether dead or alive. By the end of the fourteenth century, three generations after Ockham, the dividing line between *antiqui* and *moderni* had not moved forward. Ockham remained a *modernus* and his immediate predecessor Duns Scotus remained an *antiquus*.[19]

During the early fifteenth century, therefore, when the contrast between the *via moderna* and the *via antiqua* began to be made, 'modernism' had already been born, for the word *modernus* signified the new mentality. The *moderni* were those for whom the traditional view which holds that we best understand the world in terms of universal essences or forms that are the expression of divine ideas in the mind of God, was no longer valid. It was no longer valid because what really exist are individual things in the world, and these are physical through and through. The *via moderna* with its emphasis on the reality of singular things on the one hand and its focus on terminist logic on the other, radically broke with the old tradition. Because it carried more and more people along with it, it belonged as much to the future as to the 'present' of the fourteenth and fifteenth centuries. The spirit of the age was behind nominalism, and thus the medieval world slowly perished from within.

The Impact of the Reformation

While the radical fourteenth and fifteenth century *moderni* in many respects prefigured the early seventeenth century experimentalists and mechanistic philosophers who spearheaded the scientific revolution, they were nevertheless separated by a considerable interval of time. Before the scientific revolution could take place, the medieval world had inwardly to disintegrate. Long established habits of thought and deep-seated assumptions had to be abandoned, and for this to happen the passage of many generations was needed.[20] An entirely different kind of awareness, which would affect the individual's sense of self, relationship to nature and to the religious life, had to come into existence. This could not happen overnight, nor could it happen without an immense spiritual crisis. The crisis came in the form of the Reformation, a period of violent passions and excruciating upheavals.

Of all the many highly charged events of the Reformation, one of the most drastic was the dissolution of the monasteries in Britain, under the reign of King Henry VIII. As William Cobbett remarked, 'It was not a *reformation,* but a *devastation.*'[21] The Reformation also proved catastrophic for monastic life in Germany, Holland, Bohemia, Hungary and Scandinavia, as hostility to the very idea of monasticism was the hallmark of the reformers. While monasticism survived in countries which remained

Catholic, it was nevertheless shaken to its roots by the Wars of Religion (in France) and by the not infrequent assertions of secular authority (in Italy, Spain and Portugal).[22] As we have seen, the monasteries upheld the value of a life lived in prayer and contemplation, which throughout the medieval period provided an important counterbalance to the emphasis placed on logic, rational analysis and disputation in the universities. The assault on the monastic houses had a devastating effect on the pursuit of the contemplative life, within which not only prayer but also a discipline of the imagination, which involved meditation on the symbolic meaning of Biblical imagery, was practised.[23] The decline of such interior practices in turn affected the wider society, which lost the high regard for introversion that once it used to have and, in so doing, lost its spiritual bearings.

One factor of special relevance to the change in consciousness underway was the Protestant insistence on a literal reading of the Bible and a literalistic understanding of nature. In the monasteries, there had existed a long tradition of Biblical commentary and exegesis, which was pursued by monks in the spirit of elaborating different levels of meaning and symbolism contained in Biblical texts.[24] According to Origen, just as the human being is composed of body, soul and spirit, so all Scripture has three senses corresponding to these levels of being so intimate to our own nature.[25] By the twelfth century it had become common to think of four rather than just three levels of meaning. And so it became one of the principles of medieval Biblical exegesis that sacred texts will not properly be understood unless we move from their literal to their deeper allegorical, moral and spiritual meaning.[26] This led to commentaries becoming highly elaborate expositions of Christian teaching based on an imaginative penetration of a biblical text, sometimes in extraordinary detail.[27]

This interpretative approach to Scripture was also applied to the natural world, which according to a much-loved metaphor coined by St Augustine, was seen next to the Bible as the second of God's books, the so-called 'book of nature'. It, too, was replete with spiritual truths and symbolic meanings.[28] This meant that the depths of nature were seen as best plumbed by the contemplative imagination, rather than by a literalistic interpretation, for all creatures in nature have manifold meanings.[29] The medieval reading of the 'book of nature' was to see it as belonging to the same coherent web of meaning as could be found in the Bible. For this reason, when Luther and Calvin insisted that only the literal sense of the Bible was valid, the same stricture would inevitably be applied to the way in which nature was approached.[30] The new stance was, of course, the upshot of the nominalist denial of spiritual content to the natural world, and it served to further dry out and harden the experience of nature. Just as in the churches, images were smashed and the beautiful murals whitewashed, so too the 'book of nature' was stripped of all symbolic resonances in favour of a purely physical description of natural phenomena. The old approach to nature was regarded with suspicion, as fostering idolatry. Thus the way was cleared for Galileo to assert that mathematical analysis rather than the religious imagination should be the guide to understanding the order of nature. Turning his back on religious symbolism, he sought to reassure his readers by declaring that the 'book of nature' is, after all, 'written in the language of mathematics'.[31]

The Distinction Between Primary and Secondary Qualities

Let us briefly step back. It was the medieval consensus that the sense-perceptible qualities of things are expressive of their inner nature. Our senses reveal attributes (colours, textures, sounds, tastes and smells) that belong to a given substance, and are regarded as actualizations of its innate potentialities. In the traditional understanding, knowledge gained through sense-experience supported and contributed to knowledge of the inner nature of things. The sense-perceptible qualities are not added on to, but are rather manifestations of, a substance. And so they contribute to an understanding of its essence, and form the bedrock of its religious symbolism. But as nature's inwardness became less and less accessible to the consciousness of the *moderni*, the notion that anything of objective significance could be revealed by manifest qualities in nature became correspondingly problematic. For Ockham and the *moderni* who thought like him, what is objectively real in nature is characterized primarily by external dimensions and shape, to which the category of quantity is ultimately more applicable than the category of quality, to which belong the more capricious colours, sounds, odours and tastes.[32]

During the Reformation, the passionate insistence on a literalistic approach to nature, which sought to cut out symbolic associations, increased the pressure on qualities. A number of thinkers argued that non-measurable qualities (colour, sound, taste, odour) should be regarded as secondary to those qualities that can be subjected to measurement and quantified, thereby reversing the traditional view.[33] At the beginning of the seventeenth century, both Kepler and Galileo took up the cause, arguing that only the quantifiable characteristics of the world provide a reliable basis for knowledge. Kepler argued in *Harmonice Mundi* (1619) that quantity is the fundamental feature of things, prior to all other categories. The primary attribute of any substance is not in fact qualitative at all: it is quantitative and therefore expressible in mathematical form. It follows that the real world is the world revealed by mathematics, and the changeable sensible qualities belong only to the deceptive surface of things—a lower level of reality altogether.[34]

In Galileo's *Il Saggiatore*, ('The Assayer'), published in 1623, this new way of viewing things is forcefully driven home. The title of the work refers to someone who tests the amount of metal in an ore. For Galileo the ore is scientific knowledge and the metal-content of this ore is that part of scientific knowledge that can be expressed with mathematical precision. Galileo was convinced that the traditional approach to understanding nature was unable to lead to objective knowledge without mathematical demonstration. Mathematical certainty, based on identifying regularities in nature to which the behaviour of things conforms, effectively replaces the traditional aim of gaining insight into their inner nature or meaning. Galileo was a nominalist who didn't believe in nature's inwardness, but thought of it as a mechanism. And so the first aim of science was, for him, not to seek for essences but to establish regularities of behaviour and express them in terms of mathematical abstractions.[35]

But Galileo went much further than the mere description of mathematical regularities in nature. As a mechanist, he saw that it was imperative to analyse things into their constituent parts in order to take hold of them mathematically. He came to believe that matter must be resolvable into infinitely small, indivisible atoms which, because of their size, possess none of the 'sensible qualities' of colour, odour, taste or sound, but only those qualities of number, weight, figure and velocity that are susceptible of measurement.[36] Galileo was thereby able to reconstitute the world in such a way that the quantitative took ontological priority over the qualitative. The *real* world is a world of tiny atoms that are beyond the range of normal sensory experience. And the sensory world of our daily experience is an illusion created by the interaction of our sense organs with the colourless, tasteless, soundless, odourless world of atoms. The following statement from *Il Saggiatore* (1623), nicely sums up his credo:

> I do not believe that external bodies, in order to excite in us tastes, odours and sounds, need anything other than size, figure, number and slow or rapid movements; and I judge that if the ears, the tongue, and the nostrils were taken away, the figure, the numbers and the motion would indeed remain, but not the odours nor the tastes nor the sounds which, without the living animal, I do not believe are anything else than names.[37]

Not that Galileo had any evidence that atoms exist. He did not, as some might suppose, observe them through a microscope. Indeed, the closer physicists came in subsequent centuries to observing atoms, the more evanescent they became. Atomism, therefore, was a *philosophical theory* that Galileo introduced in order to replace the older view based on experience, which prioritized the sense-perceptible qualities so resistant to quantification.[38] Through atomism, he was able to propose a new way of explaining the world based on the category of quantity. In consequence, the qualities that previously had beckoned the observer to contemplate the inwardness of nature were now to be understood as subjective responses to quantifiable atoms (or 'corpuscles' as they were then termed) in motion. A very great number of tiny atoms, lacking the qualities of colour, odour, taste, texture and sound, were supposed simply through their numbers and motions to produce the world of sense-perceptible qualities. What we today would recognize as a description of the 'virtual world' stems from this historical moment. For Galileo, atoms in motion are the building blocks, the ultimate component parts, of the world-machine. In making their 'figure, number and slow or rapid movements' the cause of our experience of colours and sounds, etc., he establishes for the first time the model upon which the subsequent development of machine-generated virtual experience would be founded.

Galileo was convinced that prioritization of the quantifiable attributes of phenomena combined with philosophical atomism was the key to a new exact science. This conviction was rapidly and enthusiastically taken up by other natural philosophers. Descartes, who was some 30 years younger than Galileo, hit upon the same idea at about the same time.[39] It was as if a veil was lifted from a hitherto concealed truth that, once revealed, seemed to be self-evident. By the second half of the seventeenth century, the new distinction between *primary qualities* as the measurable and therefore truly objective qualities, and the non-measurable and therefore subjective *secondary qualities*, had become more or less universally accepted.[40] Underlying it was the belief that the former resolve into imperceptible atoms, while the latter resolve into the effects of the motions of these atoms on human consciousness.[41] The atoms themselves lack colour, sound, taste and smell, and yet through their number, their presumed shape and their slow and rapid movements they are able to induce in us the perception of a world alive with colours, sounds, tastes and smells. And so all educated people came to believe that the rich and varied sensations by which we orientate ourselves in the world do not really tell us anything about the world *as it is in itself*. Thus they were persuaded to distrust their own spontaneous experience of, and feelings for, nature.

The significance of the distinction between primary and secondary qualities is that it laid the foundations for the conception of the computer as an instrument through which qualities can be generated from quantitatively determined inputs that bear no similarity to the qualities that are outputted. As we have seen, the re-envisaging of nature as a machine was closely associated with the rise of nominalism, reinforced by the literalistic view of nature that was such a prominent feature of the Protestant reformers. But what made the mechanistic worldview viable was the quantification of the world by the new mathematical philosophy promulgated by Kepler, Galileo and Descartes, based on the primary/secondary quality doctrine. Re-conceived in this way, nature became susceptible to simulation by machines. The mechanical analogue of the motions of the stars in the first great astronomical clocks is an early example of this possibility being realized.

Under the sway of the quantitative worldview, the path was opened for human beings to conceive of machines that might generate new worlds of experience. The prospect of *recreating the world* mechanistically became conceivable as a direct consequence of the reversal of the traditional worldview, which had never thought to doubt that the qualitative aspects of things really inhere in a substance and disclose its essence. The mechanistic and quantitative worldview at once foreclosed the possibility of communion with the inwardness of nature and at the same time fired the mechanistic philosopher-scientists with an inebriating sense of power, for the new mechanistic, quantitative science was first and foremost a *technology*. One could do things with it. As Descartes noted, unlike the traditional contemplative natural philosophy, this was 'a practical philosophy', and by means of it we could 'render ourselves as the masters and possessors of nature'.[42]

Chapter Eleven

THE RENUNCIATION OF THE UPPER BORDER

The Upper and Lower Borders of Nature

In Chapter Nine, we touched on the view held in the Middle Ages that some bodies are intrinsically heavy and others intrinsically light, depending upon the preponderance of elements in the body. We saw that in medieval physics (which was based on Aristotle), gravity and levity were understood to be complementary principles. We now need to go into this a little further, for the denial of levity would become one of the causes célèbres of the early years of the Scientific Revolution, and its denial would prove to be an important stepping stone on the path that would eventually lead to computer technology.

In the medieval understanding, heaviness and lightness were in the first place inextricably linked to the balance of the four elements in bodies. In heavier bodies the balance is tipped towards earth or water, while in lighter bodies it is tipped towards air or fire. The four elements tend to arrange themselves above each other in nature, with earth occupying the lowest place, then water above it, then air and then fire. For this reason, heavier bodies (like rocks and stones) with a preponderance of earth will tend downwards towards the natural place occupied by the element earth, while lighter bodies (like hot ash or smoke) with a preponderance of fire will tend to rise towards the natural place occupied by the element fire. Bodies with a preponderance of water or air in their makeup will take their place between the earth beneath and the fire above (Fig. 11.1). According to Aristotle's *Physics*, the natural tendency (in Greek, *rhopē*) of heavy bodies is to move downwards and light bodies to move upwards, through a kind of innate propensity that the body has to find its rightful place in the order of elements. In relation to the strength of this propensity a body will fall either faster or slower, or equally rise faster or slower.[1]

There was a further aspect to the medieval understanding of heaviness and lightness, which it is easy for us to overlook. For the medieval mind the directions 'up' and 'down' were charged with metaphysical import. The 'upper border', which the element fire occupies, is in this understanding a *spiritual* border, beyond which one encounters the realm of

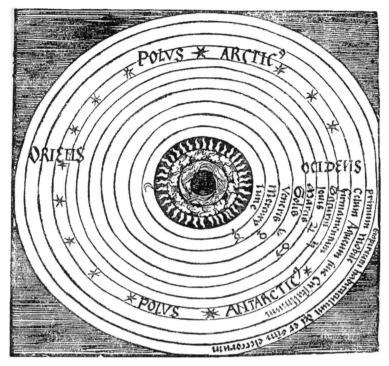

Figure 11.1

The medieval cosmos, showing the four elements arranged at the centre, one above the other. Earth occupies the lowest position. Surrounding it is the element water, and surrounding this are clouds representing the element air. Beyond the clouds are flames representing the element fire, and beyond the element fire are the seven planetary spheres. Note that the Arctic and Antarctic poles are typically located in the sphere of the fixed stars rather than on Earth. Sixteenth century diagram.

spiritual archetypes. Likewise, beneath the 'lower border', which the element earth occupies, there exists the sub-natural realm of *materia prima*—the 'first matter' that has the potential to take on form but of itself has neither visible form nor actual existence in nature, because it only becomes actual when it is 'informed' (Fig. 11.2).[2] By far the greater focus of medieval science was on the upper border, for it was believed that the explanation of the essence of things could only be found in the realm of spiritual archetypes, not in the realm of *materia prima*. From this heavenly world, which was conceived as pre-eminently a world of radiant light, spiritual archetypes and angelic beings bestowed their powers and exerted their influence upon the world below. Indeed, all forms in the world were understood to be derived from this sphere of heavenly light.[3] Other forces of an entirely negative character issued from the sub-natural world

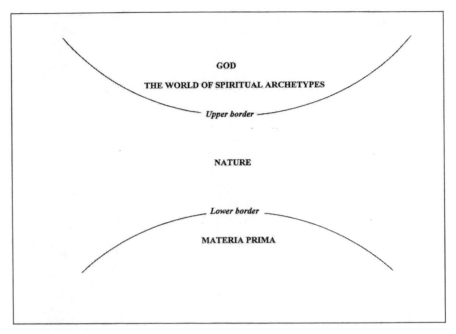

Figure 11.2
The upper and lower borders of nature, as conceived in the Middle Ages.

beneath our feet, which was the imaginal location of the realms of Hell, so vividly described by Dante in his *Inferno*. For medieval science, explanations of the phenomena of nature had to appeal to the upper border for their ultimate validation, because not only did this denote the realm of spiritual archetypes, but also because this realm was in the mind of God, the 'light which is the source of all light'. Without the all-sustaining power of God, all of nature would fall into nothingness, just as without the sun the air illumined by its light would fall into darkness.[4]

We need to dwell a little further on the medieval view of light, as this has a bearing on why so little attention was paid to the phenomena of electricity and magnetism during the Middle Ages. The medieval understanding of the nature of light was strongly influenced by the Biblical account of creation in the Book of Genesis, according to which light was brought into being before the sun, moon and stars. It was also influenced by the later identification in the New Testament of Christ with the 'light of the world'.[5] Another important influence was the writings of Dionysius the Areopagite, for whom the originating source of the light we experience in the world is the divine 'Father of lights'. The relationship of sunlight to the divine is, for Dionysius, that of image to archetype, for sunlight is an image of the divine goodness. According to Dionysius, 'Light comes from the Good, and light is an image of this archetypal Good'.[6] More than this, it actively works in material bodies to make them whole—nourishing, perfecting, purifying and renewing them.[7] To understand how Dionysius could attribute to light such an intimate role in material creation, we have to see that for him the light that makes the world visible is the outermost manifestation of the spiritual light that emanates from the divine mind. This spiritual light characterizes the heavenly world of spiritual archetypes or archetypal forms, which stand behind and within all created beings. And it is, furthermore, this light that illumines the minds both of the angels and of those humans who are able to participate in it, recalling them to the source of that radiance which is 'the goodness of the transcendent God'.[8]

Light, then, is a unifying factor in the universe, illuminating the human mind that seeks to know, as much as it inwardly informs the objects of which the mind seeks knowledge, for all creatures are so many reflections of the divine light.[9] Here we see how religious piety necessarily informed the natural philosophy of the Middle Ages. The writings of Robert Grosseteste, who lived in the early part of the thirteenth century, well exemplify this interweaving of religious piety and natural philosophy. For Grosseteste, the whole physical universe is made up of light: all its features are different forms taken by this single, fundamental energy that is 'the first corporeal form' and indeed 'corporeity itself'.[10] At the same time, for Grosseteste, as for St Augustine and for all of Christendom, God is light, a light that is intrinsically spiritual and which far excels in purity the light that emanates from the sun. But the sun's light nevertheless shares in the nature of this spiritual light, and in so far as each creature bears an intrinsic likeness to the creator, each creature too mirrors this essential light in its own limited form.[11]

Such a perspective was widely held in the Middle Ages, and it stands behind the outlooks of thinkers as diverse as John Scotus Erigena and Thomas Aquinas.[12] At the same time, the language used to describe light was far more nuanced than what we have grown used to today. Just as spiritual or 'primary' light (*lux prima*) was distinguished from visible light, so visible light was differentiated into *lux* (the light source) and *lumen* (the light emanating from a given source). Rays of light (*radii*) were a species of *lumen* rather than of *lux*, and the light that was reflected from a body (such as the moon) was referred to again with a different word, *splendour*.[13] The widespread regard for light during the medieval period was due to the focus on the upper border, from which by far the greater influence was believed to be exerted on the world of physical forms, while the lower border of material causes was seen as lacking formative power, having a receptive role in relation to the realm of spiritual archetypes.

Galileo: The Assault on the Upper Border

Galileo attacked this view with a fierce antagonism. For him the heavenly world was not different in kind from the world we live in here on Earth. In other words it is material, made of the same physical substances that exist on Earth. This was something he felt able to demonstrate with his telescope, for did he not observe through it mountains on the moon and sunspots on the surface of the sun? The heavenly bodies were physical bodies, and to the extent that Galileo considered the nature of light, this too he conceived as entirely physical, lacking any transcendent dimension.[14] There was no upper border for Galileo. How could there be if the universe was physical through and through? He believed it was precisely over-attention on the supposed upper border that had held back the progress of human knowledge since antiquity. For Galileo, the real world is the world of atoms in motion, that is to say the world that previously would have been located *beneath* the lower border of sense-perceptible qualities.

Galileo believed, furthermore, that there are no intrinsic tendencies in things to go up or down: rather, all things are subject to a primary force, which operates upon them externally and which is called gravity.[15] Levity, understood as the tendency of things to rise, should have no place in the new science. Gravity acts on bodies in such a way that there is just one Law of Fall that applies to everything and if some things fall faster or slower than others, this is due solely to the resistance of the medium that they are falling through, not to any innate propensity in the falling body. For Galileo, all objects would fall at the same speed if placed in a vacuum. Ash would fall at the same speed as lead, a feather would fall at the same speed as a stone.

Galileo's denial of the property of 'lightness' or levity (in Latin, *levitas*) to substances was simultaneously the denial of the existence of the upper border, beyond which lie the spiritual archetypes. His affirmation of the universal applicability of the law of fall was likewise an affirmation of the primacy of the lower border, on the other side of which an atomic realm was hypothesized—a realm of tiny particles of matter in motion, characterized only by 'size, figure, number, and slow or rapid movements', but nevertheless the source of all the varied phenomena of experience.[16] This atomic realm, essentially theoretical, was defined by its supposedly having no qualities save those susceptible to measurement. It was into this realm that Galileo wished to drive the sense-perceptible qualities of things (their colour, sound, taste, warmth or coldness, etc.), for only in so far as an object of experience can be measured does it possess the requisite 'weightiness' to be taken seriously. But the consequence was not simply that objective reality could only be known through mathematics, for it also inevitably led to the prioritization of the calculative faculty over the contemplative as the true bastion of reliable knowledge.[17] Galileo steered an unwavering course towards the lower border of nature precisely by rejecting contemplative insight in favour of the quest for mathematical exactitude.

The Middle Ages, under the influence of Aristotle, had avoided such an approach to nature because, as Aristotle had explained, the mathematician bypasses the real world of experienced qualities.[18] And so, for Aristotle, mathematics cannot lead us to the essential nature of a thing, which requires the contemplation of its different qualities perceived through the senses.[19] In adopting the mathematical method, Galileo gives his allegiance to a method of investigation that abandons the sensory world. He dismisses the perceived qualities of things, and the very notion that they might express an inward essence, as subjective illusions. If one really takes this worldview to heart, then one must resign oneself to accepting that human beings are nothing more than alien observers of a world bereft of meaning.[20] In the seventeenth century, it is as if a bargain is struck between those embracing the new physico-mathematical sciences and their own conscience, in its deeper sense of the inner guiding voice that keeps faith with the truly human. What was to be lost as meaning, and as connectedness with the soul of nature, would be made up for by far greater exactitude of knowledge, which would give to humanity a greater ability to control nature.

This new physico-mathematical mode of thinking, based on an ideological renunciation of the upper border, lies at the foundations of computer technology. As we saw at the end of the last chapter, once it is conceded that the world of qualities can be reduced to a world of quantities, and that the quantitative world is not only epistemologically but also *ontologically* prior to the qualitative world, it then follows that it should be possible to reconstruct the qualitative world out of the world of quantities. After all, the world of experienced qualities is, according to this view, merely subjective, a by-product of the world of quantities, which is the true locus of the real. And this locus is the other side of the lower border, from which material causes operate that can best be monitored by calculative reason, which alone provides the basis of objective knowledge. By contrast, the contemplative faculty—oriented towards the sphere of spiritual archetypes and allied with the experience of the world of qualities—must be disregarded for it is unable to lead us beyond our own subjectivity. As we shall see in later chapters, it was soon realized that the calculative faculty is susceptible to mechanization. In 1642, the year that Galileo died, the first mechanical calculator was invented. The creation of a technology dedicated to the reduction of qualities to quantities, and a corresponding reconstruction of qualities out of quantities, would become not only an epistemological goal but also an *ontological* goal. It would eventually result in the creation of a machine-generated 'virtual' world to rival the world of nature.

Electricity and Magnetism in the Middle Ages

During the medieval period, there was a notable lack of interest in electricity. No new thinking was directed towards furthering an understanding of electrical phenomena. So, for example, the manifestation of lightning in the thunderstorm was explained in the same way that it had been explained by the Romans—as fire, produced either by the friction of clouds rubbing against each other, or else by the wind violently tearing through the clouds.[21] This view, inherited from the Romans, was itself derived from earlier Greek philosophers such as Theophrastus.

Similarly, the phenomenon of static electricity was not explored any further than it had been in antiquity. Throughout the Middle Ages, the phenomena of magnetic attraction and repulsion were regarded as deriving from the unusual propensities of the lodestone, and static electricity from the equally unusual propensities of amber and jet. In scholastic terms, these phenomena were caused on the one hand by the substantial form (or inherent virtue) of the lodestone, and on the other hand by the accidental form (or acquired virtue) of the rubbed amber and jet.[22] Because lightning was regarded as a type of fire, only the electrostatic effects of amber and jet were seen as 'electrical', due to a peculiarity of amber (*electrum* in Latin), which jet also happened to share. Medieval writers on natural philosophy therefore merely noted the attractive properties of amber and jet towards chaff, accounting for them in terms of their acquiring an attractive 'virtue', similar to that of lodestone to iron. The difference between lodestone and amber or jet lay in the fact that whereas lodestone had an innate virtue which caused it to attract iron, amber and jet acquired their attractive virtue only when rubbed.[23] There was no sense that magnetism and electricity might be intimately connected with each other, or that they might be forces in their own right extending far beyond certain particular minerals with their curious propensities.

This lack of interest in electricity, which meant that its real nature went unnoticed, is in marked contrast to the keen interest shown in light during the medieval period. Medieval science, with its emphasis on the interplay of the four elements, and on the perceptible qualities, virtues and forms of things, was not equipped to really take hold of forces such as magnetism and electricity. This is because they lie beneath the realm of qualities, to which the medieval mind was so obstinately loyal. The medieval unwillingness to relinquish the realm of qualities was because of this realm's inherent openness to religious and symbolic interpretation. To move away from the qualitative would have been to move away from the meaning-infused spiritual universe, in which the religious consciousness of the time basked.

However, towards the end of the thirteenth century, there was a first awakening of interest in magnetism. In Chapter Nine, we traced the origins of the mechanical clock. We saw how the development of the early clocks depended not just on a mastery of the complex internal mechanism of cogs and wheels and the invention of the verge escapement, but also on the utilization of the force of gravity as the motive power of the clock. In medieval terms, this would have been understood as the utilization of the innate tendency of heavy objects such as lead to fall towards the earth, in other words to fall towards the lower border. Efforts towards constructing such a gravity-powered clockwork mechanism can be traced as far back as the 1260s.[24] The new interest in gravity, and the pull towards the lower border in the quest for a more accurate quantification of time, was paralleled by a new interest in magnetism in the latter part of the thirteenth century.

At almost exactly the same time that the gravity-driven clock was being worked on towards the end of the 1260s, Petrus Peregrinus composed the first systematic and empirically accurate treatise on the magnet, entitled 'Letter on the Magnet' (*Epistola de Magnete*). This short treatise was written in 1269, and established certain important facts about the magnet or lodestone.[25] In his experiments, Petrus Peregrinus used a lodestone made in the shape of a globe. By placing iron needles at various points on the surface of the globe, a series of lines could be drawn that indicated the meridians of a magnetic field (although the concept of a magnetic field was not made explicit). It is typical of his medieval sensibility that this spherical lodestone represented for him not the sphere of the earth but the sphere of *the heavens*.[26] The lodestone was thus a kind of microcosm of the greater macrocosm. For Petrus Peregrinus the source of magnetism must be in the heavenly world, for the heavens are the source of the 'virtues' of all things and of all creatures terrestrial.[27] Petrus Peregrinus was by no means the only medieval natural philosopher dedicated to painstaking empirical observation and experiment, but his research into magnetism was conducted within a sensibility that was medieval through and through. His world is still an animated world and he still has a sense that the phenomena of the earth are subject to and directed by heavenly, that is to say spiritual, forces.[28]

William Gilbert: The Emergence of the Electric Force

Petrus Peregrinus' 'Letter on the Magnet' was the major authoritative source on magnetism for over 300 years. But right at the beginning of the seventeenth century a new study entitled De Magnete (1600) by William Gilbert produced a significant breakthrough.[29] Prior to this, during the sixteenth century, a number of thinkers and experimenters had been drawn to the phenomena of magnetism and static electricity, but none of them went very far beyond the experiments already conducted by Petrus Peregrinus and others in the thirteenth century.[30] It was Gilbert's treatise that decisively broke new ground. Petrus Peregrinus had explained the phenomenon of magnetism with reference to celestial influences streaming down to Earth from the heavens. If it was the innate 'virtue' of the lodestone to have two magnetic poles that aligned north and south, for Peregrinus this was because the lodestone was receptive to the influences emanating from the stars in the corresponding parts of the heavens. What Gilbert proposed was that, far from the virtue of the lodestone being derived from the heavens, it was derived from the Earth itself. The Earth, he argued, should be reconceived as a giant magnet, and the lodestone would then appear as a miniature replica of the Earth.

Gilbert had spherical magnets made to emulate the sphere of the Earth, and he called them terrellae ('little worlds'). On these terrellae he conducted his experiments, proving to his own satisfaction that the Earth is a huge magnet and all the phenomena of magnetism could be explained without recourse to heavenly influences. Figure 11.3, a diagram from Gilbert's De Magnete, depicts a spherical terrella with its north pole at A and its south pole at B, thus representing the Earth 'lying on its side'. The magnetized needle placed on top of the stone orientates itself north–south, pointing towards the north pole, and keeping perfectly horizontal, whereas a magnetised needle placed at E assumes an angle of approximately 45° to the surface of the terrella. When placed directly over the north pole, the needle stands perpendicularly erect. By placing needles or thin wires on the surface of the terrella in this way, Gilbert was able to mark out lines of magnetic force on its surface with precision.[31]

In denying this heavenly cause of the phenomenon of magnetism, Gilbert joined forces with Galileo, who was equally intent on breaking

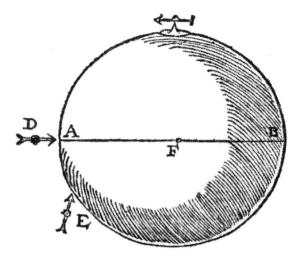

Figure 11.3

Gilbert's spherical lodestone or 'terrella', representing the Earth 'lying on its side', with A and B representing the Earth's poles, and F the Earth's centre. D and E represent pivoted compass needles applied to the lodestone ball.

the traditional pattern of thought that would always bring the explanation of things back to a spiritual cause. Gilbert was an older contemporary of Galileo, being born exactly 20 years before him in 1544. We know that Galileo read and admired Gilbert's treatise on magnetism, and subsequently conducted many experiments with magnets himself.[32] The reason for Galileo's interest was that Gilbert's treatise, by describing the Earth as a great magnet, suggested an answer to the question of why bodies fall. With the source of magnetism now relocated from the upper to the lower border, the prestige of the lower border was greatly enhanced.

But Gilbert's treatise was important for a further reason. It included a whole chapter dedicated to a discussion of electricity.[33] Because of his rejection of the upper border in his explanation of magnetism, Gilbert was also able to break the spell of scholastic thinking that held electricity confined as a property of amber and jet. He was able to show that the attractive property of static electricity characterizes many diverse substances, including sulphur, glass, diamonds, sapphire, resins, crystals and so on.[34] Far from being a peculiar property of amber and jet, Gilbert proposed that there are many 'electrics', and the electric virtue is in fact a material 'effluvium' which is released or 'exhaled' from electrics when they are rubbed, and then draws other bodies to the 'electric'. This effluvium he referred to as the 'electric force', and it characterizes those substances which attract in the same way as amber (*electrum*).[35] Gilbert believed that both the magnetic and the electric force are to be understood in purely material terms. He went so far as to describe the effluvium exhaled by an electrified substance as a 'material ray' (*'tanq materiales radii'*) that emanates from it.[36] The phrase is interesting because the term *radii* was normally used of emanating light (i.e. *lumen* rather than *lux*). Gilbert was, in effect, suggesting that the electric effluvium is a kind of materialized light.

Gilbert also proposed another radical idea. Not only is the Earth a giant magnet, but also the globe of the Earth coheres *because of electricity*. It is the electric force that cements together all its disparate parts:

> The globe of the Earth is aggregated and coheres by itself electrically. The globe of the Earth is directed and turned magnetically; at the same time also it both coheres, and in order that it may be solid, is in its inmost parts cemented together [electrically].'[37]

The full significance of this statement can only be appreciated when we compare it with the earlier medieval view that understood the coherence of all material bodies as due to formal, that is to say *spiritual*, principles, which are heavenly in origin. As we have seen, light specifically was regarded as a principle of coherence, preserving and maintaining all things in their essential unity.[38] Light was apprehended as having a transcendent dimension that reaches into the realm of spiritual archetypes in the mind of God. All things are manifestations of this divine light and all are maintained in existence by the power of God, for as Thomas Aquinas put it, 'they would instantly be reduced to nothing if God should abandon them'.[39]

In Gilbert's statement there is the first intimation that electricity could take on the role hitherto assumed by the light, conceived both as the inward essence of the divine and as the underlying energy of spirit that makes the world cohere. Just as for Gilbert it is not necessary to look to heavenly influences in order to explain magnetism, so it is not necessary to look in that direction either for the inward coherence of the world, for this is taken care of by the electric force. Thus at the same time as Gilbert forgoes the 'upper border' in his explanation of magnetism, he introduces the electric force as the internal sustaining power of the world. He thereby opens the way for a secular theology of electricity, which in subsequent centuries would challenge the traditional conception of the divine.[40] It was a pivotal moment in the prehistory of the computer.

Part Three

THE EARLY MODERN PERIOD

Chapter Twelve

THE SEVENTEENTH CENTURY
PENETRATION OF SUBNATURE

The New Scientific Societies

The abandonment of the upper border of spiritual archetypes, combined with the desire to explore the material basis of natural phenomena and the embrace of mathematics as the tool of the new science, together made up the milieu in which electricity began to reveal itself to the post-medieval consciousness of the seventeenth century. During Galileo's lifetime, scientific societies were beginning to spring up. Galileo himself belonged to one such society, called the Accademia dei Lincei ('Academy of the Lynx'), which published his book *Il Saggiatore* in 1623. It was succeeded by another scientific society called the Accademia del Cimento ('The Academy of Experiment'), founded in Florence in 1657 and dedicated to pursuing the Galilean approach to nature. Just five years later, in 1662, the Royal Society was founded in England, the fruition of a proposed 'College for the Promoting of Physico-Mathematical Experimental Learning'. French and German societies followed shortly thereafter.[1]

The shared agenda of all these societies was to develop rigorous standards of measurement, to create proper laboratory instruments, and to conduct well-documented, systematic experiments. Through these scientific societies, the centre of European culture shifted away from the Church and the traditional religious philosophy. They represented a new and quite different locus of knowledge and authority, based neither on spiritual insight nor on the contemplation of sacred Scripture, but on applied reason, calculation and experiment. Thus the latter part of the seventeenth century saw a consolidation of the new 'physico-mathematical experimental learning', in each instance under royal or princely patronage.

One of the most influential publications of the period was a volume of essays and reports of experiments conducted by the Accademia del Cimento, entitled *Saggi di Naturali Esperienze* ('Examples of Experiments in Natural Philosophy'), published in 1667. Here we read of abortive attempts to measure the speed of light, which were based on the assumption that light is a physical object like any other.[2] We also read of two experiments specifically designed 'to prove that there is no positive levity'.[3] These latter experiments, completely baffling in themselves, demonstrate (if nothing else) the deep-seated opposition to the notion of levity and the correlative 'upper border' of nature. In the publication, they are immediately followed by experiments on magnetism and electricity.[4] It is as if in order to clear the ground for the study of magnetism and electricity, it was considered imperative both to physicalize light and to discredit the concept of levity first. The rejection of the concept of levity was certainly something that William Gilbert had felt the need to do more than 60 years previously, almost as the price to be paid for the new vision of a cosmos in which everything is seen under the aspect of the forces of magnetism and electricity.[5]

The Creation of the Vacuum

It is at around this time of the burgeoning of interest in the new science that further strides were made towards revealing the nature of electricity by the German experimenter Otto von Guericke. Although von Guericke was an independent researcher, for the German Society of Sciences had not yet been formed, he presents a good example of the new scientific consciousness at work. Like the experimenters of the Accademia del Cimento, von Guericke was fascinated not only by electrical phenomena but also by the possibility of creating a vacuum. As we shall see, these two areas of interest were closely connected.

Von Guericke was as much an engineer as a scientist, and his attitude towards both the vacuum and electricity was in stark contrast to the attitudes of the earlier age. We shall look briefly at the creation of the vacuum first. For the medieval natural philosophers, following Aristotle, a vacuum could not exist in nature because a vacuum is by definition an empty space, and all space in nature is occupied by the four elements. In nature everything has its allotted place, and an empty space would in effect be a negation of place, in which nothing could exist, for where there is no place there can be no thing existing 'in' it.[6] Von Guericke understood that if he were to show that a vacuum could exist in nature, he would himself have to produce it. This he did by first of all designing and then constructing an air–pump that was able to suck air out of a glass vessel or 'receiver'. His design enabled him to place in the receiver various objects and small creatures before the air was evacuated (Fig. 12.1). Von Guericke then noted the effects. He observed how as the air was evacuated from the glass receiver a flame would slowly contract and turn blue before being entirely extinguished, how a bell would lose its ability to sound, and how a little sparrow was brought to a standstill as it gasped for breath with its beak wide open, until finally it fell over and died. All of

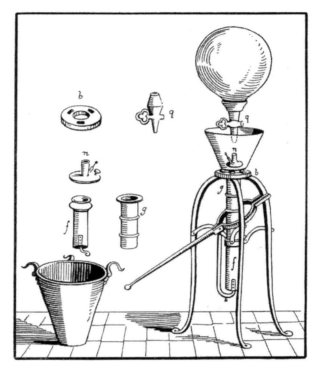

Figure 12.1

Von Guericke's improved air-pump. mounted on a tripod. The conical end of the receiver (top right) is thrust into a tube n, under which is a leather valve, which is opened at the downward stroke of the pump, allowing air to pass into the barrel below. The valve closes again at the upward stroke.

these things he recorded with unflinching objectivity and cold dispassion.[7]

The creation of the vacuum was of enormous symbolic significance, for a totally new kind of space was created through human technological prowess, which dared to challenge the order of nature. In one of his more spectacular experiments, Von Guericke made a receptacle composed of two brass hemispheres pressed together. When the air was evacuated, the hemispheres were, as a result, bound together with such great force that two teams of eight horses each were unable to pull them apart.[8] He demonstrated this at Magdeburg in 1656, an event that became legendary and was illustrated countless times over the next two centuries. The illustration below (Fig. 12.2) was made in 1850, nearly 200 years after the event, a clear sign of the lasting fascination that the ability to create a vacuum exercised. But what was it that so fascinated people? The

Figure 12.2
Von Guericke's demonstration of the pressure of air on an evacuated receptacle. From a woodcut, 1867.

scientific explanation seemed straightforward. The resistance of the hemispheres to being pulled apart was due to the pressure of the surrounding air on the evacuated receptacle: because it had been evacuated, there was nothing to compensate the external pressure exerted on it, which as a result was extraordinarily powerful.[9] But the focus of attention was inevitably not on the air surrounding it, but on the receptacle itself and what seemed to be the *power of the vacuum* within the receptacle to resist its being pulled apart. It was not the ambient air but the airless vacuum that gave every appearance of being greater in strength than the combined muscle-power of 16 horses. Here, then, a force of a quite different quality to that normally experienced in nature seemed to have become present within the natural world. This was something new and uncanny that had never before been witnessed. The negative space of the vacuum gave every appearance of possessing a gigantic, preternatural strength. It was as if it belonged to a wholly different order of existence, as if a hole had been punched through nature to a subnatural realm that lay beyond the reach of God.[10]

As a result of von Guericke's work, Robert Boyle and Robert Hooke in London made a series of air-pumps of increasing sophistication. They conducted a range of experiments in receivers that showed that, as the receivers were evacuated, candle flames consistently turned blue and went out, sounding bells became mute, and small animals and birds placed inside them did indeed die of suffocation. Furthermore, the growth of plants was arrested, bladders filled with air burst, and feathers fell at the same rate as stones, just as Galileo had predicted. On the other hand neither magnetism nor electricity were inhibited, as similar experiments conducted by members of the Accademia del Cimento and the French Académie des Sciences confirmed.[11] Indeed, as we shall shortly see, electricity in particular seemed to have a certain affinity with the vacuum, for the latter provided the conditions under which electricity seemed to thrive.

It is important that we dwell a little on the meaning of the creation of the vacuum, for it would later play a crucial role in the functioning of the first electronic computers of the Second World War. ENIAC, the first fully operational computer, was totally dependent on its 18,000 diode vacuum tubes, without which it would have been quite useless. Here we see how what was done in the mid-seventeenth century had long-term consequences, making it possible for other things to be done 300 years later in the mid-twentieth century. The creation of the vacuum was a breakthrough of enormous significance in the prehistory of the computer, and this significance lies precisely in the fact that it broke through the natural order to make manifest something that had never before been known, the very idea of which was regarded by those living in the Middle Ages with horror, for it was so obviously 'against nature' (*contra naturam*). No one then had any doubt that 'nature abhors a vacuum' (*natura abhorret vacuum*).[12]

Through the air-pump, which was perfected by Francis Hauksbee in the early years of the following century, this new kind of 'abhorrent' space became an accepted feature within the experimental laboratories of scientific researchers. It was a kind of anti-world in the midst of the real world, in which the powers of life were rendered impotent, unable to resist the forces of gravity and death. The snuffed flame and the falling feather speak as eloquently of the character of this chamber of death as the numerous small animals gasping for breath and the songbirds falling forever silent. Within the evacuated receiver, the natural order was overturned and conditions that belonged outside the natural order were allowed to take up residence within it. During the eighteenth century experiments with the vacuum proved remarkably popular and were repeated time and again, as if people were drawn by a morbid fascination to pry into a space that did not belong to the natural world, but was like a protrusion of the realm of death into the world of living nature.

The First Electric Machines

At the same time, and very often by the same people, new headway was made into the generation of electricity. Included in the same book in which von Guericke published his experiments on the creation of a vacuum, there is a chapter devoted to his experiments with electricity.[13] Inspired partly by William Gilbert's work on magnetism, von Guericke made a globe of sulphur, a sort of 'terrella', by pouring molten sulphur into a glass sphere and then breaking the glass when the sulphur had cooled. He then mounted this sulphur globe on an iron axle, which rested on two supports so that it could be rotated and at the same time rubbed with a dry hand. This was the first ever 'machine' constructed specifically for the generation of electricity (Fig. 12.3).

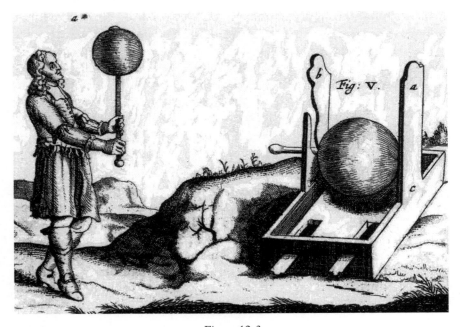

Figure 12.3
Von Guericke's sulphur globe and electric machine.

Von Guericke noticed not only that the charged globe attracted light objects, like paper and feathers, and carried them around with it as it rotated, but that it also repelled similarly electrified bodies, only to re-attract them once they had themselves come into contact with a non-electrified body. Thus a charged feather placed between the electrified globe and the ground beneath was first repelled from the globe towards the ground, but as soon as it touched the ground it was attracted back again to the globe. On the basis of this attractive power of the globe, von Guericke concluded that it was a model of the Earth, just as was Gilbert's lodestone, but for von Guericke the attractive virtue of the Earth was due to the electric force, which he equated with the force of gravity. The electrified Earth, however, conveniently kept certain bodies like the moon at a safe distance through its power of repulsion. The cause of the Earth's electrification was, he thought, due to the friction of the sun's rays on the surface of the Earth as the globe of the Earth turned.[14] Thus von Guericke came to conceive of the Earth as a vast electric machine.

Von Guericke was the first to clearly observe that electricity both attracts and repels. He was also the first to observe that the electric charge could travel from one end of a linen thread to the other, and that bodies could be electrified simply by being brought into the vicinity of the charged sulphur globe, without need of actual contact. These two as yet unnamed capacities of electricity were later termed 'conduction' and 'induction'. They both pointed towards electricity being a phenomenon that is less a quality that characterizes certain substances than a force that becomes manifest through them, a view that Gilbert had already begun to grope towards. This intimation of the nature of electricity, which von Geuricke was still not quite able to articulate, would become much sharper in time. Von Guericke also noticed the generation of light (flashes) and sound (crackling noises) as phenomena accompanying electrification. As soon as his book, *Experimenta Nova*, recording his experiments was published (in 1672), others—notably Robert Boyle— repeated the experiments.

Some time later, in the early years of the eighteenth century, Francis Hauksbee, Fellow and Curator of the Royal Society, conducted a series of experiments in which the vacuum and electricity were brought into conjunction. He made a machine that could rotate an evacuated receiver very rapidly, into which he placed amber and wool. As the amber and wool rubbed against each other, he observed an accentuation of the light phenomena caused by their friction as com- pared with the friction of amber and wool in an unevacuated receiver. This led him to a simpler and more dramatic experiment, in which he rotated an evacuated glass receiver with nothing in it at all (Fig. 12.4). When he applied his bare hand to it as it spun round, sparks an inch long shot out and an eerie purple light with bright streaks was pro- duced, sufficient to read by. As soon as air was re-introduced into the receiver the electrical effects diminished.[15] If the vacuum was inimical to life, Hauksbee showed that it was an environment in which, by contrast, electricity flourished.

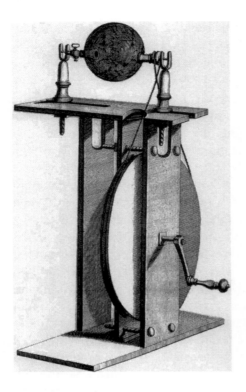

Figure 12.4
Francis Hauksbee's electrical machine,
consisting of an evacuated glass globe about
nine inches in diameter, which could be
rotated rapidly about an axle.

The Second Fall

In all of these early experiments with electricity and the vacuum, the
phenomena that are studied had first of all to be produced by human
agency. There was a need for human beings to become actively and
wilfully involved in order for these phenomena even to become manifest.
In this respect, the new experimental science differed radically from the
conception of human knowledge that had prevailed from ancient Greek
times through to the Middle Ages. Previously, the bedrock of scientific
activity consisted in the contemplation of already existent natural
phenomena through bringing concepts to bear on them that illuminated
their essential nature. Scientific activity was above all contemplative
activity, and this was felt to be appropriate because the ultimate cause of
all natural phenomena was divine, and so science itself was pursued within
a sacred context. Quite different in character again was the relationship to
nature that characterized the ancient civilizations prior to and including
archaic Greece. Then nature was not so much the subject of intellectual
contemplation as of strong emotional participation. As we saw in earlier
chapters, people had a strong emotional bond with nature such as we can
hardly imagine today. We need only consider the strength of feeling with

which the ancients participated in their seasonal festivals to glimpse the very different consciousness that prevailed in deep antiquity. The seasonal festivals were at the same time divine dramas in which the passions, deaths and resurrections of gods were enacted. The sacred rites of Isis and Osiris in Egypt, of Inanna and Dumuzi in Sumeria, and of Demeter and Persephone in archaic Greece lived deeply in people's souls.

It is incumbent on us today to place the scientific consciousness that emerged in the seventeenth century into this wider historical perspective if we are to understand its significance not only for our own times but also for the future. The experimental method enabled human beings to become aware of new phenomena that had hitherto been hidden from human consciousness, but because these phenomena were generated by human activity it is not strictly correct to refer to them as *natural* phenomena. The vacuum and the electrostatic effects produced by the experiments that we have been considering actually brought human beings to the threshold of a world that, strictly speaking, is not 'natural', for these phenomena only became manifest through technologies requiring the active involvement of the human experimenters. However, the approach to this threshold of a non-natural world was largely unconscious. Few considered its metaphysical implications. Caught up in the extroverted activity of the new experimental method, most of the experimenters were unable to grasp the meaning of what it was that they were really conjuring forth.

It is for this reason that Ernst Lehrs, in his profound study of the birth of scientific consciousness, *Man or Matter*, refers to the scientific revolution as a 'Second Fall'.[16] In the Garden of Eden, as described in the Book of Genesis, Adam and Eve represent primordial humanity living in communion with God, who shares the garden with them, converses with them, and is inclined to stroll in the garden 'in the cool of the day'.[17] The description in the Book of Genesis indicates a condition not simply of human innocence but also of an extraordinary intimacy between human beings and the divine. According to Lehrs, in the First Fall, when Adam and Eve ate of the Tree of Knowledge they succumbed to the temptation to acquire knowledge prematurely. This led to a certain illumination of human consciousness, but at the devastating cost of a separation from the original state of participation in the divine world. The Second Fall, by contrast, was due to human action outrunning knowledge, so that powerful forces came to be grasped hold of and utilized by human beings, who were unable to penetrate the real nature of what it was they had so cleverly discovered. Lehrs remarks concerning the early exploration of electricity that:

men were exploring the electrical realm as it were in the dark; it was a realm foreign to their ordinary ideas and they had not developed the forms of thought necessary for understanding it.[18]

While this was clearly the case in the seventeenth century, it was also, as we shall see, just as much the case in the eighteenth and nineteenth centuries as well. Human understanding fell far short of the moral maturity and spiritual insight needed in order to proceed wisely in relating to this realm. We still live today in the era of the Second Fall, characterized by a consciousness not only separated from the divine but also perilously overconfident in its ability to bend to its own use forces that it does not understand.

The observations of Ernst Lehrs are borne out by the fact that the sense of excitement that the new scientific consciousness engendered in intellectual circles throughout Europe was counterbalanced by a less articulated but no less real sense of a Second Fall, which was the implicit theme of Milton's great poem *Paradise Lost*. This hugely popular poem spoke to the age. First published in 1667, it was reprinted repeatedly over the next 200 years, often in illustrated versions, which vividly portrayed its archetypal themes.[19] While ostensibly about the Fall of Adam and Eve from Paradise, at the fulcrum of the poem is the Promethean figure of Satan, chief of the rebel angels, at once both chilling and inspiring, who dominates Book One.

It is an interesting synchronicity that the date of publication of Milton's *Paradise Lost* was exactly the same as the publication of the Accademia del Cimento's *Saggi di Naturali Esperienze* ('Examples of Experiments in Natural Philosophy'). Milton had a lifelong predilection for the scientific discoveries of his age, and his great admiration for Galileo led him as a young man to seek a meeting with Galileo when he visited Florence in 1638.[20] In *Paradise Lost* he refers obliquely to Galileo as 'the Tuscan artist', in a passage where the poet describes the shield of Satan as 'like the Moon, whose Orb through optic glass the Tuscan artist views ... rivers or mountains in her spotty globe'.[21] This shield of Satan, then, is not simply round like the moon. It is like the moon *as revealed through the optic glass or telescope*, through which Galileo first observed 'rivers and mountains' on its surface. Unlike poets of an earlier age, Milton does not associate the moonlike shield with silver or with the goddess Diana. Rather he associates it with the way it appears to the scientist viewing it through his telescope: specifically the Tuscan 'artist' Galileo who challenged the medieval worldview and who, when Milton visited him, was a prisoner of the Inquisition. In other words, the shield with which Satan defends himself against the spiritual powers of Heaven is the objectifying scientific consciousness that literalizes what it observes.

Milton's description of the shield comes just after Satan's rousing speech to the fallen angels in which he urges them to accept the new situation which their rebellion has resulted in: namely, that they have all fallen into the infernal world, the world beneath the natural world. 'Hail Horrors', says Satan, 'hail Infernal world, and thou profoundest Hell receive thy new Possessor . . .'[22] In the complexity of Milton's psyche one never can be sure where his deepest sympathies lie. His Satan appears in the early part of *Paradise Lost* as a magnificent anti-hero in rebellion against the tyrannical old religious order. In his rousing speech Satan speaks proudly of embracing the severance from the divine:

Figure 12.5
Satan contemplates the infernal regions, 'deep underground, materials dark and crude of spirituous and fiery spume'.

... Here we may reign secure, and in my choice
To reign is worth ambition though in Hell;
Better to reign in Hell, than serve in Heaven.[23]

Perhaps the most telling episode in the poem is that in the great battle that follows, Satan embarks on a search for a weapon that might be able to defeat the heavenly hosts. His quest takes him 'deep under ground' where 'materials dark and crude of spirituous and fiery spume' are buried (Fig. 12.5). Only gradually do we realize what this weapon could be, that would have the power to 'dash to pieces and overwhelm whatever stands adverse'. Familiar to us from the ancient mythologies of the thunder gods, from Ishkur and Baal Hadad to Zeus, Milton reveals that it is the primeval forces of electricity that Satan seeks to take possession of, so his adversaries 'shall fear we have disarmed the Thunderer of his only dreaded bolt'.[24] In this way, Milton consciously or unconsciously identifies Satan as a representative of the new science. Both the shield with which he defends himself, and the infernal weapon with which he intends to inflict a final devastating blow against the heavenly hosts, are placed in his hands by the new scientific mode of knowing.

Chapter Thirteen

FRANCIS BACON'S 'NEW INSTRUMENT'

Francis Bacon and the Seed Idea of the Computer

In the last three chapters, we have followed up the legacy of Galileo's radical new approach to nature, through his renunciation of the upper border of spiritual archetypes and essences, his embrace of the atomistic worldview and his pioneering the 'quantitative way of seeing', which demanded a regimen of experiment and observation that would continually subject itself to calculation and measure.[1] One of the key features of this new approach was the interrogation of nature through the experimental method. This required the fabrication of new scientific instruments. Contraptions such as the air-pump created vacuums that had never before existed, while the electric machine likewise produced electrostatic phenomena that no one had ever before witnessed. Before the seventeenth century, no one had believed that a vacuum in nature was possible, nor had anyone imagined the kind of electrostatic effects produced by the new machines with their evacuated glass globes. During the seventeenth century a new world was beginning to be discovered, but the discovery of this new world was totally dependent on the instruments that scientists invented in order to discover it. The phenomena were not already there, like animals, plants or stones: the conditions had to be created by human technologies in order for them to become manifest.

Just as Galileo stands at the threshold of the new technological interrogation of nature that would lead to the harnessing of electricity for human benefit, so his contemporary Francis Bacon stands at the threshold of the radical narrowing of the scope of reason in order to gain the kind of knowledge that would have practical use for human beings. If Galileo's legacy was the penetration of subnature and the awakening of electricity from its state of dormancy, then the chief legacy of Bacon was the mechanization of the process of thinking, which would eventually result in the automation of logic. Destined to be born within three years of each other, Galileo (born in 1564) and Bacon (born in 1561) may be regarded as founding grandfathers of the computer, which uses electricity in the service of the automation of logic.

It was one of the hallmarks of the new philosophy of the seventeenth century that the purpose of human knowledge was seen less as being to further our spiritual understanding of the mysteries of existence than to increase our ability to control and exploit the world about us. Lacking the experience of nature's inwardness, the pursuit of knowledge of nature now needed to be shown to be *useful*: the aim of knowledge as the spiritual contemplation of God's creation was no longer sufficient. The traditional knowledge of nature, based on the contemplation of spiritual essences and their religious symbolism, could only seem utterly pointless once the experience of nature's inner dimension was lost. Francis Bacon's thinking well complemented Galileo's approach to nature. Bacon wrote in justification of his project for the overhaul of human knowledge that he was 'labouring to lay the foundation, not of any sect or doctrine, but of human utility and power'.[2]

To achieve this kind of practically useful knowledge, Bacon believed that a 'new instrument'—a *novum organum*—was required. This instrument was to be the human mind stripped of the contemplative propensities of the *intellectus*, so as to focus all the more effectively on the analytical and discursive thinking of the *ratio*.[3] In his book bearing the title, *Novum Organum* (1620), he described how in order to create the 'new instrument' the mind needed to be 'hung with weights to keep it from leaping and flying'.[4] It had, in other words, to be subjected to the mental equivalent of the force of gravity if it was to be disciplined into becoming an efficient and reliable tool for processing, analyzing and cataloguing data. Like Galileo, Bacon was concerned that knowledge should above all be reliable, and so utmost rigour was needed in order to avoid error. Rules had to be formulated and rigidly adhered to. Bacon sought to forge an instrument, or method, that would 'proceed in accordance with a fixed law, in regular order, and without interruption'.[5]

The method was essentially a program of instructions that Bacon sought to apply to the human mind operating in a purely mechanical way. In order to acquire reliable knowledge, he believed that it was necessary that:

> ...the entire work of understanding be commenced afresh, and the mind itself be from the very outset not left to take its own course, but guided at every step, and the business be done *as if by machinery*.'[6]

For the first time, the image of the machine is applied to the inner life of the human being. It is proposed as an ideal to which human beings intent on the acquisition of knowledge must conform. For Francis Bacon, the more we can model our inner life on the machine, the more effective we will be in attaining useful knowledge that gives us power over nature.

We have already considered the enormous impact of the introduction of the mechanical clock on the way people thought about the nature of the world, from the fourteenth century onwards. By the early seventeenth century, small, portable, spring-driven clocks were increasingly common in the houses of the wealthy. Bacon would undoubtedly have possessed one or more of these mantelpiece clocks. However in 1620, the year of the publication of the *Novum Organum*, the kind of machinery that Bacon was thinking of was not so much the clock as the mill. During the fifteenth and sixteenth centuries, large geared machines powered by the energies of wind, water, horse, or human beings—windmills, grain mills, sawmills and treadmills—became widespread throughout Europe. Of these, it is the treadmill, which operates through the harnessing of human energies to the machine and its geared drive, that was probably in Bacon's mind when he wrote of the mechanization of the inner life of the human being in the service of knowledge. For while he was inclined towards a mechanistic view of the universe, what interested Bacon more was the prospect of *mechanizing the human mind*. For this reason, the treadmill certainly seems the most appropriate symbol of the harnessing of human mental labour to the new method that Bacon sought to inaugurate in the sciences (Fig. 13.1).

It is in the idea that the human mind can be disciplined to operate like a machine that we find the seed idea of the computer. For if it is really possible, by drilling the mind to think purely mechanistically, to achieve advances in knowledge hitherto unattained, then by implication a machine could in due course be designed that would function like the human mind. For Bacon, and thinkers subsequent to him enamoured of the mechanistic view of the universe, it could only be advantageous for human beings to think and act like machines: the more machine-like the better. Why? Because if the world is itself nothing but a machine, then we shall understand it best by conforming our thought to the mechanical. The machine is thereby raised up as the model both of the universe and of the process by which human beings may best attain knowledge of it.

If we look for the origins of the computer in a physical machine, we are likely to be misled into identifying the simple mechanical calculator as the earliest prototype of the computer. But the prototype of the computer had first to exist *as an idea*, before it could become physically embodied. And it is in Bacon's notion that the human mind can be 'guided at every step, and the business be done as if by machinery' that we can identify the 'seed idea' of the computer. What this means is that the originating idea of the computer was not just that of a very useful instrument—a mechanical data-processing system—that would serve humanity's needs. That is only part of the idea. The other part of the idea, which existed less as a rational

Figure 13.1

Sixteenth century chain-pump and treadmill. The treadwheel on the left is worked by two men, one of whom is partially obscured by one of the wheel's beams. A third man, apparently resting, sits on his own away from the wheel, in the centre of the picture. The wheel was used for draining water from a mine, through a series of cogwheels and pulleys.

conception and more as a fantasy, in the sense of an archetypal image that grips the mind and casts a spell on the soul, was *a re-imagining of the human being in the image of the machine.*

Once we see that this fantasy is what actually inspires the thinking of Francis Bacon, we will understand better the true nature of the project that he inaugurated. To bring the new instrument into existence, Bacon saw that it was necessary 'to level men's wits' and to leave 'but little to individual excellence'.[7] Individual genius, spiritual intuition and poetic imagination were the enemies of Bacon's ideal of the mechanization of the human mind. To realize this ideal, any orientation of consciousness towards spiritual knowledge had to be excluded. Bacon's aim was to confine human thought to the material level. Hence the need for the mind to be 'hung with weights to keep it from leaping and flying'.[8]

Having 'purged and swept and levelled the floor of the mind' and reduced it to purely mechanical functioning, it would then prove possible 'to extend more widely the limits of the power and greatness of man', so that humanity would attain dominion over the universe.[9] Here is a further key idea that belongs to the originating fantasy of the computer. It is that in so far as the human soul can be conformed to the image of the machine, and made into a 'new instrument', humanity will gain *dominion* over the material world. The kind of knowledge that Bacon envisages is less about gaining understanding for its own sake, and more to increase our ability to control and use the material world for our own benefit. It is, in other words, *technological* understanding that is foremost in his mind. In this Mephistophelean bargain, the pursuit of the spiritual life is surrendered and human thinking is reduced to a soulless, mechanical activity of gathering, sifting and analyzing data, in order that we may thereby reap the reward of extending human power over nature.

The way Bacon's new instrument functioned was by reducing human experience of nature to the collection of objectified data, which he referred to as 'particulars'—we might say 'information packets'. These particulars could then be arranged in Tables of Discovery, 'drawn up and marshalled' so that the mind could 'be set to work upon them'.[10] By analyzing and comparing the particulars, it would become possible to arrive at general axioms to which a mass of data would conform. Negative instances that contradicted the assembled mass of data would serve the function of either narrowing or widening the validity of the axiom. The strength of this method lay in the exhaustiveness of the data-collection process combined with the rigorous analysis of all instances. And the promise of the method was that, once axioms had been established, it would then become possible to apply them to good practical effect. The ultimate aim was to move beyond the formulation of general axioms to what is *practically useful*, that is to 'works'.[11] As Bacon put it: 'Truth, therefore, and utility are here the very same things.'[12] By this he means that a proposition or formula should not be regarded as true because it gives us a greater insight into nature, but because it gives us the ability to *use* nature. If it gives us the ability to manipulate and control nature, then we should be content that we have arrived at the goal of scientific understanding.

Bacon's avowed aim was to improve the conditions of human life, rendering obsolete physical hardships, which at that time were regarded as having been caused by the Fall. In this way, through the combined

application of the new scientific method and the technological domination of nature, Paradise could be restored on earth:

> For man by The Fall fell at the same time from his state of innocence
> and from his dominion over creation. Both of the losses, however, can
> even in this life be in some part repaired; the former by religion, and
> faith, the latter by arts [i.e. technology] and sciences.[13]

Bacon was one of the first modern thinkers to locate the notion of a
Golden Age *in the future* rather than in the past, promoting the concept of
progress where before people were inclined to view human history as a
process rather of decline.[14] But the Golden Age which he conceived was
that of a new 'scientific humanity', in which human society is organized
and governed by a scientific elite who, in exchange for the 'new
instrument', have dispensed with contemplative imagination and metaphysical insight, and whose minds are closed off from any access to the
spiritual.

In his Utopian novel, *New Atlantis*, Bacon describes an imaginary
island somewhere in the Pacific Ocean, which is organized according
to his ideal. It is governed by a scientific research establishment called
the 'House of Salomon', which is essentially a human data-processing
system. In other words, the House of Salomon is a computer whose
components consist entirely of human beings.[15] The House of Salomon sends forth explorers to collect data, which are then recorded,
stored and classified in precisely the way that Bacon's method sets
out.[16] Elaborate experiments are devised by researchers working in
laboratories underground or high up on top of towers, to investigate
the nature of matter under diverse conditions, in order to amass information based on 'particulars'.[17] General axioms are formulated from an
analysis of the information accumulated by these experiments and,
working from these general axioms, an extraordinary degree of control
over nature is achieved. For example, we find described in *New Atlantis*
the artificial manipulation of the weather, the practice of genetic engineering applied to plants and animals, and the creation of entirely new
species of creature. We also find described incendiary bombs, aeroplanes, submarines, robots and telephones 'that convey sounds in tubes
and pipes, in irregular lines and distances'.[18] Bacon foresees with
uncanny accuracy the new technological age that it was his mission to
inaugurate. And the driving force behind it was the 'new instrument'
by means of which human mental faculties, successfully deprived of
any inclination towards knowledge of the spirit, are harnessed to the
machine, and thereby made extraordinarily powerful.

The Invention of the Binary Code

This, however, does not exhaust Bacon's contribution to the seed idea of the computer. Bacon also anticipated the means by which data in a modern computer would come to be stored, manipulated and transmitted: namely, the Binary Code.[19] Early in his career, when he was employed by the government both at home and abroad to gather intelligence, Bacon learnt the use of ciphers. It was while he was in Paris in the late 1570s that he invented a cipher, which he called the 'Biliteral Alphabet', for the purpose of sending coded messages. He only revealed this to the world towards the end of his life when he published it in the Latin edition of *The Advancement of Learning* (*De Augmentis Scientiarum*, 1623). The alphabet is shown in Figure 13.2, where we can see how the letters of the Latin alphabet (which leaves out '*J*' and '*U*') are each represented by a sequence of five of the letters '*a*' and '*b*'.

Exemplum Alphabeti Biliterarii.

A	B	C	D	E	F	G
Aaaaa.	*aaaab.*	*aaaba.*	*aaabb.*	*aabaa.*	*aabab.*	*aabba.*
H	**I**	**K**	**L**	**M**	**N**	**O**
aabbb.	*abaaa.*	*abaab.*	*ababa.*	*ababb.*	*abbaa.*	*abbab.*
P	**Q**	**R**	**S**	**T**	**V**	**W**
abbba.	*abbbb.*	*baaaa.*	*baaab.*	*baaba.*	*baabb.*	*babaa.*
X	**Y**	**Z.**				
babab.	*babba.*	*babbb.*				

Figure 13.2

Francis Bacon's Biliteral Alphabet of 24 letters transposed into a series of different combinations of the letter 'a' and 'b'.

From this sequence of five letters, which of course could equally well be ones and zeros, 32 different combinations can be made. Thus the 24 letters of the alphabet could be translated into a binary sequence with eight combinations to spare (*bbbbb, bbbba, bbbab, bbbaa, bbabb, bbaba, bbaaa* and *bbaab*)—to represent punctuation marks or other signs. Not only did the biliteral alphabet enable the encoding of information, but it also made the transmission of it particularly easy. All that was needed was a medium that could be in one of two possible states, for example a light that is either shining or obscured. In the era of electricity, long after Bacon's time, an electric switch that is either 'off' or 'on' would become the perfect transmitter.

Bacon's intended use of the code, however, was not in a simple substitution of sequences of 'a's and 'b's for their corresponding letters, as this would have been relatively easy to decode. His idea was rather to conceal the biliteral alphabet within a medium that no one would suspect contained it. Any medium that could be divided into two types, an 'a'-type and a 'b'-type, would do. So, for example, a script with letters alternating between two kinds of font, or between bold and normal in the same font, could act as the carrier of the biliteral alphabet. If we take the second example, we could designate the letters in bold as 'a' and the normal as 'b'. When they are both employed in the same text the actual meaning of what is written is totally different from the apparent meaning, and if this is handwritten it may simply look as if the pen is writing unevenly. Bacon gives the example of a message, '*Manere te volo donec venero*' ('Stay until I come for you'), which, when deciphered, means '*Fuge*' ('Flee!'). The word '*Fuge*' spelled biliterally is 'aabab.baabb.aabba.aabaa', so all that is required is to encode this into the text as follows, here with bold type representing 'a' and normal type representing 'b':

'**Manere te** vo**lo** do**nec ven**ero'.
aababb aa bbaa bbaaa baa

Thus the text actually reads:

aabab.baabb.aabba.aabaa
  f   u   g   e

As only 20 letters are required to encode the message, the final three letters of 'venero' are superfluous, and merely add to the deceptiveness of the cipher.[20]

A series of five '*a*'s and '*b*'s was sufficient to meet Bacon's needs. What he was concerned with was creating a way of disguising meaning. His was the art of concealment and deception, so that behind the surface appearances another meaning would lie hidden. But the principle of translating each letter of the alphabet into a particular sequence of alternating and mutually exclusive options is one that can be applied to any information whatsoever. As well as to words, it could equally be applied to numbers, for instance, and Bacon used it in this way too. It could, in theory, also be applied to colours, with gradations of tone within a particular colour being represented by incremental changes in the combination of '*a*'s and '*b*'s. It could likewise be applied to sounds, and so on. There is in theory nothing in the world of perceived qualities that could not be expressed in binary form. As Bacon put it, the biliteral alphabet enables one to signify 'anything by means of anything' ('*omnia per omnia*').[21] The addition of more digits to the sequence would have vastly increased the carrying capacity of the biliteral alphabet. If Bacon had added just three digits more, so the sequence had eight 'a's and 'b's, the number of possible combinations would have increased 16-fold to 256. It is this number of digits that subsequently proved the most easy to work with in computers, in which a sequence of eight digits (known as a 'byte') and multiples thereof, were to become the basic unit for the binary code of ones and zeros.

Bacon's biliteral alphabet was an extremely flexible and cunning device for reducing the world of qualities and meanings to a non-qualitative and apparently meaningless series of digits. By curious coincidence, Bacon's binary alphabet was published in the very same year that Galileo published *The Assayer* (1623), in which Galileo put forward the view that the sense-perceptible qualities we experience in nature (through our senses of touch, smell, taste, hearing and sight) should best be understood as caused in us by strictly objective, quantifiable 'primary qualities'. As we saw in Chapter 10, the distinction between primary and secondary qualities was deeply implicated in the reconception of nature as a great machine, for the distinction enabled the world of experienced qualities to be conceived in terms of non-qualitative but quantifiable constituent parts. The distinction also contained the idea that what we experience with our senses is at one and the same time reducible to these constituent parts, and is the product of them. Therefore, given the constituent parts (that is, the atoms), one could then reconstitute the world out of them again. In other words, the realm of qualities, once decomposed into non-qualitative units, could be 'recreated' out of the realm of quantity. With his biliteral alphabet, Bacon showed how this could be done in relation to language.

We can therefore see the kinship between Galileo's atomism and Bacon's biliteral alphabet. Bacon's sequences of 'a's and 'b's out of which a meaningful text can be constructed are second cousins to Galileo's qualitiless atoms, which he believed produced in us our experience of the world of colour, sound, smell and taste. The thought that lies behind both conceptions is the same thought: whereas Galileo applied it to the world, reducing all of nature to a quantifiable substratum, Bacon saw how it was possible to reduce human language (the world of meaning) to a very simple, binary, or 'biliteral', substratum. For both thinkers, the substratum then becomes the cipher which, once decoded, enables one to recreate the world of nature (in the case of Galileo) or of meaning (in the case of Bacon). It is, however, to Francis Bacon rather than to Galileo that we owe the seminal idea of the computer, because it was he who articulated the mechanics of data-processing in the *Novum Organum* and *New Atlantis*, and it was he who saw the potential of the binary code as a means of storing and transmitting data.[22]

Chapter Fourteen

THE MYTH OF THE BINARIUS

The Binarius: The Arising of a New Archetype

Why did Bacon hit upon the binary code as the most effective for his cipher? He might, for example, have chosen a code that instead of two had three letters; but three letters would have been too many. Bacon realized that in order to accomplish the total disintegration of meaning into a system of intrinsically neutral, meaningless units, a binary rather than a tertiary code was required. The biliteral alphabet was the most flexible vehicle to hold all manner of content: it could represent numbers as easily as words, and—as we have seen—it could in principle represent qualities too: different degrees of colour or sound, and so on. Its advantage was that it combined complete lack of any inherent quality or meaning with a high degree of precision. Because its language was that of the realm of quantity, it was inherently precise, but because it was a binary system it could be the carrier for quite different kinds of content, in accordance with Bacon's maxim, *omnia per omnia*.

In Chapter Two, we considered the principle of leverage on which the *shaduf*, used in the ancient Middle East to draw water, operated. We observed that leverage in general, and the *shaduf* in particular, only works because when one end of the pole of the *shaduf* (bearing the bucket) is up, then the other end (bearing the counterweight) is necessarily down, and vice versa. There is a mutual exclusivity between the two states that the *shaduf* can be in, for the condition that one end of the pole is either up or down is that the other end of the pole is in the contrary state at the same time. There is, to express it in logical terms, an 'exclusive disjunction' between the bucket and the counterweight, which is why the *shaduf* works. As we saw, its relatively late introduction in ancient Egypt is best explained as having been due to a strong resistance to a technology that interposed a binary mechanism between the human being and the realm of god-filled nature. This resistance was only finally overcome during the exceptional political and religious circumstances of Akhenaten's reign.

In the Greek period, the principle of exclusive disjunction, already known and applied in the practical sphere of elementary technologies, was transposed into the realm of thought. This momentous occurrence coincided with the birth of philosophy in ancient Greece, founded on the recognition of certain basic principles or laws of thought, which were at the core of the new discipline of logic. We saw in Chapter Four that the principle of exclusive disjunction encompasses three closely related laws of thought formulated by Aristotle, referred to as the Law of Identity, the Law of Contradiction and the Law of the Excluded Middle. Together these laws established the hegemony of oppositional thinking, in which the binary alternatives of 'either/or' came to dominate the way in which Europeans would henceforth orientate themselves towards all questions involving truth and knowledge.

In his biliteral alphabet, Bacon now extended the principle of exclusive disjunction beyond the sphere of logic into the sphere of information storage, for his cipher in effect reduces language to no more than information: it is a means of encoding information. The biliteral alphabet has the effect of dissolving language so that it is no longer recognizable as language, for it is not a true alphabet. As we have seen, by virtue of the fact that it reduces the meaning-saturated words of ordinary language into sequences of just two letters, it is a method of quantifying both qualities and meanings. Thereby, it has the potential to absorb into itself and accurately store all manner of information, which can be retrieved with ease so long as one knows the code. Here, then, the principle of exclusive disjunction is made to serve a purpose quite different from the quest for truth, which was—despite its devastating impact on the old religious consciousness—nevertheless the *raison d'être* of this principle in the sphere of logic. By contrast, the prime motive behind the biliteral alphabet, conceived by Bacon as a cipher, was to conceal the truth.

It might be objected that all ciphers are designed to conceal the truth, and in this respect Bacon's cipher was no different from any other. Ciphers had been used by European governments since the Middle Ages, and during the fifteenth and sixteenth centuries quite elaborate methods for creating coded messages were developed. In response to this objection, we have only to point to the fact that none of the other ciphers involved using the binary code—they were all alphabetically based. What was unique about Bacon's cipher was that it was based on the binary principle. Given the historical process just outlined of this principle first entering the realm of technology in deep antiquity, then moving into the realm of logic in ancient Greek times (where it remained throughout the Middle Ages), we should not overlook the significance of its being seized upon by Bacon as a means of encoding messages. The significance lies in the fact that if Bacon pioneered the use of the binary code for information storage and communication, in this mode the binary principle was set free from human moral constraint. In so far as it was released from the arena of logic it was no longer employed to serve the quest for truth. Of course the use to which it was put by Bacon was extremely limited, but the biliteral alphabet nevertheless opened the way for what would come later.

I have referred in the previous paragraph to the binary *principle*, because in each of the three major cultural phenomena I have pointed to—the simple machine based on leverage, the laws of thought, and finally the binary code used for information processing—the same underlying principle can be identified. We might call it the principle of duality or Twoness. For those people living in the sixteenth and seventeenth centuries for whom the realm of spirit was still a reality, not only was a principle such as that of Twoness much more than just an abstract general concept, but also numbers were much more than just quantities: they had spiritual qualities. This qualitative understanding of numbers belonged to an ancient tradition, which went back at least as far as Pythagoras who, as we saw in Chapter Three, introduced such principles as the Monad and the Dyad in place of the gods.[1] In antiquity, many metaphysicians followed Pythagoras in identifying the Monad with God, the source and origin of all being and also of goodness.[2] The Dyad, by contrast, was identified with otherness, division and plurality, while from a moral point of view it was frequently associated with evil.[3] The next principle in the sequence, the Triad, was associated with the union of spirit and matter, and the perfect expression of the former in the latter.[4] In the early centuries AD, Pythagorean metaphysics was readily absorbed by both Neoplatonists and those working within the Hermetic tradition. However, with the collapse of the Roman Empire many important texts were lost to the West. They were only rediscovered again during the second half of the fifteenth century, following the fall of Constantinople in 1453. As a result of their rediscovery and subsequent translation into Latin, there was a resurgence of interest in these perspectives during the Renaissance and early modern period.[5] Furthermore, these newly rediscovered Pythagorean-Hermetic-Neoplatonic writings could readily be related to esoteric teachings contained in Kabbalistic, alchemical, astrological and magical texts.[6] It was as if a lost key to the West's spiritual heritage had been rescued. And one consequence was that during the sixteenth and seventeenth centuries a new interest arose in the qualitative significance of numbers. One alchemist in particular became fascinated by this subject, and seemed almost to have developed an obsession with the Dyad, which translated into Latin as 'the Binarius'. His name was Gerhard Dorn.

Gerhard Dorn and the Myth of the Binarius

By a curious synchronicity, it was just at the time when Bacon first conceived of the binary code in the form of his biliteral alphabet, whilst in Paris in the 1570s, that Gerhard Dorn wrote a kind of spiritual biography of the Binarius. Such was Dorn's fascination with the archetype of the Binarius, one has the impression that he kept returning to it in his writings because he had caught sight of it behind the rising new world order which, at the end of the sixteenth century, was being ushered in by Francis Bacon and others.[7] In the Book of Genesis, on the first day of creation God created light, saw that it was good and differentiated it from darkness. According to Dorn, the Binarius originated on the second day of creation when, as the Book of Genesis further records, God separated the upper waters from the lower. In this act, duality was introduced into the world. For Dorn, this was the reason why in the Book of Genesis it is stated that, unlike on all the other days of creation, on the evening of the second day God failed to say that what he had done was 'good'. It was Dorn's view that the division of the upper and lower waters on the second day allowed confusion, division and strife to enter the world.[8]

This view concerning the origin of confusion, division and strife was not original to Dorn. He was well-versed in the writings of two Renaissance sages, both of whom were steeped in the magical and esoteric teachings which were so much part of the cultural ferment of the time: Johannes Trithemius of Sponheim (1462–1516) who was old enough to be Dorn's great-grandfather, and Henry Cornelius Agrippa (1486–1535). Agrippa was a generation younger than Trithemius, who deeply influenced him, and was 44 years older than Dorn (who was born in 1530). In Agrippa, we read the following statement concerning the number Two, which Dorn was evidently influenced by when he made his observations concerning the second day of creation:

> This [i.e. the number Two] is also sometimes the number of discord, and confusion, of misfortune, and uncleanness, which is why Saint Jerome in his book *Adversum Jovinianum* ('Against Jovinianus') said that for this reason 'and God said, that it was good' was not spoken on the second day of the creation of the world, because the number Two is evil.[9]

According to Agrippa, the tradition concerning the number Two can be traced back to Saint Jerome, but of course it can be traced back further to various pre-Christian Platonic and Neo-Pythagorean thinkers, including Porphyry, Philo and Plutarch, and ultimately to Pythagoras himself.[10]

While Dorn, then, was writing within this Pythagorean tradition, he also took it further than his predecessors by casting it into the form of an unusual Christianized geometrical creation mythology. Pictured figuratively, the divine origin and source of creation was envisaged by Dorn as a geometrical point, at one with itself. This is to be understood as representing the principle of unity, the Unarius. From this point, Dorn pictures a line extending beyond it, which becomes the radius of a circle drawn around the point. The circle symbolizes the cosmos, which revolves around its creator. But in creating the cosmos, God has necessarily introduced the principle of duality, symbolized by the radius, which measures the 'distance' between the divine source and creation. The radius line introduces division and separation from source, and it represents for Dorn the Binarius. The circle, in symbolizing the cosmos also symbolizes the perfect image of the divine in creation, turning around the centre point of origination, and represents the Ternarius (Fig. 14.1).[11]

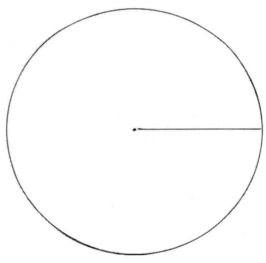

Figure 14.1
Point, line and circle, representing the three principles of the Unarius, Binarius and Ternarius in Gerhard Dorn's geometrical creation mythology.

So far in this description, Dorn is quite closely following John Dee's *Monas Hieroglyphica*, which exercised a strong influence over his thought.[12] But now he departs from Dee as he sets forth his further teaching concerning the Binarius. According to Dorn, the Binarius belongs neither to the world of the One (the point), nor to the Three (the circle). It therefore tries to create its own rival circular system, its own rival universe, by bending itself around in imitation of the Ternarius. But in this attempt it fails. All it is able to produce is the 'figure of a twofold serpent lifting up four horns'.[13] Here, then, we see the origin of the serpent that tempts Eve. It is the radius of the circle, full of envy for the circumference, the Ternarius. And in its envy it makes the attempt to bend itself into a rival circle, but fails miserably. From the description that Dorn gives, the figure that emerges would seem to be something like the one below (Fig. 14.2).

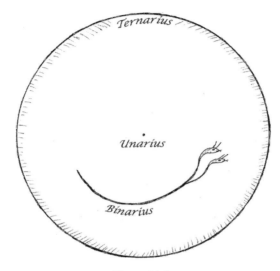

Figure 14.2
An imaginative representation of the origin of the Binarius, according to the description in Dorn's De Duello Animi cum Corpore.

Dorn himself, however, represents 'the twofold serpent lifting up four horns' in a more abstract way—in another geometrical figure. It shows two interlocking semi-circles, suggestive of the falling apart of the original unity of the circle, into a duality of two incomplete circles that have succumbed to fourfoldness and lack any centre (Fig. 14.3).[14] Dorn's caption to this image is 'Dyad', and underneath he wrote:

> The Devil is variable,
> And has no repose.[15]

Thus for Dorn, there is no doubt that the Binarius (or Dyad) is the devil in serpentine form. And it is characterized by variability and lack of repose, because it has no centre. Dorn ascribes to each of the horns a particular negative attribute, each of which reflects the restless discontent of the Binarius: ambition, calumny, brutality and separation.[16]

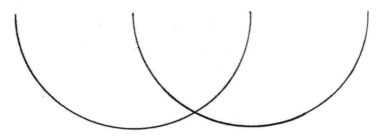

Figure 14.3
Gerhard Dorn's geometrical representation of the Binarius as the 'twofold serpent lifting up four horns'.

Intent on setting up its own alternative cosmos between the One and the Three, in Dorn's mythology the Binarius now launches an attack on the world created by God. Realizing that Adam, the first man created in the image of God, is an image of the Unarius, and would therefore not be susceptible to its influence the Binarius, now in serpentine form, approaches Eve. It insinuates itself by 'crawling into the integral mind' with 'deceptive and cunning questions'.[17] The reason for its choice of Eve is that, being divided from her husband, she is a 'natural binary' and is therefore more vulnerable than Adam to the Binarius's innate duplicity. Thus it was that the Binarius chose to tempt Eve first.[18]

In Dorn's depiction of the Binarius, we cannot help but be reminded of the medieval iconography of Dialectic and her symbolic beast, which was so close to the serpent in the Garden of Eden as to be almost indistinguishable from it (see Chapter 7). The serpent, or serpent-dragon, is an age-old symbol of duality, and hence of duplicity, and the image of the double serpent with its four horns arises in Dorn's mind with all its Christian mythological associations, as the symbolic manifestation of the Binarius. Having brought us to this point in the story, Dorn evidently felt that this represented the current human situation. The Binarius has crawled into the integral mind, and his readers are faced with the question as to what to do about it. Dorn proposes that it is only when the Binarius is rejected that the Ternarius can return to the simplicity of Unity, and by implication it is only then that the soul can return to God.[19] In taking this stance, however, he departs from the view more usually held in the Christian Hermetic tradition within which he was writing. Other authors, such as Trithemius of Sponheim, Henry Cornelius Agrippa, John Dee and later Thomas Vaughan, were all of the view that the path to the restoration of the Unarius is not through the outright rejection of the Binarius but through conjoining it with the Ternarius.[20]

Whichever attitude is adopted by those concerned with the ingress of the Binarius into the historical trajectory of Western culture, neither presents a particularly easy path to tread. But it is less in his recommendations on how to deal with the problem of the Binarius than in his having given us a mythological framework within which to imagine it that Dorn's best contribution lies. Dorn created a mythology that was able to incorporate binary thinking, and the emerging mechanical consciousness based on it, into a greater cosmological framework. In the myth of the Binarius, he took hold of binary thinking and projected it into the imaginal realm, where it could be clearly visualized, and there both its character and value were made plain. First and foremost, Dorn saw that the binary principle was to be understood as an objective reality. More than a metaphysical principle, it is a mythological *being* that those who are able to open their spiritual eye can perceive as really operating in the world. This being is characterized by an intractable antagonism to the divine. Having infiltrated 'the integral mind', it has introduced tendencies into human thought-processes that have the effect of blocking awareness of the spirit. In Dorn's myth, the Binarius is intent on setting up its own alternative world, that would exist apart from the divine and natural orders, and which would radically undermine the spiritual orientation of humanity. What Dorn wrote was prophetic. He saw the tendencies within his own times, and he saw what lay behind them and how they were likely to unfold. In the next two chapters we shall explore the manner of thinking that developed during the middle and later periods of the seventeenth century, and which further prepared the way for the Binarius to fulfil its ambition of establishing its own independent domain.

Chapter Fifteen

THE DRIVE TOWARDS MECHANIZATION

The Mechanization of the Mind: Descartes
During the period when Francis Bacon was advocating his new method and Galileo was pioneering the view of the universe as a vast mathematical machine, the mechanistic understanding of both universe and human being became increasingly influential. But it was the next generation of thinkers, born in the 1580s and 1590s, who were able to successfully establish the mechanistic philosophy, both in the sphere of cosmology and in relation to the inner life of the human being. Foremost amongst these thinkers were Thomas Hobbes (1588–1679) who, in his youth, served as Francis Bacon's secretary, and his slightly younger contemporary René Descartes (1596–1650).

The new understanding about the nature of things was revealed to Descartes at the age of 23, in the autumn of 1619, and confirmed to him in a series of dream-visions. In the latter, the 'Spirit of Truth' appeared to him and affirmed his consciously held conviction that mathematics is the key to unlock the secrets of nature. Whatever or whoever this spirit was, it had the effect of infusing Descartes with an almost religious ardour for his undertaking. It is worth briefly considering these dreams for the light they shed on the spiritual source of his inspiration.

The dreams occurred after a momentous day in November, in which Descartes felt 'filled with enthusiasm' and 'completely absorbed by the thought of having that day discovered the foundations of a "marvellous science" '.[1] The dreams that visited Descartes that night, however, had a markedly sinister tone. According to the detailed account of Descartes' early biographer, Adrien Baillet, in the first dream he had terrifying visions of ghosts and was assailed by a violent wind, which rendered him unable to hold himself upright; when he awoke, he had a strong sense of the presence of 'evil spirits, bent on leading him astray'. In the second dream the atmosphere of evil continued, culminating in a great thunderclap that made him wake in terror. This atmosphere was dissipated in the third dream, which consisted of his discovering a 'very useful' dictionary on his table, which he interpreted as signifying the connection between all the sciences. This episode was followed by the appearance of another book containing a collection of poems, which fell open on a page where a poem (written in Latin) began with the words 'What sort of life should one choose?' A man he did not know then appeared, who showed him a second poem beginning with the words 'Yes and no'. As if this binary alternative was of profoundest significance, Descartes interpreted the words as 'the truth and error of all human knowledge and profane science'. Descartes regarded this third dream as inspired by 'the Spirit of Truth that wished, through this dream, to reveal to him the treasures of all the sciences'. Why he should put quite such a positive gloss on it is less clear.[2]

The 'marvellous science' that Descartes' day-consciousness had discovered was based on the revelation that all of nature—plant and animal organisms as well as the inorganic world—could be understood in mechanistic terms. All of nature, that is, except for human consciousness, which he regarded as a thing apart and not subject to mechanical laws. But the human body (including the human brain) was conceived in the image of a machine, with parts external to other parts, and all parts interacting with each other in accordance with mechanistic principles. In his 1632 *Treatise on Man*, Descartes wrote, 'The body is nothing else than a statue or machine of clay,' and continued:

> All the functions ... follow naturally in this machine from the arrangement of its parts, no more nor less than do the movements of a clock, or other automata, from that of its weights and its wheels.[3]

Descartes frequently used clocks as his guiding metaphor for understanding the human body. In his *Meditations* (1641), for example, we read:

> Now a clock built out of wheels and weights, obeys all the laws of 'nature' no less exactly when it is ill-made and does not show the right time, than when it satisfies its maker's wishes in every respect. And thus I may consider the human body as a machine fitted together and made up of bones, sinews, muscles, veins, blood, and skin in such a way that, even if there were no mind in it, it would still carry out all the operations that, as things are, do not depend on the command of the will, nor, therefore, on the mind.[4]

Descartes assures us that if we do not actually perceive the mechanical apparatus of the bodily organism, this is due merely to the fact that the parts of the body are so small that they elude our senses, whereas the apparatus of actual, man-made machines is large enough to be perceptible by the senses. In just the same way as a clock composed of so many cogs and wheels will tell the time, so will an apple tree produce apples—a thought readily extendible to the behaviour of both animals and humans.[5] Descartes is so enamoured of the clockwork metaphor that we have the uneasy feeling that he is not simply using this image to support his argument, but that he is in some degree *possessed by it*, and that his arguments devolve from the image. Jung pointed out on many occasions that the psyche consists essentially of images, and that these images derive from a relatively limited number of archetypes.[6] By the early seventeenth century the image of the clock had come to embody just such an archetype: the archetype of the Machine—applicable to all of nature, including the human being. It was this archetype that lived in the mind of Francis Bacon, and now gripped the soul of Descartes.

In Francis Bacon's *Novum Organum*, we met the idea that there was a compelling need to reduce human thought to a purely mechanical process, in order to secure the foundations of scientific knowledge. Like Bacon, Descartes too was greatly concerned with formulating a reliable method for the acquisition of knowledge. In his *Rules for the Direction of the Mind* (written circa 1629, but not published until 1701), the extent of his desire to limit the mind to a purely mechanical way of functioning is laid bare. The *Rules* effectively reined in the free-floating mind—the *res cogitans*— that in other writings he felt obliged to regard as lying outside the mechanistic universe. Through these rules, he set out how the mind could be made to conform to the unbending laws of rigid mechanical thinking.

In Rule 13, Descartes tells us that in order to solve problems they must be reduced to their simplest terms, or split up into their smallest possible parts. Then we may come to apprehend, step by step, the inter-dependence of the terms (Rule 17). Needless to say, the process of thought that Descartes advocates is reminiscent of taking a clock apart, but with the added finesse of making our thinking conform exactly to the operations of arithmetic. In Rule 18 we are told that ultimately only four operations are needed for problem solving—addition, subtraction, mul-tiplication and division. The 'rules for the direction of the mind' thus resolve into a kind of applied mathematics (Rule 19), in which all unknown terms are treated as magnitudes, and reasoning itself becomes a matter of formulating equations (Rules 19–21).[7]

In this respect, Descartes' thinking is matched by that of Thomas Hobbes. In Hobbes' masterwork, *Leviathan*, first published in 1651—the year after Descartes' death (Descartes died at the relatively young age of 54)—reason is redefined as simply a way of doing maths:

> When a man reasons, he does nothing else but conceive a sum total, from addition of parcels, or conceive a remainder, from subtraction of one sum from another . . .

That is to say, for Hobbes, reasoning is nothing more than a kind of calculating or, as he put it, a kind of 'reckoning':

> In sum, in what matter so ever there is a place for addition and subtraction, there is a place for reason; and where these have no place, there reason has nothing at all to do. For Reason, in this sense, is *nothing but reckoning* (that is, adding and subtracting) . . . [8]

The Reconstruction of Language

When we read such a statement, it is hard not to conclude that it would only be possible to reason in this way if language itself were reconstructed on the model of mathematics. That in fact is precisely the view that Descartes came to at the time he composed his *Rules for the Direction of the Mind*. In a letter to Mersenne in 1629, Descartes put forward the radical proposal that language itself be reinvented so as to conform to the orderliness and precision of mathematics. He wrote:

> Order is what is needed: all the thoughts which can come into the human mind must be arranged in an order like the natural order of the numbers. [9]

Such an artificially constructed language, in which numbers would be correlated to simple ideas drawn up in lists, and so become susceptible to arithmetical reasoning, was the obvious corollary to his *Rules*. It would ensure that the risk of error in the pursuit of scientific knowledge would be minimized:

> The discovery of such a language depends upon the true philosophy. For without that philosophy it is impossible to number and order all the thoughts of men or even to separate them out into clear and simple thoughts, which in my opinion is the great secret for acquiring true scientific knowledge. If someone were to explain correctly what are the simple ideas in the human imagination out of which all human thoughts are compounded, and if his explanation were generally received, I would dare to hope for a universal language very easy to learn, to speak and to write. The greatest advantage of such a language would be the assistance it would give to men's judgement, representing matters so clearly that it would be almost impossible to go wrong. [10]

Descartes never got further than this brief formulation of the idea of creating a purely rational, mathematical language, but in the course of the seventeenth century other thinkers—notably Leibniz, as we shall shortly see—would pick up the idea and run with it. This idea that we need to

create an artificial language follows closely on the heels of the notion that
the human intellective faculty extends no further than the *ratio*, or calculative thinking. In fact it is integral to it, for how can the *ratio* function
effectively with words that are ambiguous, or with phrases that are open
to many different kinds of interpretation? Bacon had already come to the
conclusion that words themselves cause endless confusion, and he
regarded ordinary language as one of the four 'idols' against which science
had to fortify itself.[11] Although Bacon did not have the same predilection
for mathematics as Descartes and Hobbes, he shared their dislike of
ordinary language, which he called the 'idol of the marketplace'.[12]
Because they all wished to subject nature to the ordering and calculating
logic of the *ratio*, it was inevitable that they would also want to clean up
language so that it became an effective medium of exact thought.

All three thinkers believed that humans must learn to think
mechanically if they are to master the complex and intricate mechanism
that is the world. Precisely because they understood the world to be a vast
machine, only that type of thinking that is able to be simulated by
machines—namely computation—was seen by them as valid for
acquiring genuine knowledge. For them, the mind functions correctly
when it functions mechanically, because only in this way can it penetrate
the logico-mechanical structure of the universe. But the vehicle of all
thinking is language, and if the new mechanistic science demanded that
thinking be constrained by the need for clarity, mathematical exactitude
and logical precision, then language itself inevitably became a target in the
war against all that lives in the human soul in immeasurable and logic-
defying ways.

In these thinkers we see how the mechanistic worldview cannot—as
Descartes, at least with one part of himself, hoped—be separated from our
inner life. To subscribe to this worldview involves more than just the
intellect; one soon finds one's whole self dragged into its orbit, body, soul
and spirit. This is because underlying the worldview is something far
more powerful than an idea. It is an archetype, or from a strictly esoteric
point of view a spiritual *being*, which Gerhard Dorn identified as the
Binarius. Its influence takes hold of the human imagination, insinuating
itself into the unconscious, or as Dorn put it 'the integral mind'.[13] And,
from the unconscious, or integral mind, which has succumbed to this
influence, there surfaces the conviction that if we are to attain true
knowledge, then we must relinquish feeling, intuition, imagination and
contemplation, and surrender our inner life to the cold dictates of
mechanical reasoning.

The curious illustration of a stamp-mill turbine in Figure 15.1, pub-

lished in 1629 (the year of Descartes' letter to Mersenne and his composition of *Rules for the Direction of the Mind*), portrays one of many
machines invented by the Italian engineer, Giovanni Branca. Despite its
precise depiction of all the different components of the turbine, so one
can clearly see how it works, it also presents a kind of symbolic snapshot of
the encroachment of the machine archetype into the human domain.
The artist shows the human being ensnared in a mass of cogs and wheels,
so one sees illustrated not just the mechanism of the stamp-mill turbine
but at the same time the inner predicament of the human being in relation
to a huge, sprawling machine.

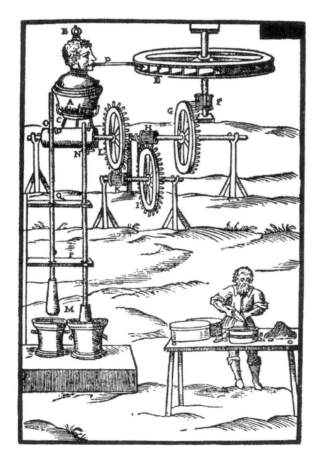

Figure 15.1
Branca's steam-driven
stamp-mill turbine, 1629.

The illustration shows a large metal vessel or cauldron *A* (top left), the
lid of which is in the shape of a human head and shoulders. Underneath it
is a large fire. The form of the cauldron's lid is a bizarre choice, for it gives
the impression that the man in whose shape it has been made is himself

slowly being cooked in the cauldron. At the same time, he is enmeshed in the mechanism of the machine through a tube inserted in his mouth (D), and unable to make his escape. As the water in the cauldron is heated over a fire, it turns to steam, which escapes through this tube or pipe held in the man's mouth. There we see that the conscious intention of the image is humorous, for the man is smoking a pipe at the end of which, pre-sumably, puffs of steam can be seen escaping. The steam drives round the wheel *E*, which, acting through a train of cogs, works a stamp-mill *M*, in which ore would be crushed before being put into a smelting furnace. To the right of the mill is a table where a 'real' man, overshadowed by the vast mechanical apparatus, labours as if in a dream. On the level of symbolism, the image of the human lid over the cauldron speaks of the increasing pressure exerted on the human being within 'the cauldron', or unconscious mind, which is now in servitude to the unrelenting drive towards mechanization, represented by the impersonal mechanics of the turbine. Thereby, the picture accurately portrays the tensions and pres-sures within the European psyche in the middle of the seventeenth century.

Pascal's Mechanical Calculator

It was not by chance that at just the time when the machine archetype was being applied to the inner life of human beings by Bacon, Descartes and Hobbes, the first mechanical calculators were invented. In 1642, one year after the publication of Descartes' *Meditations* (1641), his fellow countryman Blaise Pascal produced one of the earliest calculating machines. As people became more open to the idea that the human mind functioned, or at least could be made to function, like a machine, it became more possible to conceive of a machine that could carry out human mental functions. The capacity of thought that distinguishes the human being from the animal gradually came to be seen as precisely the same capacity that brings us closer to the machine. Referring to his mechanical calculator, Pascal noted in his *Pensées*:

> The arithmetical machine produces effects which approach nearer to thought than all the actions of animals.[14]

With the spread of the mechanistic philosophy at the beginning of the seventeenth century, the course was set for humans and machines to draw closer to each other.

At the time when Pascal invented his calculator, people whose job was to work with figures used to be called 'computers'. The word 'computer' derives from the Latin *computare*, 'to calculate'. Pascal (1623–1662) was a young man, not yet 20 years old, when he designed his arithmetical machine. He made it specifically to assist his father (who was a tax collector) in adding up sums of money. Thus it was to alleviate the burden of purely mechanical calculation on human computers that Pascal's calculating machine was invented. About 20 years after Pascal invented his machine, Samuel Morland invented machines capable of performing not only the addition and subtraction, but also the multiplication of sums of money, as well as certain other arithmetical calculations. Both Pascal and Morland conceived the purpose of their inventions as being to aid human computers in very specific and limited mathematical tasks (Fig. 15.2). But in the process they showed that there is a purely mechanical aspect to the act of calculation, and this could therefore be replicated by a machine. It should be borne in mind, however, that—in contrast to Hobbes—neither Pascal nor Morland regarded the act of *calculating* as the same as the act of *reasoning*. Pascal was very clear that there are higher levels of thought than mere calculation, which a machine could not replicate.

Figure 15.2
A human 'computer' with his calculating machine.

Pascal's machine was a table-top device, enclosing six or more cylinders with numbers 0 to 9 inscribed on each, corresponding to units, tens, hundreds, etc., each number becoming visible through a sight hole in the cover (Fig. 15.3 top row; Fig. 15.4 on the left). By means of a stylus, one of the six horizontal wheels (Fig. 15.3, bottom row; Fig. 15.4 top right) could be advanced from one-tenth to nine-tenths of a complete turn, and

Figure 15.3
Pascal's Mechanical Calculator.

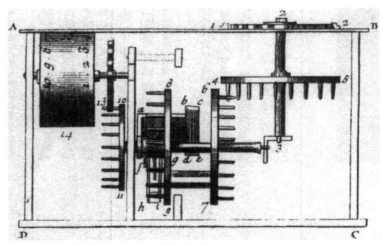

Figure 15.4
The internal mechanism of Pascal's mechanical calculator.

this movement would then be transmitted via pinwheel gearing to the corresponding numbered cylinder (Fig. 15.4). If, for example, the number *3* is visible on the far right cylinder through its sight hole and the stylus is inserted against the number *4* on the horizontal wheel in front of it, and then moved round to zero, the number on the cylinder will move correspondingly and the figure 7 will appear, thereby accomplishing the addition *3* + *4* = 7. A carrying mechanism enabled the addition of higher numbers, involving the deployment of more and more wheels and cylinders (to the left of the first cylinder).

Pascal's machine was capable of providing answers to arithmetical problems requiring addition and subtraction, and (through repeated additions and subtractions) to multiplication and division. What was so groundbreaking about his calculator, as can be appreciated when we contemplate its internal mechanism, was the fact that it substituted the rotation of cogs and wheels for the human mental effort required to perform mathematical calculations. The design shown in Figure 15.4 was published in Diderot's famous *Encyclopaedia* (1767), more than 100 years after the machine's invention. It became a new icon of the age, rivalling the clock, for one could *actually see* how a physical mechanism could be made to perform a human 'mental' function. The mechanical calculator thus confronted human beings with the deeply challenging fact that the interaction of physical cogs and wheels could be arranged in such a way that they could simulate the process of human thinking (at least in so far as this was restricted to arithmetical calculation). Did they therefore truly present a picture of how the human mind actually works?

As we have seen, Pascal himself did not think so. Arithmetical operations were not, for him, the basis of genuinely *human* thinking. The human mind has other capacities than merely the ability to calculate arithmetical sums—specifically it has the capacity to think with concepts. Pascal never dreamed that his calculating machine could develop into a machine capable of conceptual thought. Crucially for Pascal, there was in the human being an organ of knowing far superior both to mathematical calculation and logical reasoning. He called it 'the heart', by which he meant not so much an emotional sense as an organ of spiritual perception, by which we know truth, and through which we become aware of transcendent realities. 'The heart,' he said 'has its reasons of which reason knows nothing.'[15] In this statement we hear an echo of the traditional distinction made in both ancient and medieval philosophy between the *ratio* and the *intellectus*, the latter enabling us to apprehend meaning and giving us the possibility of gaining spiritual insight.[16] As we shall see, the conception of a computer in its modern sense arose in the minds of thinkers for whom, almost without exception, this distinction was no longer an experienced reality, and for whom all thinking was just the operation of the *ratio*.

Chapter Sixteen

LEIBNIZ: THE QUEST FOR AN
ARTIFICIAL LANGUAGE

Leibniz: The Art of Combination

Like all enlightened thinkers of the late seventeenth and early eighteenth centuries, Gottfried Leibniz (1646–1716) was a mechanist. He fully subscribed to the clockwork model of the world, which represented the universe as a machine, and living organisms as automata.[1] But, like Descartes, Leibniz drew back from fully embracing materialism, and ingeniously—though it has to be said not very convincingly—managed both to place God in the centre of his philosophical system, and to instil souls into animals, and spirits into human beings.[2] In his thinking we can trace the further development of the seed idea of the computer, as envisaged by Bacon, towards the point at which the attempt could be made to embody this idea in an actual machine.

Leibniz was an extremely clever man. A brilliant mathematician, jurist, philosopher and theologian, he was a man in love with reason. And it was his love of reason that led him at an early age to the notion of creating a completely new, artificial language, a *lingua characteristica*, the grammar of which would conform exactly to the rules of logic.[3] This was a project that he repeatedly returned to during the course of his life. We have already seen that Descartes conceived of a universal mathematical language back in 1629. Philosophers of the next generation, such as John Wilkins (1614–1672) and George Dalgarno (1626–1687), subsequently put forward proposals for a new, artificially constructed language more than 30 years later in the 1660s, at almost exactly the same time as Leibniz, who was then a young man.[4] But Leibniz's conception was more far-reaching than theirs, and closer in spirit to that of Descartes. Everything that was not logically necessary—for example distinctions of gender, declensions, conjugations and most nouns—Leibniz believed could be eliminated. The alphabet itself should be reinvented so that it represented not sounds but concepts. That is to say, Leibniz's artificial language would be mute. It would be an alphabet of soundless concepts, consisting of signs, or ideograms (Leibniz initially thought geometrical figures would be most suitable) that expressed simple ideas. And because the grammatical rules of the new language would be none other than the rules of logic, these simple ideas could only be linked together in a logical way.

Crucial to his project of creating a new 'alphabet of concepts' was the notion that the signs that represented them would 'perform the same task as arithmetical marks do for numbers and algebraic marks do for magnitudes considered abstractly'.[5] This would be a language from which imprecision would be excised. It would be without any opacity, without any irrational substratum, and without any possibility of ambiguity. It would be a language that would stop poetry dead in its tracks, because its conceptual alphabet would necessarily exclude metaphor and multiple levels of meaning. It would have no etymology, because it would have no past; there would be nothing organic about it, because it would be purged of all historical and human content. Its content would be determined by pure, computational reasoning alone, and in its abstract alphabet of concepts there would be no place for the unconscious to hide.

Leibniz worked out the principles in some detail in his seminal essay, *The Art of Combination* (*De Arte Combinatoria*), published in 1666 when he was only 20 years old. For Leibniz, as for Thomas Hobbes, the key principle is that thinking is no more than a form of computation. Early on in this essay he expresses his admiration for Hobbes in the following words:

> Thomas Hobbes, everywhere a profound examiner of principles, rightly stated that everything done by our mind is a computation, by which is to be understood either the addition of a sum or the subtraction of a difference. So just as there are two primary signs of algebra and analytics [i.e. logic], + and −, in the same way there are as it were two copulas, 'is' and 'is not'; in the former case the mind compounds, in the latter it divides.[6]

Leibniz believed that it should be possible to re-express all concepts as numbers, and thereby literally to apply Hobbes' computational model to the act of thinking. In his *The Art of Combination* (and in his later essay, *Elements of a Calculus*, 1679), he argued that every idea is either simple or complex, depending on whether its definition can be reduced to either one or more than one predicate. Since numbers, too, are either simple or complex, Leibniz reasoned that if ideas could be re-expressed as numbers, then we would have the key to computational thinking. A simple (or prime) number is one which can only be divided by the number *1* or itself; for example *2, 3, 5, 7, 11, 13*. A complex number is one which can be divided by numbers other than the number *1* or itself; in other words, it can be produced by simple numbers multiplied together, for example

$6 = 2 \times 3$; $12 = 2 \times 2 \times 3$; $35 = 5 \times 7$, and so on. A complex number therefore *contains*, or is composed of, simple numbers. But so too does every true proposition 'contain' the ideas out of which it is composed.[7] It occurred to Leibniz that simple ideas could be represented by simple numbers in the range *2, 3, 5, 7, 11, 13* . . . while complex ideas could be represented by the numbers in the range *4, 6, 8, 10, 12, 14* . . ., the number for a complex idea being the product of the numbers that represent the simple ideas contained within it.[8]

So, for example, Leibniz takes the statement 'The human being is a rational animal' and explains that if we ascribe to the simple ideas 'animal' and 'rational' the numbers *2* and *3* respectively, then the number for 'human being' will be the same as 'animal' × 'rational', namely *6*. Here the composition of concepts is regarded as equivalent to multiplication in arithmetic. The statement could be expressed in the formula $ar = h$, which now has the same certainty as $2 \times 3 = 6$. Just as the number *6* is a product of the numbers *2* and *3* multiplied together, so also is the concept 'human being' a product of the two predicates 'rational' and 'animal' combined (i.e. multiplied) together. For just as *6* arithmetically equals 2×3, so 'human being' contains, or is composed of, the predicates 'animal' and 'rational', and so proving the truth of the proposition is analogous to the arithmetical operation of division.[9] If we were to ascribe to the ape the number 10, Leibniz argues, then it is clear that the concept of ape cannot contain the concept of human being, since *10* cannot be exactly divided by *6*.[10]

In this example we can see where Leibniz is trying to go, even if he himself was unsatisfied with the attempt. In the same year that he wrote *Elements of a Calculus*, he also tried a different tack, writing another paper in which he attempted to set out the conditions for the arithmetization of propositions, entitled *Rules from which a decision can be made, by means of numbers*.[11] Now he tried to represent each term of a *proposition* by two numbers, qualified by a plus or minus sign, and argued that it should be possible to see whether or not a line of reasoning is true or false simply by arithmetical means. What Leibniz was reaching for was the translation of the old Aristotelian logic, which operates with words and sentences, into arithmetical formulae based on numbers and arithmetical operations.

Leibniz came to think that logic should be like algebra. He conceived of an 'algebra of logic' that would specify the rules for manipulating concepts in just the way that algebra specifies the rules for manipulating numbers. This would ensure that, through simple calculation, we could judge whether a sequence of thoughts is correct or not. So, for example, if two philosophers are in dispute, they could cast their arguments into the formal language of the *lingua characteristica*, and then simply say, 'let us calculate' (*calculemus*), and they would be steered to an indisputable conclusion through the very structure of the formal language itself.[12] The intention behind Leibniz's *lingua characteristica* was to create an artificial language that could wrest human thinking away from the soil of ordinary language in which it is naturally rooted, and to replant it in a medium less susceptible to error, namely mathematics.

Logic, as we have already seen in Chapter 7, had traditionally been conceived as belonging to the Liberal Arts: it was the basis of Dialectic—the art of correct thinking, complemented on the one side by Grammar (the correct composition of sentences) and on the other side by Rhetoric (the art of expressing oneself elegantly and persuasively). Thus conceived, it was an important element in the training of the inner faculties of the human being, but it was also embedded in human language. Leibniz wanted to separate logic both from ordinary language and from the human soul, in order that it might function in the autonomous and entirely objective sphere of the *lingua characteristica*, to which all propositions could be reduced, so that they could then be subjected to a calculus of reasoning, a *calculus ratiocinator*. Those instances of language that are not propositional in form, or in which metaphor, paradox and ambiguity are utilized in order to convey a deeper meaning, as in most of poetry, devotional and religious writing, mysticism and much metaphysics would, in a sense, be presented with an ultimatum: either re-express yourself in purely rational terms or else fall outside the new system of knowledge.

Leibniz wanted to secure the foundations of knowledge in rationality, specifically in the exercise of calculative thinking. Despite the loss of access to the trans-rational realities that this would entail (for such realities cannot be merely calculated nor can they be brought within the grasp of a *lingua characteristica* such as Leibniz conceived without being distorted beyond recognition), this was Leibniz's fundamental choice. For him, if access to this trans-rational level of reality was lost, the gain would be a new certainty in knowledge and the end of error, 'for the characters and the words themselves would give directions to reason, and the errors—except those of fact—would only be mistakes in calculation'.[13] Great, then, would be the benefits. The reduction of words to quasi-arithmetical signs, and of the process of reasoning to simple calculation, he believed would inevitably lead to immense advances in human knowledge. His mathematical language would give humanity a new kind of tool that would increase the power of the human mind 'much more than optical lenses have helped our eyes':

By using these numbers I can immediately demonstrate through numbers, and in an amazing way, all of the logical rules and show how one can know whether certain arguments are in proper form. When we have the true characteristic numbers of things, then at last, without any mental effort or danger of error, we will be able to judge whether arguments are indeed materially sound and draw the right conclusions.[14]

The phrase 'without any mental effort or danger of error' is especially pertinent. Leibniz understood that if thinking could be reduced to arithmetic, it would be susceptible to mechanical treatment. It is for this reason that his invention of a mechanical calculator (Fig. 16.1) falls into a different category from the machines invented by Pascal and Morland. Leibniz's *machina arithmetica* itself did not greatly differ in its capacity from their machines—it was (like Morland's calculator) able to multiply and divide as well as to add and subtract. Although it was a marginally more complicated device, for it utilized a 'stepped reckoner'—a cylinder with nine cogs of varying length for the digits *1* to *9*, which was to play an essential part in later calculating machines—it was still basically a set of cogs and wheels in a box.

And yet the conception behind Leibniz's machine was quite different from that of Pascal and Morland. Having redefined reasoning as a species

Figure 16.1
Leibniz's Calculating Machine

of computation, Leibniz could not but regard his machine as capable of physically carrying out, albeit on a limited scale, the process of reasoning.[15] It was in his eyes the first step towards a fully-fledged 'thinking machine'. Indeed, Leibniz's goal was ultimately to build a more complex machine on the same lines that *would* be capable of carrying out all the functions of logical reasoning.[16] Since he believed that all concepts, and thereby everything that the mind conceived, could be quantified and then manipulated by a reasoning that was no more than a purely arithmetical process, his calculating machine promised far more than just to ease the labours of human 'computers' doing their business accounts. The *machina arithmetica* may have been to all appearances just a calculating machine but it was, in Leibniz's underlying conception, potentially capable of conducting logical operations, and of arbitrating all disputes concerning the truth and falsity of claims to knowledge.

In this respect, Leibniz's machine came much closer than those of Pascal and Morland to the modern conception of a computer and should be regarded as its precursor, for fundamental to the idea of the modern computer is that it is able to go beyond mere arithmetical calculations and can store, analyse and manipulate *data of any kind*, leading to its becoming an encyclopaedic source of knowledge. This requires, however, that the data be converted into a form that is amenable to the machine: a 'language' that the computer can 'read'. Had Leibniz succeeded in developing his *lingua characteristica*, it would have had to have been 'machine compatible' in order to fulfil his grand vision of a computationally based universal science. The *lingua characteristica*, as an incipient 'programming language', was in the end both too underdeveloped to be used in this endeavour, and too overdeveloped for the crude calculating machines that existed at the time, which were tied into the decimal system.

Leibniz and the New Theology of the Binarius

Leibniz came closest to conceiving of a 'machine code' in his repeated toying with the binary number system. He independently discovered the binary system in 1679, probably quite unaware of Bacon's 'Biliteral Alphabet'. His first thoughts on binary numbers appear in some notes entitled *De Progressione Dyadica* (March, 1679), in which he outlines how the binary system works. He then remarks that owing to its simplicity it has the potential for automation, explicitly stating: 'This type of calculation could also be carried out by a machine.'[17] In his notes he describes in some detail the way in which this could be done.[18] But he does not pursue this initial insight into the potential of binary arithmetic for automation. Later in the same year he wrote an essay, 'Of an *Organum* or

Ars Magna of Thinking', in which he more formally introduced the binary system and referred to its 'immense advantages' but without specifying what these were. The title of the essay, however, betrays the fact that he thought of binary arithmetic as a 'new instrument' (*organum*) and a 'great art' (*ars magna*) of thinking.[19] Unlike Bacon, it seems that it did not occur to Leibniz that the binary number system could be used as the carrier of information other than numbers. This is despite the fact that Leibniz was convinced that concepts and logical operations could be reduced to arithmetic. Perhaps the problem was that he could not relate the binary system to his quest for a *lingua characteristica*, and so it receded into the background of his interests.

Many years later, however, the binary system resurfaced once again in his mind, stimulated by a conversation with Duke Rudolph August in 1696. It was not the potential usefulness of the binary system that now aroused Leibniz's enthusiasm, but rather its theological symbolism. In a letter to the duke, written in January 1697, Leibniz expounds a kind of binary mysticism, in which the binary system represents for him an image of the origin of the universe. He sees Zero as signifying the void before Creation, and One the work of creation itself. In his letter he writes:

> One of the chief tenets of the Christian faith, one of those which have met with the least acceptance on the part of the worldly wise and are not easily imparted to the heathen, relates to the creation of all things out of nothing by the all-power of God. It can be rightly claimed that nothing in the world better represents this, indeed almost proves it, than the origin of number in the manner represented here, where, by the use simply of unity and zero or nothing, all numbers originate. In nature and philosophy it will hardly be possible to find a better symbol of this mystery . . .[20]

Leibniz tried to persuade the duke to have a medal struck, which depicted his binary vision of the creation of the world, entitled, *Imago Creationis* ('Image of Creation'). The design was surmounted by the words *Omnibus ex nihilo ducendis sufficit unum* ('To draw out all things from nothing, unity suffices'). Although the words on the medal proclaim that 'unity suffices', the overall message seems rather to highlight the cosmic importance of duality, with all the emphasis going on the binary number system in a design featuring contrasting areas of darkness and light. The medal was never struck but the design for it has survived and is reproduced below (Fig. 16.2). It shows rays of light, shining down from a unitary source, the One, onto a square divided into two vertical rectangles placed side by side, in which the binary numerical scale of 0s and 1s is correlated with its

Figure 16.2
Leibniz's design of the Binary Medallion for Duke Rudolph August.

decimal equivalents. At either end of the square are zones of darkness, representing the void or Zero, in which further Ones and Zeros are immersed.

In the rectangle on the left, the seven days of creation are symbolized, with the first Zero at the top representing the condition of Nothingness preceding the creation of Heaven and Earth. The One on the second line symbolizes God, becoming active on the first day of creation. By the second day both Heaven and Earth have been established, and are respectively symbolized by a One and a Zero (here Leibniz is influenced by Chinese thought, in which he was greatly interested, and to which he sought to relate his binary system). So we see how on the second day, the two numbers of the binary system appear for the first time side by side: the principle of duality, the Binarius, here makes its appearance. But Leibniz interprets this in terms of the creation of Heaven and Earth (which actually occurs on the first day) rather than the separation of the lower from the upper waters, concerning which God (as Gerhard Dorn pointed out) failed to say 'it was good'. The work of creation then proceeds through various admixtures of Ones and Zeros. Finally, on the seventh day, when all is completed, the Trinity appears in the form of three Ones at the bottom of the left-hand column.[21]

In 1703, Leibniz's essay 'Explanation of Binary Arithmetic' was published, arguing that the binary system affords us a far more economical way of calculating, because it requires very little thought, by cutting out all that is excessive (that is, all numbers beyond Zero and One).[22] For this reason, Leibniz never let go of the idea of a binary calculating machine. In 1716, during the last months of his life, he wrote that through the binary system he had essentially automated calculation.[23] In other words, he saw that of all arithmetical systems, the binary brings us closest to one that can be put into action in a purely automatic way.

The curious combination of mystical theology and mechanistic thinking reveals the extent to which Leibniz came under the spell of the Binarius. We have seen that the Binarius had assumed a central place in Gerhard Dorn's creation theology as the origin of evil, but there is no equivalent moral dimension in Leibniz's binary creation theology, for the Binarius is essentially God at work in the world. This positive gloss on the Binarius reflects Leibniz's underlying attitude towards rationality, specifically the function of reason carried out by the *ratio* alone, in relation to which he feels an almost religious sense of devotion. Just where Dorn, the pious alchemist, views the Binarius as an image of the devil, and seeks the Ternarius as the means by which the human soul returns to God, Leibniz, the pious rationalist, exalts the Binarius to the status of an image of God's activity in creation, and has no place at all for the Ternarius. In this contrast of attitudes between the two thinkers, separated from each other by 100 years and by a totally different relationship to the religious sphere, we are able to perceive how during the seventeenth century the pall of rationality descended on European spiritual life, driving to the margins of acceptable philosophical discourse a tradition of thought that had understood not only the inherent opposition of the principle of twoness to the divine unity, but also the redemptive role of the principle of threeness.[24]

In Leibniz's thought the Binarius is exalted to an epiphany of the divine as it manifests in nature as *imago creationis*. And yet, as he well understood, it is not to the world of nature that the Binarius belongs, but to the world of the machine. Ignoring the earlier, age-old tradition concerning the moral aspect of the principle of twoness, Leibniz bestows upon it a new positive theology that enables it to step out into the open, freed from the ancient stigma that for so many centuries had attached to it. This is Leibniz's spiritual deed, which prepares the way for what is to come.

Chapter Seventeen

THE POWER OF THE ELECTRICAL DISCHARGE

Electricity Begins to Reveal Itself

Leibniz died in November 1716, convinced that the binary system was the key to the automation of logic. But the future would prove that the automation of logic could only be fully realized when the binary system was used in conjunction with electric circuits, the basis of the electronic computer. The electric circuit was not recognized until the mid-eighteenth century, almost 30 years after Leibniz's death. At the beginning of the eighteenth century, for the newly established mechanistic consciousness the science of electricity was confined to electrostatics. Electricity was generated by friction, and the phenomena of electricity were essentially the phenomena of static electricity, i.e. of bodies that had become electrically 'charged'. Nevertheless, a good deal had been learned this way about the 'electric virtue': notably, that it had, like magnetism, the property of both attracting and repelling; that it involved certain sound and light effects; that it had the ability to travel away from an electrified body through being transmitted along a rod, a thread or a piece of string; that it could also jump from an electrified body to a non-electrified body close to it; and finally that it had a peculiar affinity with the vacuum. Hauksbee also discovered that lines of force emanated from an electrified body, comparable to magnetic lines of force, so that woollen threads placed near an electrified rotating glass cylinder would go rigid and assume a radial position in relation to the cylinder's axis.[1]

These initial findings, especially those relating to the transmission of the electric virtue, were confirmed and developed by Stephen Gray in the 1720s and Charles Dufay in the 1730s. Gray discovered that the distance the electric virtue could be transmitted through the medium of a length of thread was far greater than anyone had previously thought. He succeeded in getting it to travel 765 feet, and when he substituted wire for thread it travelled a distance of 850 feet. In France, a few years later, Dufay succeeded in conducting electricity 1256 feet (just under a quarter of a mile) using the medium of wet string.[2] Both Gray and Dufay also experimented further with the induction of an electric charge in a substance simply through its coming into proximity with an electrified body, but without the need of physical contact. What slowly but surely was being revealed was that the electric virtue did not 'belong' to any single body. It was not the virtue of a specific substance or group of substances. Dufay showed that any number of substances were capable of being electrified, thereby confirming what Robert Boyle had already suspected: that the electric virtue is not simply a property of amber, and a small number of other substances, alone. Electricity is not a 'virtue' of any particular substance, but rather is a power in its own right, which certain substances are more or less capable of absorbing and transmitting.[3]

What Dufay discovered was that those substances most difficult to electrify (for example metals and wet objects) are excellent transmitters or conductors of an electric charge, whereas those substances that are easiest to electrify (for example amber, glass and silk) are extremely bad at conveying it. Both Gray and Dufay understood that the transmission of an electric charge via a conductor over any distance would require that it be insulated so that the charge was not drawn off or dissipated by coming into contact with another conductor. While Gray had already employed silk as a non-conducting support for the threads and wire, which he used to transmit the electric charge, Dufay utilized glass as a solid insulator and was the first to coin the term 'insulator'.[4] So just those substances that had previously been thought to possess the electric virtue, because of the ease with which they were charged, were now understood to be insulators. By contrast, the very substances that seemed to have no affinity with electricity turned out to be the ones that best transmitted it.

In 1730 Gray demonstrated that even the human body can conduct electricity. By suspending a boy horizontally with clothes lines from the ceiling of a room, and holding a charged glass tube to the boy's feet, he showed that feathers and brass flakes were then attracted to the boy's face and hands (Fig. 17.1). The experiment was repeated by Dufay, and in the 1740s became a set piece for electrical researchers. Many of these researchers realized that this and other demonstrations of electrical phenomena could be exploited as a marvellous source of entertainment.[5] Electrical demonstrations soon became extremely fashionable, with so-called 'electricians' performing all kinds of wondrous tricks with their electric machines and electrified wands. An atmosphere of joviality came to surround the varied electrical displays that it had now become possible to produce. Serious investigation of electricity was caught up in a playful mood, in which magical effects, such as lighting candles with electric sparks, setting fire to alcohol, electrifying cutlery and producing electric halos around people's heads, all contributed to an amusing night out.[6]

Figure 17.1
Gray's 'suspended boy' experiment. A charged glass tube is brought close to his legs causing hands and face to attract brass flakes.

But then in 1745, a hitherto unknown characteristic of electricity presented itself, announcing in no uncertain terms that it was not a force to be played with. Up until this time, experimenters had experienced certain physical effects when brought into proximity with electrically charged bodies: Von Guericke reported tingling sensations, while Hauksbee described feeling a current of 'wind' as he brought an electrified tube near to his face. Hauksbee also described how when the electrified rod touched him he had the sensation of a multitude of fine hairs pressing against his skin, a sensation also noted by Dufay, who had described it as being like a spider web being drawn over his face.[7] Dufay also described the pricks and burns of the electric sparks that flew from his fingers when, suspended by silk like the boy in Figure 17.1, he was electrified.[8] In the early 1740s, Bose in Germany was producing stronger and stronger electrical discharges, powerful enough to give minor shocks that left a wound on the skin.[9] But in 1745, a quite different order of direct experience of electricity occurred. It was announced by Pieter van Musschenbroek at Leyden in the Netherlands. Van Musschenbroek experienced something truly terrifying.

Van Musschenbroek was particularly interested in the question of whether it might be possible to store an electric charge. Since glass was known to be a good insulator and water a good conductor, it occurred to him that the electric charge could be conveyed from a charging machine via a gun barrel (which acted as a conductor) to a piece of wire let into a jar of water, and there the charge would gradually build up. Holding the

glass jar in his right hand, after some minutes van Musschenbroek reached
with his left hand towards the gun barrel in the hope of drawing some
sparks from it (Fig. 17.2). This is his description of what happened:

> Suddenly I received in my right hand a shock of such violence that my
> whole body was shaken as by a lightning stroke. The vessel, although of
> glass, was not broken, nor was the hand displaced by the commotion:
> but the arm and body were affected in a manner more terrible than I
> can express. In a word, I believed that I was done for.[10]

Figure 17.2
Van Musschenbroek's experiment. An iron tube is suspended from silk lines and charged by
rubbing a swiftly rotating globe attached to it by a metal chain. At the other end of the tube, a
wire is hung, leading into a glass jar, partly filled with water and held in the experimenter's
hand. When the experimenter tries to draw sparks with from the tube with his left hand, he
receives a violent shock in his right hand.

Van Musschenbroek warned other experimenters not to try to replicate this 'new but terrible experiment', for he only 'survived by the grace
of God' and would not do it again 'for all the kingdom of France'.[11]
Needless to say, his warning went unheeded. Johann Winkler, for
example, conducted the same experiment and gave a somewhat more
detailed account of its effects:

I found great Convulsions by it in my Body. It put my Blood into great Agitation; so that I was afraid of an ardent Fever; and was obliged to use refrigerating Medicines. I felt a Heaviness in my Head, as if I had a Stone lying upon it. It gave me twice a Bleeding at my Nose, to which I am not inclined.[12]

Others reported symptoms of pain in the arms and the chest, nosebleeds, temporary paralysis, convulsions, dizziness and even concussion.[13] Nollet, who had earlier collaborated with Dufay, now worked more systematically, noting the effects of the discharges from van Musschenbroek's so-called 'Leyden jar' on plants and animals. He discovered these discharges could kill small birds and mammals. And yet the attitude that electricity was an amusing source of entertainment continued. Nollet was also a showman, and he arranged an extraordinary spectacle before the king at Versailles, in which 180 royal guards held hands, with the first in the chain holding a Leyden jar. When the last soldier in the chain touched the prime conductor they all involuntarily jumped at the same instant into the air as the charge passed through them. This demonstration was followed by a second in Paris in which 200 Carthusian monks formed a line 900 feet long, with each monk connected to the next by an iron wire. Once again the effect was that the whole company suddenly sprang into the air in response to the shock when the circuit was closed.[14] In the United States, Benjamin Franklin likewise could not resist seeing electricity as a source of amusing diversions, despite his otherwise serious research. On one occasion, for example, he organized an 'electric picnic' with his electrician friends, which began by a spectacular firing of spirits with a spark sent across the River Schuylkill.[15]

Lessons of the Leyden Jar

Along with its entertainment value, the real importance of the Leyden jar, which was the earliest form of 'condenser', is threefold. First of all, it demonstrated that electricity 'travels' in a circuit. This had not been understood before. Experimentation revealed that the electricity confined in a charged jar would not only travel beyond the jar (this had been understood) but would *return to it* if a circuit was made between the electricity that had accumulated inside the jar and the exterior of the jar. Simply to bring an object near to the jar or to make contact with the jar would produce no result unless that object itself was connected to the electricity that had accumulated *inside* the jar, for only then was the circuit established. The circuit required that the wire going into the jar was linked to the outside of the jar. Initially this linkage occurred when the person holding the outside of the jar in one hand touched the wire going into the jar with the other hand (Fig. 17.3). But it was quickly realized that in order to set up this circuit an iron wire could be substituted for the human being. Once the circuit was made, then the electricity contained in the jar would immediately become present outside it, either in the human being (experienced as a shock) or in the iron wire. It was not until the invention of the battery in 1800, when the electric current was effectively tamed and brought under greater control, that the extraordinary usefulness of the electric circuit was realized. But van Musschenbroek's experiment was a highly significant milestone.

Figure 17.3
The Leyden experiment demonstrates the electric circuit for the first time.

A second reason for the significance of the Leyden jar was that it showed the potentially lethal power of electricity. If it could kill small animals then, with bigger jars or through linking several jars together to enable storage of a higher charge, it could surely be lethal to human beings. Electricity had shown itself to be a potentially deadly force.

The third reason why the Leyden jar was important is that it showed that previous explanations of electricity were clearly inadequate. Before Gray's discovery of electrical conduction, the standard theory was that by rubbing an 'electric' (such as amber), certain subtle but nevertheless physical emanations or 'effluvia' were evoked from it that either impinged upon or could be transferred to another body, and this accounted for both electrical attraction and repulsion. With the collapse of the distinction between electric and non-electric bodies, the idea of electrical effluvia continued to be held, but they were now conceived as having a semi-independent existence, and thought of as a kind of fluid, somewhat like the element water. Because electricity generated by friction required two different kinds of substance to be brought together, two kinds of electric fluid were postulated, one produced by rubbing glass ('vitreous') and one produced by rubbing resinous bodies like amber, sealing wax, paper or silk ('resinous'). Bodies with vitreous electricity repelled other bodies with vitreous electricity, but attracted bodies with resinous electricity, and vice versa. The Leyden jar experiments undermined this two-fluid theory and led to Franklin putting forward the view that there was only *one* electric fluid, which all bodies naturally possess and which exists in a neutral state of equilibrium within any given body. According to Franklin, when a glass tube is rubbed with silk cloth, this equilibrium is upset. The glass tube acquires a positive charge, and the cloth becomes negatively charged. The former charge he referred to as a 'plus', the latter a 'minus'. Both objects then seek to neutralize their respective charges in order to regain their equilibrium.[16]

Crucial to Franklin's theory was the principle that an electric charge is not created or destroyed, but rather is 'collected' by friction.[17] This meant that if a positive charge is produced by an influx of electric fluid, then it must have produced a deficit elsewhere. In the case of the Leyden jar, Franklin viewed each unit of charge collected on the inside of the jar as positive or 'plus' but, in order for it to accumulate, the equivalent unit of charge had to be driven out of the jar to accumulate on its external surface as negative or 'minus'.[18] As we have seen, the jar would only regain its equilibrium when, as a result of external contact made between its inner and outer surface, electricity passed from the inside to the outside of the jar, sending convulsions through the person holding it.

Franklin's theoretical conceptions proved extraordinarily influential and remain with us today, although the 'negative' charge is now regarded as a surplus of electrons and the 'positive' charge as a deficiency of electrons. They opened the way to a quantitative approach to the study of electricity, enabling the emphasis of research to shift from the question of *what* electricity is to considering how to measure it.[19] Some modern commentators have suggested that Franklin's quasi-mathematical conception of 'plus' or 'minus' electricity most probably derived from his experience in business, where an account balance is formed by the sum of receipts (+) and expenditures (−). Thus a body could, from an electrical point of view, either be in credit or in debit, in the black or in the red.

Electricity was thereby conceived in terms reassuringly familiar to Franklin as being like money in the bank.[20] Thereby—consciously or unconsciously—Franklin drew on concepts from the mercantile sphere of banking and business and imposed them on a realm where they could hardly be expected to reveal the inner nature of what was being investigated. While this mode of thinking may have served to organize the phenomena so that they could be grasped by human thinking, at the same time this conceptual scheme trivialized the uncanny and potentially awesome power of electricity, deflecting future research from understanding electricity from a deeper, metaphysical perspective.

A Failure of Vision

One further fruit of the Leyden jar experiments was the identification of atmospheric lightning with electricity. Several researchers had already suggested that lightning was an electrical phenomenon, but it was Franklin's description of how this might be experimentally proved that led to its finally being demonstrated. This took place in France in 1752, when French researchers, following Franklin's directions, were able to conduct atmospheric electricity down a 40-foot high iron rod during a thunderstorm.[21] A month later, Franklin himself successfully charged a Leyden jar from electricity conducted down the string of a kite that he flew in a thunderstorm in Philadelphia.[22] For Franklin, the ability of lightning to charge a Leyden jar was incontestable proof that electricity and lightning are one and the same. But this did not lead him to reconsider the phenomena connected with the Leyden jar from a greater, cosmic perspective. Franklin's explanation was, in accordance with the spirit of the times, entirely naturalistic and mechanistic. He argued that water vapour rising from the sea already carries an electric charge because particles of salt and water rub against each other in the sea, which could be regarded as a giant electrical machine. When these sea-generated, electrically charged clouds meet either a less electrically charged mountain or fresh-water clouds, then a discharge of electricity occurs with a sudden flash and a loud noise.[23]

When we set such an explanation beside the perception in the ancient mysteries of mighty sub-natural forces, wielded by death-dealing gods, we can see the extent to which, by the middle of the eighteenth century, human beings had become incapable of understanding electricity in anything remotely corresponding to the terms which, in ancient times, it was felt necessary to approach it. Franklin's charging of his Leyden jar from the kite flown into the thunder clouds both literally and symbolically reduced lightning to the puny proportions of mid-eighteenth century technology: atmospheric electricity was seen as different only in scale from the electrical discharges produced by earthbound Leyden jar experiments. Appropriately enough, Franklin was hailed as a new Prometheus snatching lightning from the heavens.[24] Once snatched, however, it was then cast into mechanistic modes of thought supplemented by concepts borrowed from the world of eighteenth century mercantile capitalism. We are reminded of a passage from Shakespeare's *All's Well That Ends Well*:

> We have our philosophical persons to make modern and familiar things supernatural and causeless. Hence is it that we make trifles of terrors, ensconcing ourselves into seeming knowledge when we should submit ourselves to an unknown fear.[25]

Thanks largely to Franklin's contribution, the focus of research in the second half of the eighteenth century shifted towards inventing instruments that could accurately measure electricity, so as to further a purely quantitative approach to it. The aim of quantitative researchers was less to understand *what* electricity was and more to formulate quantifiable laws that would capture its behaviour. Just a few years after Franklin's kite experiment the German mathematician Franz Aepinus followed up Franklin's work with a treatise on electricity and magnetism employing a sophisticated mathematical approach that included algebraic analysis.[26] Already by the 1760s Joseph Priestley was speculating that the attraction of electrically charged bodies was subject to the same laws as gravitation, varying with the inverse square of their distance. This was confirmed in the next decade by Charles Coulomb, who also invented a highly sensitive instrument—the 'torsion balance'—that was capable of measuring electric charge.[27] But while Coulomb was able to measure electric charge, he was unable to define what electricity is. Although his preference was for the older two-fluid theory, he insisted that this was just a congenial theoretical framework and could not be taken as giving us a true insight into its nature.[28]

The nature of electricity was too enigmatic for people to grasp hold of. Many were unsatisfied even with the notion of electricity being a fluid. Influenced by Newton's conception of the ether—which he regarded as the most subtle matter pervading all bodies—it seemed that electricity could be understood as a kind of ether.[29] According to the English researcher, William Watson, the electric ether was both omnipresent like air and 'elastic', so it could become more dense in one body and less dense in another, but it always sought equilibrium, and this explained the electric discharge. Crucially for Watson, electricity consists in the atmosphere *around* an electrified body.[30] In his conception of an electrical ether, Watson sowed a seed that would germinate later in the more fruitful insights of Michael Faraday. Meanwhile, the ether theory appealed especially to those seeking a theological interpretation of electricity as the life-spirit in nature or *Spiritus Mundi*. Just as Leibniz had been moved by a curious religious enthusiasm to see in the binary system a theology of creation, so now a kind of religious zeal for electricity encouraged people to interpret it as a fundamental spiritual principle active in all of creation. Some even chose to identify it with the First Light of the Book of Genesis, spreading out over chaos as a stimulating, warming and form-giving life-principle, manifesting again and again in living shapes, and infusing all matter as an innate 'electrical fire'.[31] Thus a spurious theology of electricity arose to contest the traditional metaphysics of light, twisting both the Bible and the older spiritual understanding of light beyond recognition, while conveniently ignoring the death-dealing power of the electrical discharge.

But if some religious enthusiasts were inclined to apotheosize electricity, the stronger tendency amongst researchers was to pursue a quantitative approach that turned its back altogether on the vexed question of the actual nature of electricity. As the eighteenth century drew to a close, few people could honestly claim to understand what it really was that they were dealing with. As Joseph Priestley wrote in his compendious *History of the Present State of Electricity* (1775):

> By electricity, I would be understood to mean, only those *effects* which will be called electrical; or else the *unknown cause* of those effects.[32]

It was an honest admission of ignorance. From a spiritual standpoint, the difficulty in understanding electricity was in part due to the atrophy of the old awareness of the upper and lower borders of nature, in the absence of which an accurate theological orientation towards it was rendered impossible. As a result, the way was opened to the propagation of illusory notions such as the identification of electricity with the First Light or *Spiritus Mundi*. But on the scientific side, the gradual displacement of the qualitative approach by a mathematically based approach, while having the advantage of grounding knowledge of electricity in exact, quantitative thinking, diminished the chances of a deeper contemplative insight into its nature. Something had been discovered that seemed to promise much more than anyone was able to articulate, and as with Aladdin's lamp, the genie remained largely hidden from view. There was little more to do than to keep rubbing.

Part Four

THE ERA OF THE MACHINE

Chapter Eighteen

THE DIABOLICAL MACHINE INVADES
THE PSYCHE

La Mettrie: Man in the Image of the Machine

Between van Musschenbroek's experiment with the Leyden jar and
Franklin's flying his kite into thunder clouds, a book was published that
caused a furore throughout Europe. Ostensibly, it had nothing to do with
electricity, and yet its publication was as historically significant for the
emergence of the computer as was the growing awareness of the potential
of electricity in the mid-eighteenth century. The book was written by a
French physician, Julien Offray de La Mettrie, who was born in 1709, just
three years after Franklin, and so an almost exact contemporary. But La
Mettrie and Franklin were very different characters. While Franklin was a
man of utmost moral probity, La Mettrie rather prided himself on causing
scandal with his radical opinions and hedonistic lifestyle, and yet both of
them embraced the same mechanistic worldview. For Franklin, this
meant that 'God Almighty is himself a mechanic', but for La Mettrie it
dispensed with the need for God altogether.[1] He shamelessly advocated a
thoroughly materialistic outlook on life. In his most famous book
L'homme machine, or 'Man a Machine' (1748), La Mettrie denied the
existence not only of God but also of any spiritual aspect to human
nature, and sought to show that since humans are animals, and animals (as
Descartes and other mechanists upheld) are no more than machines, then
human beings too are merely machines.

La Mettrie's significance in the historical development of the computer
lies in the fact that he was the first person to propound—and to propound
unflinchingly—the view that the human being, like any other animal, has
no soul or spirit. In La Mettrie, we meet a man whose inner life is so
thoroughly permeated by the mechanistic world-picture that he feels a
passionate need to apply it to the human being. For him, the human
being is a physical machine, and nothing but a machine. The reflexive
core of human consciousness, the ability to think, which for Descartes was
an inner experience that could not be accounted for merely in material
terms, is explained by La Mettrie as nothing more than a property of the
'organized matter' of the brain. Thereby human freedom is denied.
Interestingly, La Mettrie compares thought to electricity, seizing the

opportunity presented by the flurry of experiments being conducted at the time to enlist electricity to his cause.[2]

La Mettrie was a practical man, an empiricist who believed that our ideas should be based on experience and observation. He also appreciated, however, that the forming of mental images played an important part in the way human beings reason and make judgements.[3] The emphasis that he gives to metaphor in *L'homme machine* is important, because it is above all the *image* of the human being as a machine that La Mettrie is most keen to promulgate. He tells us over and over again that we are just clocks or large watches, a collection of wheels and springs, and so on, but he clearly does not mean this literally. He is talking figuratively, as if seeking to persuade us to welcome into our fantasy life this conception of human nature as mechanistically determined. Consider the following passage from *L'homme machine*:

> Man is but an animal, or a contraption of springs, each of which activates the next without our being able to tell which one nature used to start the merry-go-round of human society ... These springs differ among themselves in location and strength but never in nature. Thus the soul is only a principle of movement or a sensible, material part of the brain, which one can regard as the machine's principal spring without fear of being mistaken.[4]

This passage is typical of La Mettrie's style. What he seeks to impress upon us is an image, rather than an argument. He gives voice to the fantasy that the way the human body works is mechanical, and since everything we experience as belonging to our inner life is generated by this 'contraption of springs', the human being—body, soul and spirit—is simply a complex physical mechanism.

La Mettrie urges his readers to embrace the mechanistic fantasy of the Man-Machine as an erotic image. His book is peppered with *risqué* references to human sexuality. He enjoyed playing the role of the libertine.[5] In his portrait, this side of his character is well conveyed (Fig. 18.1). In both the way he sought to present his image to the world, and also in his writings, what is essentially a cold and utterly soulless view of human nature is made to seem appealing by being given a cheeky, even rakish, veneer. The logical implications, however, remain. If the human being really is a machine, then it follows that machines can be made that will act like human beings. Not just physically, but mentally too, for if the human mind is nothing more than the physical brain, then it too is a mechanism like the heart or digestive system. Thus La Mettrie, by softening up the human moral sense, pre-

Figure 18.1
Portrait of Julien Offray de
La Mettrie.

pares the way for the creation of a machine that would simulate human logical reasoning.

La Mettrie's reduction of the human being to a machine is significant because historically the activity of thinking was considered the characteristic that distinguishes human beings from the animal kingdom. Once thinking is conceived as nothing more than an activity of calculation, it becomes possible both to redefine what a human being is, and then to set about disproving the traditional view of thinking as a spiritual activity (which, precisely because it is not mechanical, is the source of human freedom), by showing that machines can indeed replicate the act of thinking. The re-conception of the human being in the image of the machine actually becomes one of the chief motivations to create a thinking machine. For those in the eighteenth century who subscribed to the view that human beings are nothing more than machines, just as for those who subscribe to it today, nothing could better vindicate their view than to create a machine that could carry out all the functions of logical thinking. As much as it was a motivation for La Mettrie, the radical and nihilistic agenda to prove that human beings are machines was a major motivation of later computer pioneers like Alan Turing and John von Neumann.[6]

The crucial point here is that once the fantasy of the human being as a machine gains hold of the human mind, then the embodiment of this fantasy in an actual machine becomes an irresistible challenge, a philo-sophical imperative. In order for computers, and subsequently ever more sophisticated robots, to be invented, the image of human nature as a mechanism had first to be accepted into the inner life of human thought and inwardly pictured. La Mettrie's *L'homme machine* is one of the first attempts to make this image attractive, even desirable. Before La Mettrie, both Bacon and Descartes had sought to articulate a method of thinking that was intrinsically mechanical. But neither dared to deny the human spirit. With the gradual entrenchment of the new objectivizing, quanti-tative approach to knowledge during the seventeenth century, scientific method itself became ever more allied to the mechanistic ideal. Finally, in La Mettrie the Machine-Man appears as a fully-fledged archetype, arising from the depths of a psyche that has turned away from what was pre-viously regarded as essential to our humanity.

La Mettrie's *L'homme machine* sold well, and was swiftly translated into English. It resonated with a rising tide of thought that was slowly colouring the way more and more people were coming to view both nature and themselves. This was, however, more than just a movement of thought, for it was psychically highly charged. That is why we need to see that the idea of the Machine-Man that La Mettrie so boldly put forward was also an archetype, in the Jungian sense of an 'autonomous factor' that roots itself in the semi-conscious and unconscious world of image and symbol.[7] From this world of image and symbol it both expresses itself in, and is nurtured by, the new kinds of physical object that increasingly fill the external world—initially mechanical mills, pumps, turbines and clocks, but later, from the seventeenth century onwards, the new cal-culating machines. These machines fed people's inner fantasy lives as much as they impacted on their outer practical lives.

During the eighteenth century, there were many attempts to improve the design and capabilities of mechanical calculators, developing further the earlier prototypes of Pascal and Leibniz. None of these attempts were particularly successful, as they were generally more complicated but rather less reliable than their prototypes.[8] Nevertheless, the fact that many minds were intent on improving the ability of machines to make calculations—something that hitherto only human beings could do—impressed itself deeply on the age. We have already seen that Diderot's *Encyclopaedia*, published between 1751 (the year of La Mettrie's death) and 1772, featured Pascal's mechanical calculator (see Chapter 15, pp. 158–159). The secular and naturalistic principles on which the *Encyclopaedia* was based, its thinly concealed hostility to the religious outlook, along with its celebration of the mechanical arts, all helped to promote a new conception of the human being.

In the mid-eighteenth century, the Machine–Man fantasy was given even more credence by the increasingly clever automata that were being created by engineers like Jacques de Vaucanson. Vaucanson, who was born the same year as La Mettrie, in 1709, is renowned for having created several remarkable automata, one of the most celebrated being a mechanical duck that could supposedly flap its wings and both eat and excrete. Inside the duck, and also within the foundation on which the bird stood, was an intricate mechanism of over 1,000 moving parts (Fig. 18.2).[9] Vaucanson's duck presented a vivid example of the power of the mechanistic paradigm to deliver convincing images (most people only encountered it on the printed page) in support of the view that animals and, by implication, humans are merely sophisticated automata.

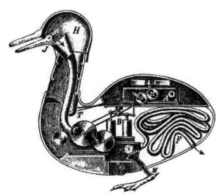

Figure 18.2
Vaucanson's mechanical duck (1739).

Dark Satanic Mills
La Mettrie died at the age of 42 in 1751. The world had to wait another 40 years before the man who would attempt to create a physical machine that would 'think mechanically' was born. That man was Charles Babbage, to whom we shall return in a later chapter. Meanwhile, although the ground was being prepared for the advent of 'thinking machines', the time was not yet ripe for this next step to be made. Opposition to the mechanistic image of the human being was still strong enough to have La Mettrie banished from France even before the publication of *L'homme machine*. After its publication he was then banished from Holland, to where he had fled, and was forced to seek refuge with Frederick the Great. Opposition was strongest within the religious establishment, and copies of the book were actually burned by the ecclesiastical authorities.[10]

Those who thought like La Mettrie were still in a minority, and Europe in the mid-eighteenth century was not yet ready to accept such a radically reduced view of what it means to be human. Furthermore, European engineering was nowhere near technically advanced enough to be able to create physical machines that could do any more than perform straight-forward mathematical calculations and, as we have noted above, these machines were not particularly reliable. Nevertheless, the mid-eighteenth century was on the verge of the industrial revolution, and this would dramatically increase the ability to create more sophisticated calculating machines than had hitherto been possible.

It was in the English textile industry that the industrial revolution really began, with a series of innovations in the machinery used for spinning and weaving that massively raised the output and quality of both yarn and cloth. The growing dominance of machinery in people's lives was one of the main social factors of the industrial revolution. Machines and human beings were brought into a much closer association, with the primary aim of increasing production. Between 1766 and 1787 there was a fivefold increase in the output of cotton due entirely to mechanical innovations.[11] Production was concentrated in large buildings, which were called 'mills' because of the dominance inside them of machinery. The machinery was initially horse-powered or water-powered, as at Richard Arkwright's mills at Nottingham and Cromford for spinning cotton, established in 1769 and 1770 respectively. But during the 1770s and 1780s the steam engines of Watt and Boulton slowly began to come into service, and an increasing number of cotton mills became steam-powered. Mills got larger—often six, and sometimes seven or even eight, storeys high, with some buildings accommodating a workforce of up to 2,000. By 1838, there were 1,600 cotton mills in England, three quarters of which were in Lancashire.[12] The cotton mills housed the machines. And thousands upon thousands of people, many of whom were women and children, now spent the best part of their lives working long hours (normally a twelve hour working day) in service to these machines.

The physician, James Phillips Kay, vividly described the desperate conditions in both the mills and the homes of the workers in 1832 in his book *The Moral and Physical Condition of the Working Classes Employed in the Cotton Manufacture in Manchester*. What comes across most powerfully is the numbing, dehumanizing effect on the workforce of their servitude to the machine:

They are engaged in an employment which absorbs their attention, and unremittingly employs their physical energies. They are drudges who watch the movements, and assist the operations, of a mighty material force, which toils with an energy ever unconscious of fatigue. The persevering labour of the operative must rival the mathematical precision, the incessant motion, and the exhaustless power of the machine.[13]

The industrial revolution must be understood as introducing a new relationship between machines and human beings, in which the human being for the first time experiences the greater-than-human power of the machine as enslaving rather than as liberating. In the mill, the traditional relationship of machine to human being is reversed: rather than the machine serving the human being, the human being must serve the machine.[14]

It was not just cotton, of course, that was subject to the new mechanical imperative. The industrial revolution introduced mechanical methods and the techniques of mass production into every sphere of manufacturing, from pottery to the production of food. Powered by the new steam engines, 'mills' sprang up everywhere. William Blake, who moved to Lambeth in 1790, would have walked past a new steam-powered flour mill in Blackfriars Road every time he went into the city of London from his home.[15] For Blake, the mills could only appear as 'dark' and 'satanic' for he knew that the mechanical reasoning (or *ratio* as he called it) they embodied would in the end turn everyone and everything into nothing more than a conglomeration of cogs and wheels—'the same dull round, even of a universe, would soon become a mill with complicated wheels'.[16]

As the eighteenth century gave way to the nineteenth, and the industrial revolution stamped its impress upon the European soul, more and more human beings were exposed to machines in their daily lives. At the same time the worldview of materialism, in which human beings were seen as mere animal-machines, slowly gained ground, in the face of the still formidable opposition of the Church.[17] It was as if the triumph of secular rationality, with its inbuilt mechanistic logic, was grimly reinforced with every new machine that was produced. The steam engines manufactured by Boulton and Watt epitomized the triumph of the logical mind, for in every cog, lever, pump, piston and valve, the power of applied logic combined with utmost mathematical precision was displayed (Fig. 18.3).

These great engines confronted people with the image of that mechanical intelligence, which in the mechanistic philosophy had been elevated to cosmic proportions in the figure of Franklin's 'God the

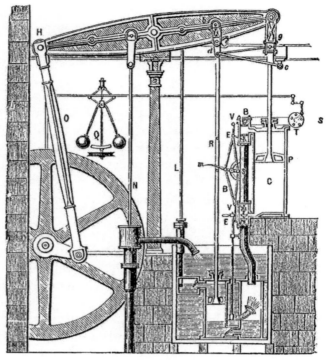

Figure 18.3
Boulton and Watt's Double-acting Steam Engine, of 1784.

Mechanic', the creator of an entirely mechanical world. This mechanical intelligence was exercising a greater and greater influence on the inner life of human beings, so whilst the invention of the steam engine and the machinery of mass production could be claimed as human in origin, its inhuman effect on people's lives revealed the extent to which human creativity and genius had in fact been captured by a mode of thought intrinsically inhuman. The encounter with the engines of industrialism brought people face to face with something within the human being that had never before been made manifest in such a terrible way.[18] It was as if the machines were being produced by an intellectual power that up until this point in human evolution had been restrained by a living spiritual awareness. Without this living spiritual awareness, the world of nature as much as that of human beings, fell inexorably under the dominion of the machine. These machines could not yet reason, but they were manifestations of thought-processes that were divorced from the human heart. They mirrored an aspect of the mind in which no trace of the divine could be found.[19]

The Enthronement of the Goddess Reason

In Chapter Ten, we saw how the change in consciousness wrought during the fourteenth and fifteenth centuries culminated in the spiritual catastrophe of the Reformation.[20] A major part of that catastrophe was the assault on the contemplative life, which outwardly resulted in the destruction or closure of monasteries and religious houses in many countries in Europe. As a consequence, the European psyche was thrown out of balance, for it lost (or at least suffered the severe weakening of) an essential countervailing force to the impulse towards the promotion of a one-dimensional, literalistic stance towards knowledge. This stance, which became deeply ingrained as 'common sense', led in the sphere of science to distrust in anything other than a quantifying and therefore merely calculative mode of thought. Put in terms of the medieval understanding of the mind, from the fourteenth to the seventeenth centuries the *intellectus* was slowly but surely displaced by the *ratio* in the esteem accorded to our mental faculties. In so far as human beings could be defined as creatures of reason, increasingly reason came to be regarded as no more than the *ratio*—the faculty of logical reasoning and calculation.

This reduced view of reason, which by the eighteenth century had become so prominent that few educated people would challenge it, was one of the hallmarks of the Enlightenment. It was typically accompanied by suspicion of the inner life of imagination, inspiration and numinous experience. The Enlightenment project was to establish an 'Age of Reason', understood precisely as excluding the supposedly non-rational capacities associated with the interior life.[21] In Catholic countries, it took the form of outright hostility towards the Church, and a deep antagonism towards monasticism, which was viewed as an excuse for laziness and an indulgence in superstition. Diderot, who was the moving force behind the *Encyclopaedia*, mercilessly attacked monasticism in his novel *La Religieuse*—written in 1760 but not published until 1796—in which he castigated the vows of poverty, chastity and obedience as debasing, and declared the life of the cloister as being for 'fanatics and hypocrites'.[22] Such was the animosity towards the monastic life that a new wave of secularization swept through Europe, beginning with the Emperor Joseph II issuing a decree in 1781 closing all the religious houses of the 'idle' contemplative Orders (i.e. Benedictines, Cistercians, Carthusians, etc.) throughout the Holy Roman Empire. Eight years later, in France the Revolution targeted the monasteries, with the state taking possession of monastery lands in 1789, and then in 1790 declaring all religious vows null and void and abolishing all religious Orders. In 1792 all religious congregations were abolished throughout France and any remaining religious houses closed.[23] Napoleon then spread secularization to Switzerland, Spain and Italy in his wars of conquest, suppressing monasteries wherever French troops gained control.[24]

For the European psyche, the new 'Age of Reason' thus involved an inner spiritual convulsion. By way of compensation, the loss of authentic religious orientation led to reason attracting to itself fanatical, pseudo-religious feelings. A new cult of reason, understood as unaided logical thought, arose. It became a substitute for the genuinely religious life that honoured the intrinsic numinosity of the authentically spiritual. Instead, the reduced reasoning process that had become the norm during the eighteenth century acquired for itself a vicarious numinosity. This culminated in a most significant symbolic event during the turmoil of the early 1790s—an event that well expressed the turbulence within the collective psyche at the time. In the year following the abolition of all religious congregations in France, a ceremony was held on November 10, 1793, in which the Goddess Reason was enthroned in the Cathedral of Notre Dame (renamed the 'Temple of Reason').[25] In place of the Virgin Mary, the Goddess Reason, enacted by a chorus girl, was installed in the Cathedral, robed in a tricolour flag and crowned with a red cap, while the crowd sang to her a hymn to Liberty (Fig. 18.4). The very idea of making

Figure 18.4
The Enthronement of the Goddess Reason in the Cathedral of Notre Dame, November 10, 1793.

reason into a goddess with her own temple (let alone her part taken by a chorus girl) was on the face of it a complete contradiction in terms. And yet it perfectly symbolized the dangerously one-sided inflation of naked rationality that had taken place during the previous centuries, which now drew to itself an unthinking mass religious zeal.

What the enthronement symbolized was the degree to which the collective consciousness was willing to project the God-image onto logical, calculative thinking—the very capacity that was embodied in machines. Although the symbolic act of enthronement occurred in France, it was a deed that issued from the depths of the European psyche as a whole—according to Jung equivalent to the hewing down of the oak of Wotan by the first Christian missionaries.[26] As we have seen, Franklin's conception of 'God the Mechanic' elevated to cosmic proportions this kind of reduced rationalism, which stood behind the newly emerging machine age. What was adored was that power of intelligence, now so much in the ascendancy, which could be directed towards any problem in order to discover the most efficient way of solving it, and whose emphasis was on practical application, on achieving practical results, rather than on the insights born of 'useless' leisurely contemplation.[27] This intelligence, despite its totally anti-religious character, drew to itself a deep-seated emotional allegiance, hardly distinguishable from religious fervour. It was, we might say, a massive psychic investment in technological thinking, in which mechanical intelligence was apotheosized.

The Diabolical Machine Invades the Psyche

The cost of the industrial revolution to psychic health can be gauged by the fact that at the beginning of the eighteenth century, in 1700, there was in the British Isles only one lunatic asylum, Bethlem, which held little more than 100 patients. By the end of the same century, in 1800, it is estimated that numbers of confined lunatics had risen to as many as 10,000. By the end of the nineteenth century, as if it were Psyche's tribute to the Goddess Reason, the number had escalated to *100,000*.[28] It is to the period of the turn of the eighteenth century into the nineteenth that the first fully documented case of schizophrenia can be traced. Significantly, this case involved a man who believed that he had come under the influence of a diabolical machine.

The case of James Tilly Matthews is one of the more poignant episodes in the prehistory of the computer. It gives us insight into the collective psyche's response to the ever-greater incursions of machines into the lives of human beings. At the beginning of the nineteenth century, from the point of view of the collective psyche, these incursions seem to have reached a kind of tipping point. Apart from the arithmetical computations of the mechanical calculators, the physical machines that existed at this time were not yet able to replicate human mental functions, but it was as if at unconscious depths people were feeling increasingly menaced by the growing sophistication of the machines that were coming to populate the world. The paranoid fantasy of Matthews—the fantasy of the 'influencing machine'—was to become widespread in subsequent decades and is today one of the most prevalent in paranoid schizophrenia.[29]

Matthews was a tea merchant who travelled to revolutionary France in the early 1790s, where he became embroiled in political intrigue, and was as a result imprisoned at the height of the turmoil of the French Revolution for three years between 1793 and 1796. On his return to England, he complained of conspiracies against his life, and he accused the Home Secretary, Lord Liverpool, of treason (he actually interrupted a debate in Parliament, shouting 'Treason!' from the Public Gallery). He was arrested and taken to Bethlem asylum, where he claimed that he had been employed as a secret agent in France by the British government, but had been betrayed by it. Matthews believed that a gang of criminals—the 'Air Loom Gang'—was torturing him with a machine called the 'Air Loom'. He claimed that many such machines had been installed throughout London, and were targeted on leading government figures. Through rays emanating from these machines, groups of French revolutionaries were able to influence the thoughts of British government ministers. Matthews himself experienced the emanations of the Air Loom physically, and described in vivid language their excruciating effects.[30] But it seems that many of those targeted were unaware of the fact that 'the intellectual atmosphere of the brain' was being influenced both during their waking life and during sleep.[31]

So intense was Matthews' delusion that he was able to describe this 'influencing machine' in detail, and even produced a design of it (Fig. 18.5). The Air Loom utilized recent developments in gas chemistry, combined with magnetic and electrical effects, to create magnetically charged air currents that could be directed towards those unfortunate souls whom it targeted. The mind of the victim was then forced to accept 'a train of ideas very different from the real subject of his thoughts'.[32] Thus the denial of freedom to the human mind by La Mettrie 50 years previously was now experienced as a nightmarish reality by Matthews.

Despite the delusional nature of Matthews' fantasies, they gave expression to a groundswell of unease in the mass of the population, as it had to face the encroachments of the machine, both without and within. Matthews' Air Loom was aptly named. It was, like the steam-powered

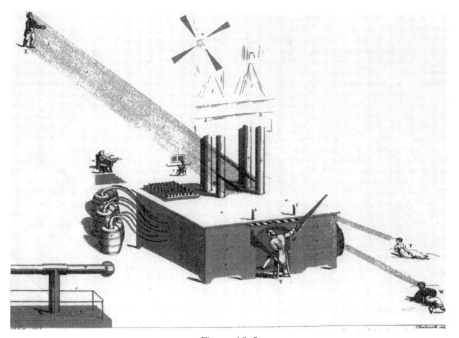

Figure 18.5
The Air Loom of James Tilly Matthews, showing its victim at the top left-hand corner, and various members of the Air Loom Gang operating it.

looms in the new mills, a weaving machine, but the fabric it wove was the fabric of thought rather than cloth. In this respect it eerily presaged Charles Babbage's Analytical Engine, which was also designed to 'weave thoughts'.[33] Unlike the Analytical Engine, the Air Loom was a nightmare vision confined to Matthews' own subjective inner life, and yet he nevertheless brought vividly to expression a notion that had taken root in the depths of the collective psyche. It lived there in the prophetic notion of a machine that had the ability *to weave thoughts*, combined with the notion of a machine that could operate directly on the material of the human mind. What Matthews experienced in his inner visions was a machine that was capable of thoroughly infiltrating, and even merging with, the human being.

Chapter Nineteen

THE PUNCHED CARD LEADS THE WAY

Air Loom, Treadle Loom and Drawloom

In the previous chapter, we have seen that, in the first documented case of schizophrenia in England, James Tilly Matthews suffered from the paranoid fantasy of the 'influencing machine'. The machine in question was a loom, which Matthews called an 'Air Loom'. One reason why the loom should have featured so strongly in Matthews' inner life was that during the late eighteenth and early nineteenth century, the cotton industry had dramatically expanded in England. The new cotton mills, with their steam-powered looms to which the workforce bent itself in unremitting labour, constituted an affront to the human being. The loom itself was an inhuman agency that undermined the soul, turning its human operatives almost into zombies. But there is also a second reason why the loom may have featured so negatively in Matthews' fantasy life. It was a *weaving* machine, and in the form of Matthews' Air Loom it had the capacity to *weave thoughts*, assailing its victims with an invisible 'pneumatic warp'. This, he disclosed, had a range of approximately 1,000 feet.[1]

At the end of the eighteenth century, looms had been in use in Europe for over 1,500 years, since at least the fourth century AD. Between the fourth century and the eighteenth century, however, the basic principles of their construction and operation remained pretty much constant, although two modifications of real significance should be mentioned. The first occurred during the Middle Ages, around the mid-thirteenth century, when treadles were introduced. These enabled the weaver to pull down alternate warp threads using his or her feet. Treadle looms were still simple machines that enabled only the simplest of patterns to be produced with ease. It is worth briefly considering how they worked, because the future development of the loom would have a profound influence on the development of the computer.

The basic principle of the treadle loom can be seen in Figure 19.1, which shows a medieval lady operating a fourteenth-century loom.

Alternate warp threads running along the length of the loom are attached to vertical draw cords that enable the weaver to raise one set of threads and lower the other set by pressing her foot down on the treadle. Each time she does this, she opens up a space between the warp threads

Figure 19.1
Simple medieval treadle loom (fourteenth century).

through which she can send a weft thread attached to her shuttle across the loom, first one way and then the other. In the first shot (at the bottom of the picture), the weft thread will lie below the first, third, fifth etc. warp threads, (because the respective warp threads have been raised). When it returns it will lie above them (because they will have been lowered). In this way the threads are woven into a fabric (Fig. 19.2). It

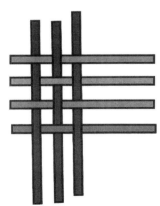

Figure 19.2
Example of plain weave, where each warp thread (dark) goes over one weft thread (light) and under the next.

will be appreciated that this kind of loom was perfectly adequate for weaving plain cloth, but it was not able to weave complex patterns. In order to weave a pattern, a different kind of loom was needed, which could select and group together warp threads. Such a loom was called a 'drawloom', and the introduction of the drawloom to Europe was the second really significant innovation in the history of European weaving.

We know that drawlooms existed in the Byzantine Empire, but were not introduced to Europe until the fourteenth century, first to Italy and Spain, then to France, and from there to Holland and Flanders and much later to England towards the end of the sixteenth century.[2] What was innovative about the drawloom was that it had the ability to gather together selected threads so they could be either raised or lowered simultaneously by a single draw-cord. This enabled more complex patterns to be woven on the machine. Every pattern could be regarded as a thought, because a lot of thought had to be given by the weaver as to how the pattern could be realized in the criss-crossing of warp and weft. First of all, the weaver would mark out the pattern on a grid of squares, each square representing a point where a weft thread and a warp thread crossed each other. In this way he or she could show which threads would need to lie on top of the others at each pass of the shuttle. This would determine the grouping of threads, how many different draw-cords were needed and which ones had to be raised or lowered at each shuttle-pass. Setting up the drawloom for a particular design was a long and painstaking process, which could take many weeks, but once set up the same pattern might then be produced by the weaver for many years. But even though the drawloom had been set up, this did not mean that the pattern was then produced automatically. Master weavers would have one or two young assistants, called 'draw-boys', who would crouch beneath or perch on top of the loom, and whose job it was to pull up the right group of threads at the right moment. With complex or unusual patterns, especially those involving images, the master weaver would make a sequence of cards, on which the squares representing the draw-cords that had to be raised for each pass of the shuttle were clearly marked. These cards were then read by a card reader (usually a woman), who would call out to the draw-boys, and tell them which cords to pull up. We may surmise that Matthews' 'Air Loom Gang' consisted of a reader and the draw-boys assisting the diabolical master-weaver of his fantasy to weave a pneumatic fabric of dastardly influencing thoughts.

In the real world, the weaving of fine patterned cloth was a complex process, that required a high degree of accuracy from the designer, and a great deal of concentration and stamina from all concerned. It was slow work, especially with the weaving of silk, which is so much finer than cotton and wool, and mistakes could easily be made. The rate at which decorated silk cloth could be produced was no more than two rows of woven fabric each minute, a mere 120 rows each hour.[3] There was therefore a very strong motivation to discover some means of automating pattern-weaving to make the whole task quicker and less prone to error. This led to the third major innovation in the history of European weaving, and the only one to have definitely originated in Europe itself. With this innovation we come to the boundary of James Tilly Matthews' prophetic fantasy, but it is precisely at this boundary that we discover the birthplace of a far more powerful kind of loom, capable of 'pneumatic' feats of thought-weaving. For this was where binary thinking was at last applied to the operation of a machine in such a way as to instil it with a kind of mechanical intelligence.

The Binary Revolution in Weaving

The heart of the silk-weaving industry during the eighteenth and early nineteenth centuries was Lyon, in France. Silk cloth was in fact France's largest export commodity at this time, and this fact added to the motivation to increase the speed and ease of production.[4] During the eighteenth century, various attempts were made to see how it might be possible to automate the selection of draw-cords. The first and most important breakthrough was made in 1725 by Basile Bouchon. This was followed by further innovations in the 1730s by Jean Falcon, and in the 1740s by the inventor Jacques de Vaucanson. Finally, at the beginning of the nineteenth century, Joseph-Marie Jacquard produced the definitive automated loom, known as the Jacquard loom. The fundamental idea that all these innovations shared in common was that it should be possible to replace the draw-boy with a mechanism that would automatically raise or lower the required threads. The question was how to ensure that the mechanism would be able successively to change the threads it raised, so as to correspond with the requirements of the pattern. The mechanism needed to be *instructed* what to do before each pass of the shuttle, just as the draw-boys had been instructed, and here was the crucial challenge: how to devise a way of expressing these instructions so that the mechanism could both 'read' and enact them.

We have seen that it was common practice for the master weaver to make a sequence of cards, with squares marked on it representing which draw-cords needed to be raised at each pass of the shuttle. Bouchon's solution was to translate the pattern of squares on the cards into a series of squares on a roll of paper, perforating holes in certain selected squares and leaving others unperforated. The paper was then sent round a revolving wooden cylinder, itself perforated by rows of holes regularly placed to correspond exactly with the squares. The cylinder was then pressed by the operator against a box that contained a horizontal row of blunt needles through which the draw-cords were threaded. When the cylinder was pressed up against them, those needles that slid through the paper remained motionless, but those that came up against unperforated squares were pushed back, along with the cords attached to them. The operator then pulled the cords down by depressing a foot-operated 'comb', thereby selecting the desired combination of threads for that shot of the weft. As the cylinder was released, it swung back towards the operator and at the same time rotated so as to bring the next row of squares into line with the horizontal needles. Once again the squares unperforated by holes would push back certain needles required for the right threads to be selected for the next shot of the weft, giving the next line of the pattern, and so on (Fig. 19.3).[5]

Bouchon's new device applied binary thinking to automate (or more accurately semi-automate) the operation of the loom. The key idea behind the perforated paper roll presented to the needle box at the loom's

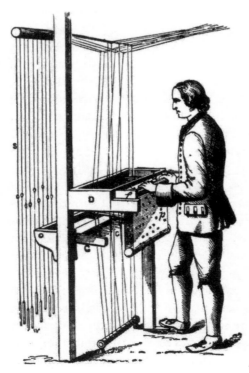

Figure 19.3
Bouchon's device for selecting draw-cords.
The perforated paper (p) revolves around
the wooden cylinder (f), which is pushed
against the needle box (D), thereby
selecting certain draw-cords, which in turn
would raise the threads needed to be raised
for the next line of the pattern.

interface was that the machine could best be 'instructed' by giving it one of two options: either the needles could pass through the hole and were unaffected, or else they met the resistance of the paper and were then pushed back. Through the distribution of the holes on the roll of paper, and nothing more, the required weaving pattern, which involved the selection of a multiplicity of draw-cords, was implemented. That is to say, it was implemented by presenting the machine with just two options. Bouchon's genius lay in his grasping the potential of binary logic, the logic of 'either/or', to become the medium through which information could be conveyed that was (at least to all appearances) neither mathematical nor logical. The paper roll with its perforations was the simplest language possible—a language of just two words, 'Yes' and 'No'. It bore precious little resemblance to the actual design to be woven, and yet the mere presence or absence of holes was sufficient to govern the actions of the machine and compel it to weave a design of potentially great complexity. The machine had become 'intelligent' in the sense that while it had not itself acquired any innate intelligence, nevertheless through the binary language encoded in the paper roll it had been made receptive to the intelligence of the designer.

Falcon's improvement to Bouchon's original design was to substitute perforated cards for the perforated paper. Instead of a perforated paper roll, each card was attached to the edge of the next card in a series, and as with the roll had to be manually pressed against the needles in order to make the required selection. Some years later, Vaucanson took the radical step of doing away with the draw-cords altogether by placing the selecting box on top of the loom, from where the sequence of perforated cards acted directly on a series of controlling hooks. It was this design that was then, in the 1790s, worked upon, adapted and developed further by Jacquard.

Jacquard's Punched-Card Loom

The 1790s was a turbulent time in France, with the revolution in full spate. It was also a time when, as we have seen in the previous chapter, the faith in reason that had been the hallmark of eighteenth century thought assumed the status of a religious cult. The process of human reasoning was accorded extraordinary status, as if it could be depended upon to solve any and every problem human beings faced. Once the revolution gave way to the dictatorship of Napoleon, this faith in the power of reason was rediscovered in a new confidence in applied science and technology, largely inspired by the personality of Napoleon himself. Napoleon had a deep respect for science, and he set out to encourage both science and industry from the time he came to power.[6]

As the eighteenth century came to a close and the nineteenth began, there was a sense of relief that the bloody days of the revolution were over, and a new optimism for the future arose. It was in this atmosphere that Joseph-Marie Jacquard applied himself to the task of automating the drawloom.

Jacquard, who came from a silk-weaving family, had as a child probably worked as a draw-boy. Now he carefully studied the previous attempts to automate the drawloom, and how to create a really practicable method of instructing the machine to raise the required threads in the right sequence. The answer that he came up with was really an adaptation of Vaucanson's method, itself based on Bouchon's original inspiration.[7] It was to use a string of cards with holes punched in them, each set of holes representing a single row of weave, and the placement of the holes acting as an instruction to the mechanism, causing it either to raise or lower the required threads. But now, unlike Bouchon's device, the loom would be fully automated, it would use rigid hooks rather than draw-cords, and it would employ many more needles, enabling much more complex patterns to be woven. The bar of the Jacquard loom, against which the punched cards would be pressed, can be seen in Figure 19.4. Each side, with its multiple rows and columns of holes, corresponds to just one pick of the shuttle, after which another side would automatically turn to present itself to the next punched card. Figure 19.5 shows three sets of punched cards, with quite different sequences of perforations, which would perform the function of instructing the machine to raise or lower different threads.

The basic principles of Jacquard's loom can be understood by looking at Figure 19.6, a simplified diagram of how the loom worked. In the Figure, the rope on the right marked 'h' is attached to a treadle below (not shown) operated by the weaver, enabling him to raise and lower the beam or griff ('g'), on which a number of vertical hooks rest. At the bottom end

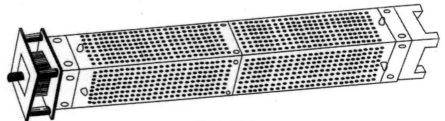

Figure 19.4

The bar of the Jacquard loom turns with each pick of the shuttle, offering another side to a fresh punched card.

Figure 19.5
A series of punched cards, each with eight rows and twelve columns of holes. On a particular card some of the rows remain un-punched, so the number of punched rows varies from card to card. With each turn of the bar (Fig. 19.4) a fresh punched card will press up against the next side of the bar.

of each of the hooks is an eye to which draw-cords are attached, tied to selected groups of warp threads below. How many hooks (and hence threads) are raised and lowered by the griff's movement up and down is determined by the horizontal movement of the rods ('b'), equivalent of Bouchon's needles, through which the hooks run. A hook will either be tilted away from the griff (and thus neutralized) or left on it depending on the horizontal movement of the rods. This horizontal movement of the rods is determined by which rods are able to enter the square-section roller ('e'), over which the punched cards ('f') pass, and which ones are not. If a rod meets resistance because no hole is present in the punched card, then it will have the effect of tilting the hook attached to it away from the griff, whereas if a hole is present in the punched card then the hook will remain on the griff and will be raised with it, thereby raising the threads that are attached to it.

With this mechanism, then, a pattern is created on the basis of the sequence of holes on the series of punched cards, which control whether or not a rod will enter or not enter the roller. The cards are pulled forward over the roller each time the weaver presses his foot down on the treadle, thereby causing the square-section roller to turn ninety degrees. The position of the holes in the cards, by controlling the movement of the rods, determines just a single row of weave, but the complete series of cards implements the whole complex pattern. In

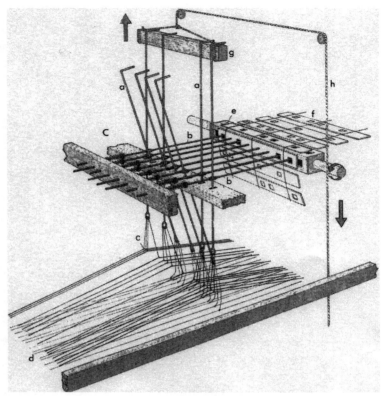

Figure 19.6
Simplified diagram of the Jacquard loom.

Figure 19.6, for the sake of clarity, only eight hooks and eight rods are shown, but the actual number would have been considerably greater. By the end of the nineteenth century, looms with 400 or 600 hooks were common.[8] This gives some indication of the extraordinary degree to which the punched card was able to embody a concentrated thought-content, which because it was machine-compatible, could be directly transmitted to the mechanism to implement, without the human operator having to exercise his mind any further than pressing his foot down on the treadle.

Figure 19.7 shows a late nineteenth century Jacquard loom in action, with a long string of cards being fed through the loom, at the top, determining the pattern that appears on the rolls below. During the first decades of the nineteenth century, the original man-powered Jacquard loom spread extremely fast. Jacquard patented his loom in 1804; by 1812, there were already 11,000 in operation in France.[9] Although the first

Figure 19.7
A mid-nineteenth century American
Jacquard loom.

steam-powered looms referred to at the beginning of this chapter began to appear towards the end of the eighteenth century, these looms were of the traditional simple type, only good for plain fabric, or fabric with the simplest of patterns, like stripes.[10] But by the 1830s, the new Jacquard looms inevitably began to be steam-powered, for the punched card had effectively made the human operator redundant. The machine had acquired its own 'intelligence' and motive power.

Jacquard's invention allowed the weaving process to speed up enormously. Decorated fabric that used to take an hour to weave could now be woven in two and a half minutes. Whereas a mere inch of brocade in a day was all that could be managed on the old drawloom, now two feet could be achieved, without any loss of quality. One of the most famous designs, made in the 1830s, was a woven silk portrait of Jacquard himself (Fig. 19.8). It took 24,000 cards to produce this portrait, and each card had over *1,000* hole positions. The portrait was so fine that it was almost impossible to tell that it was made of woven silk and was not an engraving. Indeed, most people took it to be an engraving. The portrait well demonstrated the principle that through purely quantitative multiplication of minute inputs, an increase in quality can be obtained. Here

Figure 19.8
Portrait of Jacquard, made of woven silk, circa 1839.

was a work of art that had no mistakes in it, and could be reproduced with unfailing exactitude over and over again.

In viewing this portrait today, despite all the admirable precision and technical mastery that it so clearly displays, a creeping doubt enters our minds. Is it really a work of art, or is not rather a very clever *imitation* of a work of art? The doubt arises because it is characterized by a degree of soullessness, and something in us remains unmoved. The great Victorian art critic, John Ruskin, was sensitive to precisely this kind of incursion of industrial technology into the sphere of art. Ruskin observed that the characteristic of all authentic art is that it has in it 'that spirit and power which man may witness, but not weigh; conceive but not comprehend; love, but not limit; and imagine, but not define . . .'[11] Judged against each of these criteria, Ruskin saw mechanical production fall short of those transcendent qualities which he deemed essential to true art. The portrait of Jacquard, while exemplifying all that is most admirable in mechanical production, lacks those non-measurable and indefinable qualities that are the hallmark of genuine art. It betrays itself to the sensitive observer as being precisely the product of a machine.

The Path to a Data-Processing Machine

For 30 years, the revolutionary method so successfully introduced by Jacquard, of using punched cards as a means of instructing machines, was restricted entirely to the weaving of finely decorated silk cloth. No one thought that it might be applicable to other machinery, until in 1834 the English mathematician Charles Babbage realized that punched cards could be used in mechanical computing. In the autumn of 1834, Babbage, then 43 years old, had already established himself not only as one of the leading mathematicians of his age, but also as the inventor of one of the most complex and sophisticated calculating machines ever devised. This was the Difference Engine, so called because it used the 'method of finite differences', which avoided the need for multiplication and division, and relied instead on successive additions, to calculate mathematical tables.

Mathematical tables, published in printed books, were much in use in the nineteenth century by professional engineers, surveyors, architects, bankers, navigators, and so on, who would all turn to their books of tables to find the answers to complicated mathematical questions, such as those involving multiples of fractions, the conversion of units of measurement, or—in the case of navigators—the hourly flux of tides and the position of the planets and stars. The tables needed to be accurate and reliable, but, owing to human error, they were seldom free from mistakes. Babbage saw that as tables of very many different types can be generated by the method of differences, this method of differences was eminently susceptible to mechanization through using gear wheels. And, once mechanized, the construction of mathematical tables would be free from human error.[12]

In principle, Babbage's Difference Engine was not so dissimilar to Pascal's mechanical calculator. Each gear wheel had the numbers 0 to 9 inscribed around its perimeter, and each number corresponded to an individual tooth on the wheel, so the value of a digit was represented by the amount by which the wheel was rotated. This was then transmitted to another gear wheel, whose rotation would then be transmitted to another until the required result was attained. Babbage's Difference Engine was of course vastly more complex than Pascal's calculator, as it stacked the gear wheels one on top of another in columns, the lowest representing units, the second from lowest representing tens, the third hundreds and so on up through thousands, hundreds of thousands, and millions (Fig. 19.9). As with Pascal's calculator, the operator would set the starting values on the wheels by hand, but then the Difference Engine would automatically perform the repeated additions.[13]

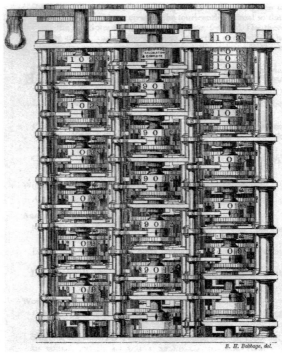

B. H. Babbage, del.

Figure 19.9
Woodcut of a portion of Babbage's Difference Engine, constructed in 1833.

Because of the complexity of the Difference Engine and the fact that Babbage's more important contribution to the development of computer technology was the Analytical Engine, which succeeded it (discussed in the next chapter), we shall not attempt to go further into the detail of how the Difference Engine worked. Suffice it to say that the first Difference Engine was designed to be made up of approximately 25,000 parts, and would have been huge—eight feet high, seven feet long and three feet deep—and would have weighed fifteen tons.[14] This great engine was, however, never completed, despite vast expenditure and years of pains-taking production of the parts.[15] Only a small hand-operated demonstration model of Babbage's machine was built during his lifetime, which survives to this day in the Science Museum in London. It is one of the finest examples of precision engineering of the first half of the nineteenth century. Babbage's improved version, the Difference Engine No. 2, was much simpler but it was not constructed until 1991, more than 100 years after his death.[16]

The Difference Engines were, however, nothing more than dedicated calculators. They were not yet prototypes of the modern computer. One of the reasons why the Difference Engines were not completed by Babbage was that towards the end of 1834, he conceived of something vastly more ambitious, which took him beyond the bounds of strictly mathematical calculation. It was in December of this year that he realized the extraordinary potential of punched cards. Babbage was greatly impressed by the silk portraits of Jacquard, and went to great pains to acquire one.[17] This was because he saw that the punched card was applicable beyond the sphere of weaving silk threads, and could equally well be applied to the 'weaving' of numbers. That is to say, the operation of a number-crunching calculating machine such as the Difference Engine could be controlled in a similar way to the Jacquard loom by employing punched cards. But Babbage saw that such a machine would have capabilities far beyond the calculation of differences, which was the main remit of the Difference Engine. The conception arose in his mind of a more 'general purpose' machine that could be instructed to perform the arithmetical operations of addition, subtraction, multiplication and division using the medium of the punched card, and on a scale of complexity never before attempted.

Babbage saw that if, as Hobbes, Descartes and Leibniz had argued, reasoning is nothing but reckoning, and if Leibniz was right that concepts could indeed be cast into mathematical form and that the logical operations of combining and separating concepts were equivalent to the mathematical operations of addition, multiplication, subtraction and division, then the punched card was the key to creating a machine that could conduct a logical analysis of the data fed into it. He would call this machine an 'Analytical Engine'. Behind Babbage's idea of the Analytical Engine there was a conception similar to that which lay behind Leibniz's mechanical calculator. The Analytical Engine would be the machine Leibniz had vaguely conceived but did not have the ability or resources to realize—a machine that would be capable of conducting sequences of logical operations not just on numbers but on concepts signified by numbers. This was the reason Babbage chose to call it an *Analytical* Engine, for by it the operations of logical analysis would be performed. In the following chapter we shall look further at the background from which Babbage's conception of the Analytical Engine arose, the basic principles on which it was based, and its wider significance in the prehistory of the computer.

Chapter Twenty

CHARLES BABBAGE AND THE
ANALYTICAL ENGINE

Babbage's Dream of a Mechanical Mind

In his autobiography, Charles Babbage informs us that as a child, he was fascinated by automata.[1] When still a young man he became 'passionately fond' of algebra, and soon became attracted to the idea of a universal language, largely through the writings of Leibniz.[2] He tells us that the idea of mechanizing calculation came to him in 1812 when, as a student aged 20, he had been labouring with tables of logarithms, and he fell into a 'dreamy mood'. A friend came into the room and seeing him with his head slumped on his desk with a table of logarithms lying open before him, said: 'Well Babbage, what are you dreaming about?' to which Babbage replied: 'I am thinking that all these Tables (pointing to the logarithms) might be calculated by machinery.'[3]

It is often the case that the most profound impulses come to human beings from a level deeper than their conscious mind. Then something is revealed concerning the direction from which the flow of inspiration enters the mind of a thinker or inventor, for we glimpse the stream of thought within which a person's inner life—not just conscious but also unconscious—is immersed. After his dreamlike reverie, Babbage felt so strongly motivated to design a machine that could, with minimal human intervention, calculate mathematical tables, that he devoted years of his life to it. And the outcome of his creative endeavours was the Difference Engine. But having designed this machine, and gone a significant way towards its construction, we have seen that he then turned his best energies to the design of the less specialized Analytical Engine that would be able to conduct not only arithmetical but also logical operations on the data fed into it.

Babbage was a man of enormous intelligence and ability. He was charming, sociable and able to relate with equal ease to royalty, government ministers, tradesmen and working men. He was a devoted husband and had lifelong close friends. At the same time, he had a passion for machinery; he loved the world of factories and the process of industrialization that was occurring in Britain and Europe during his lifetime. The practical application of exact scientific thinking was what most engaged him. He threw himself into project after project, applying his great intelligence to finding solutions to an enormous diversity of problems, including methods of storing energy through the use of compressed gasses, detailed research into how to reduce the shaking and discomfort of trains, and the design of machine tools.[4] But from the moment of Babbage's reverie, his primary ambition was to replicate human thought processes in the purely material and mechanical operations of cogs and wheels. His aim was to create the means by which a purely mechanical mode of thinking could take place as far as possible independently of the human being.

How is it that Babbage could dedicate himself to the realization of such an aim? Certainly there was a problem with the slowness and inaccuracy of human calculations and thought processes, which had obvious detrimental effects (in, for example, the compiling of navigation tables and statistics), but this was only the stimulus, not the underlying cause, of Babbage's resolve to find a way to mechanize thinking. Clearly we have to do with the nature of the man, and his peculiar destiny. If, as a child, he was especially fascinated by automata, what does this tell us of his spirit? It is interesting that as a young man materialism seemed to be a standpoint entirely obvious to him: he didn't arrive at it through a crisis in religious faith. It was as if he was born a materialist: materialism was for him simply part of his way of thinking. It was an unquestioned approach to life.

We catch sight of the young Babbage's worldview when, shortly after his reverie, he was required to defend a thesis of his choice in a formal debate at Cambridge, as part of his candidacy for his mathematics degree. The thesis he chose to defend reveals the tenor of his thought at this formative period in his life, for he set out to prove that God is not a spiritual being but a purely material agent. The choice was eminently unsuitable, given the fact that the Church regulated every aspect of university existence at the time, and his moderator was a clergyman. Babbage was failed on grounds of blasphemy before any debate took place! But the incident enables us to see that for him the concept of a world of spirit held no meaning. For Babbage the only reality was material reality, therefore God, if He is real, must be a material agent.

To hold such a view means that the life of the mind is valued to the extent that it engages with material reality. The mind of a man like Babbage must occupy itself with finding solutions to the practical problems that confront us living in the physical world. It is interesting that he once defined the human being as 'a tool-making animal', singling out the ability to make tools rather than the activity of contemplation or prayer as expressive of the human essence.[5] While Babbage had a phenomenal ability to grapple with and find solutions to extremely complex intellectual problems, these were not speculative or metaphysical problems; they were nearly always entirely practical. Philosophy didn't interest him, whereas (like Francis Bacon) ciphers and code-breaking fascinated him.[6] The only philosophical essay that he wrote (in the 1830s) puts forward a thoroughly materialistic and deterministic view of the universe, and draws heavily on his experience with calculating machines. God is conceived as the supreme scientific intelligence that set the universe going and calculated in advance all the consequences of the laws by which the universe is governed.[7] Although Babbage would never publicly admit it, this feeble notion of deity was entirely dispensable. Reality, for him, was that with which the analytical mind could engage, and if the analytical mind could be repli- cated in a machine, then not only would this enable us more accurately to perform calculations, but it would also account for the kind of intelligence at work in the universe. The Analytical Engine would demonstrate that not only human intelligence but cosmic intelligence too was essentially material and mechanical, and so could be emulated by a machine.

The New Algebra

In order to understand the operating principles of the Analytical Engine, we need to return to the idea, first broached by Leibniz in the 1670s, of an 'algebra of logic' that would specify the rules for manipulating concepts in the same way that algebra specifies the rules for manipulating numbers (see Chapter 16, pp. 162–163). It took a long time for this idea to resurface again, but at the beginning of the nineteenth century the seed sowed by Leibniz at last germinated. Up until the 1820s, algebra was still no more than a theory of equations in which letters were substituted for numbers, to which the four operations of addition, subtraction, multi- plication and division were applied.[8] But in the 1830s, mathematicians such as George Peacock and Augustus de Morgan (both of whom were friends of Babbage) began to see that algebraic signs did not have to represent numbers or arithmetical operations alone, but could equally well represent non-numerical objects and operations. They argued that the algebra of numbers was just *one type* of algebra: there could also be an algebra of entities which were not numbers. And the laws that pertained within these other algebras might be different from, but nevertheless as valid as, the laws that pertained in the algebra of numbers.

For example, in the algebraic law $a + b = b + a$, it was seen that a and b need not stand simply for quantities. Anything whatever could be sub- stituted for a and b, providing they still satisfied this law. Furthermore, the plus sign need not stand for addition only, but could signify any relation of such a kind that when it is substituted for the plus sign the law still holds. According to de Morgan, the plus sign could equally well mean 'tied to' since if *a is tied to b*, then also *b is tied to a*, thus $a + b = b + a$.[9] So, for example, let us say that *a* is a man and *b* is a woman, then the logical statement, 'if the man (*a*) is married to the woman (*b*), then the woman (*b*) is married to the man (*a*)' could be expressed in precisely this way: $a + b = b + a$.

The realignment of algebra with logic, rather than purely with the realm of number, took some time to accomplish. The most important step was taken in the next decade, in 1847, when George Boole published his short treatise, *The Mathematical Analysis of Logic*. In this treatise, the 32-year-old Boole argued that algebraic formulae might be used to express logical relations that form part of the syntax of ordinary language. Babbage was aware of Boole's work and in the margin of his copy of this early treatise, Babbage wrote, 'This is the work of a real *thinker*.'[10] What Boole was doing was taking up once again Leibniz's idea of a language of symbols (the *lingua characteristica*). Into this language of symbols (to be understood here as 'signs' rather than symbols in the true sense) all propositions could be translated, and thereby subjected to a *calculus ratiocinator*, which Boole termed a 'Calculus of Logic'.[11]

We shall return to Boole's contribution to the development of the computer in Chapter 23. At this point, it is important only to understand that it was exactly at the time when the revolution in mathematics was taking place that Babbage commenced work on his new engine. In the intellectual ferment of the 1830s, it became clear to Babbage that if a machine could be constructed that could—through the system of punched cards—function with algebraic signs, then such a machine would have capacities that extended far beyond arithmetical computation. Babbage believed that it should be possible to reduce thoughts to 'the language of algebra' and, precisely in so doing, the thoughts would then be drawn into the orbit of the machine.[12] That the Analytical Engine would be an 'algebraical' rather than a merely arithmetical machine was explicitly stated by Babbage:

> Every formula which the Analytical Engine can be required to compute consists of certain algebraical operations to be performed upon given letters, and of certain other modifications depending on the numerical value assigned to those letters.[13]

It was one of the 'great principles' of the machine that it would go beyond arithmetical operations and exercise 'entire control over the *combinations of algebraic symbols*'.[14]

Ada Lovelace, Babbage's closest and most articulate associate, went to some pains to explain this point:

> Many persons who are not conversant with mathematical studies, imagine that because the business of the engine is to give its results in *numerical notation*, the *nature of its processes* must consequently be *arithmetical* and *numerical*, rather than *algebraical* and *analytical*. This is an error. The engine can arrange and combine its numerical quantities exactly as if they were *letters* or any other *general* symbols...[15]

This being the case, it could process any data according to a given sequence of instructions, providing both data and instructions could be transmitted to the machine via the binary sequence of holes on a punched card. The emerging understanding of the potential application of algebra beyond the 'arithmetical and numerical' was of course perfectly suited to this. It meant that the Analytical Engine had the possibility of becoming—in Lovelace's words—'the executive right-hand of abstract algebra'.[16] Furthermore, Lovelace saw the possibility that the engine might engage not only analytically but also 'creatively'. She thought that its capacities might well include composing 'elaborate and scientific pieces of music of any degree of complexity or extent'.[17]

The Basic Principles of the Analytical Engine

Like Babbage, Lovelace was inspired by the extraordinary precision of pattern-making made possible by the Jacquard loom and saw the Analytical Engine as a kind of algebraical loom. She wrote:

> We may say most aptly, that the Analytical Engine *weaves algebraical patterns* just as the Jacquard loom weaves flowers and leaves.[18]

To form a picture of how the Engine would have worked, we can do no better than to return to the diagram of the Jacquard loom (Fig. 19.6), and to imagine that the differently coloured warp threads are now transformed into numbers.[19] But only a few of all the possible numbers are visible at any one time, namely the ones that have been set up for a particular operation. These numbers are engraved on gear wheels (with numbers *0* to *9* marked on their edges), stacked one above the other on columns ten feet high (not dissimilar to Fig. 19.9). They are housed in an area of the engine called 'the Store'.

In the Analytical Engine, three different kinds of punched cards are used, each controlling a different phase of the engine's work. Instead of pushing rods as in the Jacquard loom, they engage levers, which in turn cause gear wheels to move round, bringing certain numbers to the fore.

The first kind of punched card, tasked with generating the numbers to be held in the Store, Babbage called 'number cards'. Their purpose is to set up the numbers in a way analogous to the setting up of the sequence of coloured warp threads of the loom.

In the weaving of cloth, selected groups of warp threads must be attached to draw-cords so they can be raised and lowered at each pass of the shuttle so as to weave the desired pattern. The equivalent selection of numbers in the Analytical Engine is determined by a second set of punched cards, which Babbage called 'variable cards'. The function of the variable cards is to transmit to the machine what numbers are to be selected and subsequently operated on in another part of the machine called 'the Mill'. The great virtue of both the number cards and the variable cards is that they can be engaged with the mechanism and then disengaged, so that another card with different instructions can be engaged, or an earlier card re-engaged. Although the mechanism is put into a different state by each card, the information on each card remains unchanged. When not engaged, therefore, the card can be stored in readiness to be used again, if need be.

Now imagine that instead of the single operation of a griff being raised and lowered, as with the Jacquard loom, there are four possible operations that can be performed on the numbers that have been selected by the variable cards, namely addition, subtraction, multiplication and division. In the Analytical Engine, these operations are performed not by a griff but by dedicated, precision machinery housed in the Mill. The Mill is formidably complex, and we need not describe the machinery it comprises here. Suffice it to say that this machinery is controlled by a third set of punched cards named 'operation cards'. Despite the complexity of the machinery involved, these are the simplest cards, as the number of levers acted on by the operation cards is small, being restricted to performing the operations of addition, subtraction, multiplication and division. Once carried out, the results of these operations are then conveyed to a large communication hub within the Mill, through which information can then be mechanically relayed between the Mill and the Store.

The Mill, where numbers are operated on and where results are calculated, corresponds to the *central processing unit* in the modern computer. The Store corresponds to the *memory* of a modern computer. Babbage wanted the Store to consist of a series of at least 100 or more columns of rotating gear wheels, arranged in two rows facing each other, so the Store alone would have extended about ten feet (or three metres) in length. Since the selection and transference of numbers from the Store to the Mill and back again is controlled by the variable cards, not only do the variable cards specify which numbers in the Store are to be selected for any given operation, but also new variable cards can be punched with each new result and then returned to the Store. The Store, therefore, holds both the initial figures to be computed and also the intermediate results of an operation. And from there, depending on subsequent results obtained, they are either printed out or else transferred to the Mill once more to be subjected to further operations.[20] The overall design of the Analytical Engine is shown in Figure 20.1.

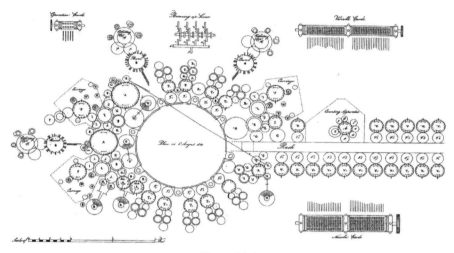

Figure 20.1

General Plan of Babbage's Analytical Engine, showing the Mill on the left of the drawing and the Store on the right. The Mill is organized around a large central communication hub, surrounded by an array of cogs and gear wheels, stacked in columns and put into operation by three rotating cylinders on the periphery, each marked 'Barrel'. The Store (truncated in this picture) comprises columns of gear wheels, positioned either side of a long toothed 'rack', which transmits information between the Store and the Mill, via the central communication hub. The bar for the 'variable cards' (which select numbers from the Store) is depicted top right, and the bar for the 'number cards' (which give the numbers to be entered into or retrieved from the Store by the variable cards) is shown bottom right. The much smaller bar for 'operation cards', which control the barrels in the Mill, is shown top left.

To give some impression of how the machine would have worked in practice, Babbage's son Henry (who took on his father's mantle after Charles's death) provides the example of the application of an algebraic formula $(ab + c)d$ to a series of numbers.[21] As we have already seen (and we shall return to consider this in more detail in Chapter 23), such a formula could as well be applied to entities other than numbers, for there is no reason why the letters a, b, c, and d should signify numbers alone, nor is there any reason why the operations of multiplication and addition should be simply arithmetical operations: Babbage was well aware that they could just as well be logical operations. And so the Analytical Engine would be as adept at handling non–numerical data as at handling numbers, even though ostensibly the data were presented entirely in numerical form.

In this example, the *number cards* first of all generate the numbers corresponding to *a*, *b c*, and *d* on four separate gear wheels on four separate columns in the Store. Then two *variable cards* select *a* to be brought from the Store to the Mill, followed by *b*. An *operation card* then causes *a* and *b* to be multiplied together, and the result is recorded as *p*. Now *p* (which represents *ab*) is returned to the Store and placed on another column there. From its new position in the Store, *p* is summoned back (by a variable card) to the Mill, followed by *c*. A second operation card causes *p* and *c* to be added together. The result is recorded as *q*, and *q* (which represents *ab* + c) is taken back to the Store and entered on another column. Now *d* and *q* are both brought to the Mill and a third operation card causes them to be multiplied together, giving the result *p2*, and the operation (*ab* + *c*)*d* is completed.[22]

What happens next? *P2* is returned to the Store where its value is entered on another column. From there the result can either be printed off, or the operation can be repeated a determinate number of times (this is called 'looping' in modern computer parlance). Another series of calculations based on this result might also be instigated, as the machine has the ability to take alternative courses of action depending on the result of a particular calculation. This ability is the mechanical application of '*If . . . then . . .*' logical judgements in human thinking, and its mechanization was one of the Analytical Engine's most revolutionary features, for if the machine could apply '*If . . . then . . .*' judgements to the results of its calculations, this surely proved that it was indeed a 'thinking machine'.[23]

An Awe-Inspiring Vision

The various designs for the Analytical Engine that Babbage produced between 1834 and 1871 (the year of his death) included all the essential features of the modern electronic computer, save the electronics. Since the development of electricity for powering machines was only in its infancy during Babbage's lifetime (the first practical designs for electric motors began to appear towards the end of Babbage's life), Babbage presumed that his engine would be steam-powered. Indeed, the scale of the machine would have been comparable to a steam locomotive, 20 feet (nine metres) or more in length. Despite the utilization of electricity for communication by telegraph from the end of the 1830s onwards, Babbage realized that electrical technology was not yet sufficiently advanced for use in the transfer of data within his machine.[24] This was a critical factor in the failure of Babbage's vision to be realized. Fully operational computers, capable of data-processing, could not function effectively until data transfer and analysis was electrified. Babbage was forced to think mechanically rather than electrically, and so his machine had to bow to the technology of the time and conduct its operations with cogs, gear wheels and rotating cylinders. This imposed a heavy limitation on its ability to function as an instrument of artificial intelligence. So, for example, although Babbage employed the binary system in his Analytical Engine, this was restricted to the punched card interface, while the machine itself operated with cogwheels based on the decimal system. Only with the incorporation of electricity into machines could they become fully binary, and only then would it be possible to push the operation of the machine beyond the engineering limitations that Babbage kept coming up against.

Babbage never succeeded in building a fully functioning computer, despite years of effort and vast expenditure of resources. Like the Difference Engine, the Analytical Engine—so promising in its conception—never saw the light of day. It was, in a sense, stillborn. Nevertheless, the significance of Babbage's designs lies in the fact that they brought the concept of the computer almost within reach of physical embodiment. Babbage's life might seem to have been a failure, since none of his designs was implemented, and yet, from a spiritual point of view, one must see that the tremendous intensity of thought that he devoted to designing his engines prepared the ground for what was to come.

The energy generated by Babbage's labours can be felt in the effusive writings of his much younger collaborator, Ada Lovelace. Unlike Babbage himself, Ada Lovelace sometimes writes with an almost visionary fervour, gripped by the enormity of the new conception of a machine able to process, analyse and 'weave' a wide variety of data in accordance with the rules of algebra. She was one of very few people at the time to fully comprehend the truly awe-inspiring potential of what Babbage envisaged. Babbage himself undoubtedly approved of what she wrote about the Analytical Engine, since he scrutinized and corrected drafts of her more definitive statements on it, and referred to her as 'my dear and much admired Interpreter'.[25]

The meaning and import of the Analytical Engine lived in Lovelace with utmost clarity and she embraced the machine with unrestrained enthusiasm, unclouded by any shadows of doubt. We detect something of the spirit of Francis Bacon in her fervour for the new engine that Babbage had designed:

> A new, a vast, and a powerful language is developed for the future use of analysis, in which to wield its truths so that these may become of more speedy and accurate practical application for the purposes of mankind than the means hitherto in our possession have rendered possible... We are not aware of its being on record that anything partaking in the nature of what is so well designated the *analytical* engine has been hitherto proposed, or even thought of, as a practical possibility, any more than the idea of a thinking or of a reasoning machine.[26]

Lovelace's careful use of language only thinly disguises her evident conviction that in the Analytical Engine the plans for a 'thinking or ... a reasoning machine' had finally been drawn up.[27]

While Francis Bacon had invented an 'instrument' that would ensure that human thinking proceeded entirely mechanically, cutting out all subjective feeling, flights of fancy, desire, emotion, inspiration and strokes of genius, Babbage had gone further. What he was aiming for was a mechanical model of the human mind that would function far more effectively than Bacon's all-too-human 'instrument', for the simple reason that it *was* a machine, and therefore could not be deflected from its assigned tasks by all those foibles that belong to an ensouled being. Like Jacquard's loom, but far more comprehensively, his was to be an instrument whose mechanical method would be pursued without danger of any interference from the human soul, which was now excluded from the analytical process as soon as the machine was set in motion.

While Babbage himself was careful to avoid directly claiming that the Analytical Engine could think, it was central to his endeavours to construct a machine that could carry out mechanically what human beings would otherwise only accomplish through a great deal of mental effort. He wrote in his autobiography that as a result of his invention of the Analytical Engine,

> the whole of the development and operations of analysis are now capable of being executed by machinery.[28]

Notice the present tense: 'are now capable'. Even though the invention was in reality no more than a highly detailed design plan that was more or less incapable of realization, the idea of it lived so vividly in Babbage's mind that he regarded his engine as tantamount to having been invented.

And it was the *idea* of it that, ironically, proved to be a far more

important legacy of Babbage than any actual machine. Hitherto, operations of analysis (i.e. logical operations) had been regarded as specifically
mental operations. Babbage demonstrated, albeit imperfectly, the possibility of constructing a *physical* machine that would perform *mental*
operations. He made the idea that abstract mental processes could be
replicated in concrete physical processes seem feasible. This is what he
intended the Analytical Engine to show. As Ada Lovelace expressed it,

> In enabling mechanism to combine together *general* symbols in suc-
> cessions of unlimited variety and extent, a uniting link is established
> between the operations of matter and the abstract mental pro-
> cesses...'[29]

Establishing the 'uniting link' between the operations of matter and
abstract mental processes was the crucial contribution that Babbage made
to the prehistory of the computer. In the conception of Babbage and
Lovelace, it was not just that the machine was, in effect, a mechanical
brain. The unspoken implication was that human mental processes are in
fact *nothing more than the operations of matter*, and that is why they can be
replicated by a machine. What Babbage and Lovelace were not able to
grasp, however, was that for such a machine to work, these operations of
matter had to be mediated by electricity. This was despite the fact that
during the 1830s, at precisely the time when Babbage threw himself into
the design of the Analytical Engine, great strides were being made in
telegraphy, which swiftly adopted binary codes (such as Morse) as the
favoured means of communication. It was not that Babbage was unaware
of these developments: far from it. He considered using an electromechanical switching system in his machine, but he rejected it because
the technology was too unreliable at that time.[30] It was Babbage's fate to
conceive a revolutionary idea, to pour his intellectual energies into the
struggle to physically realize this idea, and to have the bitter experience of
failure because the technical obstacles to its realization were too great to
be overcome at the time during which he lived.

Chapter Twenty-One

THE ADVENT OF THE ELECTRIC CURRENT

Electricity: Pseudo-Life-Principle of the Mechanistic Philosophy
We have seen that one of the main reasons why Babbage's Analytical Engine failed to be realized was that it was not possible at the time to utilize the potential of electricity for the transfer of data in his machine. Despite the sometimes faltering but frequently triumphant attempts to harness the potential of electricity in the first half of the nineteenth century, Babbage was nevertheless obliged to ignore these developments when it came to the practical design of his machine. Had he lived just 20 years later, things might have been different. The failure of his project to build the Analytical Engine highlights the fact that the practical realization of the computer was dependent upon electricity. Electricity was essential for the creation of a functioning 'intelligent' machine. In certain respects it performs the role of an animating principle, something we all may experience today in the act of 'turning on' a computer, smart phone or a thousand other more or less intelligent electrical technologies. When they are 'on', they are in an entirely different state from when they are 'off'.

This animating, or at least apparently animating, role of electricity applies very obviously to modern electrical devices. But before there were any electrical devices for domestic use, back in the eighteenth and early nineteenth centuries, the idea that electricity was indeed an animating or life-giving principle was very widely held. And the fact that it was so widely held both provided the spur for further investigations into it, and seemed to many to be confirmed by these further investigations. It was as if electricity offered human beings not only the possibility of under-standing the secret of life, but also the possibility of taking hold of this secret and gaining power over it. In this chapter we shall catch up a little with the unfolding relationship between electricity and humanity as this was played out in the drawing rooms, studies and laboratories of eight-eenth and nineteenth century thinkers and researchers.

One of the reasons for the ascendancy of the mechanistic worldview during the eighteenth and nineteenth centuries was that the original life-principle, which in the earlier Aristotelian philosophy was active as a form-giving and form-sustaining energy within the organism, had been lost sight of. As a consequence, living organisms came to be conceived in the image of machines, made up of so many physical parts external to other parts, and lacking any inherent spiritual dynamism that was both organizational and life-giving. The question then inevitably arose: How does the animal-machine or plant-machine get life injected into it? The mechanists' quest for an external, physical source of life coincided with the scientific exploration of the nature of electricity, and sure enough, electricity was seized upon as the animating principle of living organisms, mechanistically conceived. We have already noted in Chapter 17 that several theologically minded thinkers of the mid-eighteenth century identified electricity with the life-spirit in nature. In their thinking, matter was from the beginning infused with 'the electric fire' that lives concealed in all things. They argued that on the first day of creation, God mingled this fiery life-spirit into matter so that it would then bring forth all future forms into existence. Following this line of thought, the German 'electro-theologian' J. L. Fricker could write in 1765:

> There is, however, in the entire world no matter or body in connection with and in which the electrical fire does not manifest itself in one way or another.[1]

In England, Fricker's contemporary Richard Lovett, suggested in his *Philosophical Essays* (1766) that the 'electric fluid' is 'the active mechanical Agent in Nature' that gives motion to both the world and to human bodies.[2] Such ideas were fashionable throughout the later years of the eighteenth century. Thirty years after Lovett, Adam Walker, in his *System of Familiar Philosophy* (1799)—a grand survey of current scientific knowledge—went into more detail. He claimed that both the generation and growth of flowers is accelerated through electrification, so too the hatching of eggs and the growth of vegetables.[3] Walker provided no evidence to support this claim, but rather used the claim to support his own desire to see electricity upheld as the life-principle of all creatures. Referring to recent experiments on frogs (by Galvani), he argued that the influence of the brain and nerves on muscles is of an electric nature.[4] It seemed to him that all the most recent research was pointing to electricity as 'a living principle' and the 'the soul of the material world'.[5]

The Experiments of Galvani

Galvani published the results of his many painstaking (and also pain-delivering) experiments on frogs, hens and sheep in 1791, in his treatise *De Viribus Electricitatis* ('On the Effects of Electricity').[6] Of the three animals, it was the frogs that bore the brunt of his experimentation, which was mostly conducted on dismembered or (as sometimes in the case of the hens and sheep) partially dismembered parts of the animals. In *De Viribus Electricitatis* he describes how he noticed quite by chance that some amputated frogs' legs, close to his electric machine, began to twitch when his assistant set the machine going. Intrigued, Galvani conducted a series of tests, systematically observing the circumstances under which this phenomenon of twitching would occur. These tests included fastening one end of an iron wire to a point high up outside his house, while connecting the lower end to the nerves of frogs and other animals' legs, and attaching to their feet another wire that led to a nearby well (Fig. 21.1). Thereby Galvani hoped to ascertain the effect of an electrified atmosphere (before and during a thunderstorm) on the amputated limbs. He records:

> The thing went according to our desire, just as in artificial electricity; for as often as the lightning broke out, at the same moment of time all the muscles fell into violent and multiple contractions, so that, just as the splendour and flash of the lightning are wont, so the muscular motions and contractions of those animals preceded the thunders, and, as it were, warned of them . . .[7]

Figure 21.1
Galvani's experiment on the effect of atmospheric electricity on the amputated limbs of frogs.

In this macabre scenario, in the glowering sky with flashes of lightning breaking from the dark storm clouds, momentarily illuminating the amputated body-parts, which would at the same instant jerk into uncontrollable spasms, we already see a foreshadowing of the labours of Mary Shelley's Dr Frankenstein. For Galvani, this must have seemed a powerful confirmation of the view that electricity is a cosmic life-principle, which at that moment injected a revitalizing energy into the dismembered limbs.[8] However, in the next stage of his investigations he found that he was able to replicate the same spasms simply when the copper or brass hooks, by which he hung up his frogs' legs, were placed on an iron railing. No atmospheric electricity was required. He followed this observation up by systematically employing different combinations of materials, both metallic and non-metallic, to see under what conditions the spasms would occur (Fig. 21.2). He discovered that they only happened with certain combinations of metals, like copper and iron or silver and iron. But instead of concluding that the cause of the spasms lay in electricity being generated by the interaction of the two different metals with each other as well as with the animal's body, he concluded that the source of electricity was entirely internal to the animal. The central

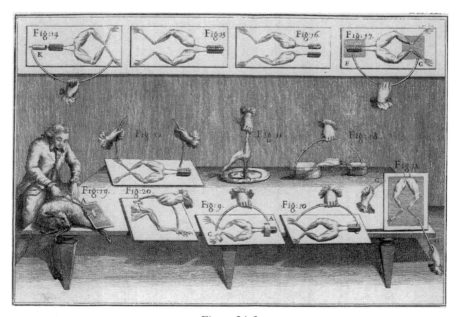

Figure 21.2
An illustration from Luigi Galvani's De Viribus Electricitatis, *1791, showing a range of experiments carried out by Galvani on the amputated legs of frogs, to which different combinations of metals have been attached, and on a sheep in the left-hand corner.*

conclusion of his treatise was that 'animal electricity' originates in the animal's brain, from which it is secreted as a 'neuro-electric fluid', which accumulates in the muscles and is forced from the muscles to the nerves. The muscle fibre could then be compared to a Leyden jar, and the nerve to the conductor.[9]

The implication is that muscular contraction in animals can be explained by the same law which governs the operation of electric machines and the accumulation of electric charge in Leyden jars. In this way, Galvani convinces himself that the principle of movement in the animal machine may be identified as electrical.[10] The fact that his explanation of the cause of the frogs' legs going into spasm was shown to be incorrect by Alessandro Volta within just a few years did little to upset the powerful picture which impressed itself on the minds of so many people at this time: namely that electricity was the animating life-force of a world that during the previous centuries had increasingly come to be experienced as a mechanism. It was not noticed that electricity has no real formative or generative power, but is capable only of provoking a brief and unsustainable semblance of life.

Instilling Life into the Man-Machine

In 1803, three years after the publication of Volta's ground-breaking work, Galvani's nephew, Giovanni Aldini, applied electricity to the corpse of a convicted murderer. Simply through sending an electric current through metal plates held to various parts of the body, he was able to cause the dead man's eye to open, his arm to rise into the air and his legs to kick violently, much to the consternation of those present (Fig. 21.3).[11] Through such experiments as this, which were repeated by others, the view that electricity is an animating, life-giving principle was widely adopted despite the fact (perhaps it would be more accurate to say *because* of the fact) that electricity was at the same time regarded as both physical and measurable.[12] The poet Samuel Taylor Coleridge, who was greatly interested in the new discoveries, knew better. For Coleridge, a life-principle cannot be observed like any other phenomenon, for it is not itself phenomenal but is different in *kind*, belonging to the realm from out of which physical phenomena arise.[13] Coleridge argued that while electricity is not corporeal or 'embodied', it is nevertheless essentially material, and its very materiality precludes it from being a life-principle.[14] The fact that no dead man was ever actually revived through applications of electricity should surely have supported Coleridge's point of view.

During the first decades of the nineteenth century, the loss of an instinctive orientation towards the real source of life became evident in

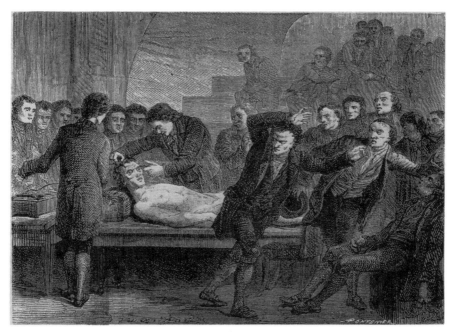

Figure 21.3
The application of electricity to the corpses of convicted criminals became quite popular in the early nineteenth century, seeming to show that electricity was the animating force or life-principle of bodies. Here an electric current applied to muscles around the eyelid causes the dead man's eye to open, to the consternation of those attending the demonstration.

the debate over 'vitalism'. Just because for the materialist orthodoxy there was no essential difference between living creatures and machines, if the life-force fell within the purview of physics, all the better. If, furthermore, the life-force could be quantified and both generated and delivered from a machine, then the Promethean impulse of science would be fully satisfied. To such a mindset, the dreadful fantasy explored in Mary Shelley's *Frankenstein* (1818) of how a living creature could be created from the parts of dead bodies retrieved from graves could then seem gruesomely possible. Like Giovanni Aldini, Victor Frankenstein turned to the dead to create life, putting together his monster from bones and body parts collected from graveyards and charnel houses. All that was required was to instil this hotchpotch creature with the electric spark of life.[15] And so, collecting the 'instruments of life' around him, Frankenstein infused 'the spark of being' into the lifeless thing that lay at his feet. And then, like Aldini, he 'saw the dull yellow eye of the creature open: it breathed hard, and a convulsive motion agitated its limbs'.[16]

It is at this moment that, in the pages of what is often described as the first science-fiction novel, the Machine-Man archetype articulated by La Mettrie comes fully into its own. It is a fiction that gives birth to a diabolical image of the human being which, through many centuries, had slowly been crystallizing, and now presents itself fully formed. The fiction is all the more powerful for the fact that Frankenstein's monster is a sensitive creature, endowed with a human soul, human longings and desires. We cannot but feel empathy for him, for he reflects back to us our own fate, should humanity continue down the path that Bacon and Descartes laid out at the beginning of the seventeenth century. The contours of this path, formed above all by the quest for power over nature, were becoming ever clearer with the passage of time. By the beginning of the nineteenth century, science was being extolled by Humphry Davy in the following words:

> Science has given to him [Mankind] an acquaintance with the different relations of the parts of the external world; and more than that, it has bestowed upon him powers which may be called almost creative; which have enabled him to change and modify the beings surrounding him, and by his experiments *to interrogate nature with power*, not simply as a scholar, passive and seeking only to understand her operations, but rather *as a master, active with his own instruments . . .*'[17]

The Advent of the Electric Current

We have seen that according to Galvani, the reason why the dismembered legs of frogs and other creatures twitch when they are brought into connection with certain metals is that there is an innate electricity within the muscles of the animal, which function like a Leyden jar. For this reason the electricity can remain in the muscles even after the animal has died. Alessandro Volta demonstrated, however, that the source of the electricity was not in the muscles, through the simple expedient of applying an electric current to the nerves and removing the muscles from the conducting circuit. He found that the same twitching resulted.[18] Volta realized that the source of the electricity was not in the animal at all, but in the metals, and furthermore that the two metals had to be different from each other in order to produce an electrical effect. If two metals of the same type were used, nothing happened. The metals, then, were not conductors of electricity as Galvani had supposed, but were responsible for the production of the electricity: the frogs' legs convulsed not because of an innate *animal* electricity but because of an external *metallic* electricity.[19]

Volta then went on to experiment with different combinations of metals in order to gauge the electrical charge generated by bringing them together. He found that, after contact, one of the two metals would become positively charged while the other would become negatively charged: so zinc after contact with copper became positively charged, whereas the copper became negatively charged. However if copper was brought into contact with silver, the copper acquired a positive charge and the silver a negative charge. Volta was able to draw up a list of substances, beginning with zinc, lead, tin and iron and ending with copper, platinum, silver, gold, graphite and charcoal, in order of their relative charges. When any two of these substances was brought into contact, the substance earlier in the list would acquire a positive charge and the substance later in the list would become negatively charged.[20] But his most significant finding was that a continuous electric current was only made possible if, as well as the two dissimilar metals being brought into contact with each other, there was also the mediating presence of a fluid, for example brine (salt water). By stacking combinations of two metals interspersed with brine-soaked cardboard in a repeated pattern, Volta was able to create a 'pile' of cells, which produced just such a continuous electric current (Fig. 21.4).

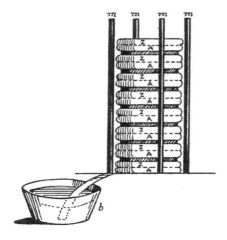

Figure 21.4
Volta's first 'pile', 1800. Discs of zinc (Z) and silver (A) are soldered together, and interspersed with brine-moistened cardboard. On touching the uppermost disc with one hand while dipping the other hand into the vessel (b), the experimenter receives a shock. The circuit can be less painfully closed by running a wire from the zinc disc at the top to the silver disc at the bottom.

While the pile's electric charge was much weaker than that of a Leyden jar, the advantage of the pile over the Leyden jar was that its charge was constantly renewed. By adding more discs to the pile, or by connecting one pile with another, the charge could be greatly increased. Volta thought that the generation of electricity was caused by the contact between the two dissimilar metals, which then induced a flow of 'electric fluid' from one metal to another, enhanced by the presence of the moist conductors. It was very soon realized, however, that a rather different account could be given for the generation of electricity in the pile, namely that it was due to a chemical reaction between the metals and the brine.[21] According to this account, the metal that acquires the positive charge (zinc) is 'attacked' by the brine, which oxidizes the zinc, and this chemical attack releases an electric current which flows to the other metal (silver) and so on down through the pile. The electric current could then be conducted back again to the zinc by the connecting wire. This chemical reaction was observed by the two British investigators, Carlisle and Nicholson, who replicated Volta's pile within just a few weeks of the publication of Volta's findings.[22]

But what especially drew the attention of Carlisle and Nicholson was another chemical reaction. They devised an experiment in which a tube of fresh water was placed between the two wires connected to the top and bottom ends of the pile, and observed the gas hydrogen being given off at one end and oxygen at the other.[23] The experiment demonstrated that the effect of electricity on water is to decompose it into its constituent chemical elements, a phenomenon which came to be known as 'electrolysis' (Fig. 21.5).

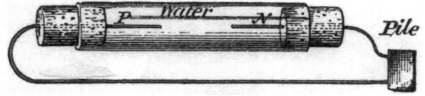

Figure 21.5

In the electrolysis experiment of Carlisle and Nicholson, the tube is filled with fresh water and both ends are plugged with cork. Through the cork at either end a platinum wire is inserted, with one wire (P) connected to the zinc end or positive terminal and the other (N) connected to the silver end or negative terminal. At the positive end bubbles of oxygen were emitted and at the silver end bubbles of hydrogen. In the actual experiment the tube was placed at an angle of 40 degrees to the table. The pile is only schematically indicated in this drawing from a nineteenth-century textbook.

Interrogating Nature With Power

Something more, therefore, was learnt about the nature of electricity through the invention of the Voltaic pile: not only did chemical action between three dissimilar conducting bodies produce an electric current, but an electric current was also seen to have a disintegrating effect on a substance to which it was applied. Very soon, Humphry Davy would show that other substances apart from water could also be decomposed into their chemical constituents by passing an electric current through them. In 1807 Davy sent an electric current through a lump of damp potash, and found that at the positive pole oxygen gas was produced and at the negative pole the chemical element potassium appeared for the very first time. Many more experiments would follow in which Davy used current electricity to reveal the hitherto unknown chemical constituents of substances. He called it 'electro-chemical analysis', and through it he discovered such new chemical elements as magnesium, calcium, strontium, barium, boron and silicon. Through the discovery of current electricity the analytical mind, which functions through analyzing wholes into their constituent parts, had found a powerful new tool.

To gain perspective on the significance of this new development, we should bear in mind that it was not until the latter part of the eighteenth century (in the generation preceding Davy's) that the ancient Greek and medieval doctrine that all matter is composed of the four elements Earth, Water, Air and Fire was finally dismantled. Joseph Priestley only comprehensively established that air is itself composed of nitrogen, carbon dioxide and oxygen in 1774.[24] Water was not analyzed into hydrogen and oxygen until Lavoisier demonstrated this eleven years later in 1785. It was Lavoisier who also proposed that fire may be explained as a rapid combination of oxygen and carbon. This decomposition of the Elements into their chemical constituents was not, however, due simply to a sudden increase in human powers of observation or the ability to perform experiments. It was due rather to a change in consciousness, which, as we saw in Chapter 10, had already begun to occur in the fourteenth century when William of Ockham declared that 'matter cannot exist without having parts distant from part'.[25] Ockham was challenging the established view that all material substance was 'informed' with spiritual content. At that time matter was generally regarded as incapable of existing in its own right. There was no such thing as purely physical matter, because it had to be spiritually informed in order to exist. Precisely to the extent that matter was experienced in this way, human beings—rather than experiencing themselves as observers of a world that was external to their consciousness of it—still felt that they had the capacity to participate in the world's spiritual dynamics. And this is what Ockham challenged.

In the Middle Ages, the Four Elements were regarded as fundamental not in the sense that they were the ultimate physical 'building blocks' of matter, but in the sense that they were seen as indicating four underlying conditions of material manifestation. They were never regarded as physical substances, but rather as four underlying and essentially dynamic matrices through which the world came into physical manifestation, and which were symbolically represented by the physical substances earth, water, air and fire. These matrices were described in terms of combinations of the four elementary qualities: moistness, dryness, warmth and coldness.[26] Each of these was an inner, soul-quality as much as it was an outer sense-perceptible quality. For this reason these qualities could also define the four temperaments of the human soul: melancholic, phlegmatic, sanguine and choleric. Even today, we still use the words 'warm' and 'cold' to describe the soul-qualities of other human beings, forgetful that in the Middle Ages it was not just physical warmth or coldness that was indicated by these terms. The same breadth of meaning spanning both the inner and the outer world meant that nature too was encountered as displaying qualities not just of matter, but also of soul.

By the seventeenth century, what Ockham had begun to experience, namely the solidification of the world into externally existing physical substance confronted by human consciousness, was philosophically enshrined by Descartes in his distinction between 'extended substance' (*res extensa*) and 'thinking substance' (*res cogitans*). By the time of Descartes, this was increasingly accepted as the new experiential norm. The old feeling that true knowledge had to entail a participation in nature's inwardness was slowly being lost, as more and more people experienced nature as simply a world of solid physical objects. This is how atomists such as Galileo and Robert Boyle wrote about nature, and it necessarily entailed the consciousness of standing fundamentally *outside* of nature, as a detached observer. But it was also this consciousness that lay behind the atomistic theory of matter.[27] As the seventeenth turned into the eighteenth century, the doctrine of the Four Elements could no longer be sustained because this mode of experience of the 'spectator-mind' had become so thoroughly entrenched. Whereas in the previous era human beings could relate to the Four Elements not only as pervading outward nature but also as determining the soul-qualities of plants, animals and humans, now the way of knowledge demanded an approach that excluded soul. By the second half of the eighteenth century, then, the time was ripe for the discovery of chemical elements such as nitrogen, hydrogen, oxygen and carbon dioxide, precisely because nothing soulful could be found in them.

It was at this historical juncture, at the end of the eighteenth century when the Four Elements had been forced to give way to the new chemical elements, that the scientific consciousness, trained in the discipline of detached, systematic observation and experiment (as Alessandro Volta was) discovered current electricity. As we have seen, within just a few weeks of the discovery of current electricity, the decomposition of water into hydrogen and oxygen (first achieved by Lavoisier 15 years previously) was dramatically confirmed by means of electrolysis. In the early years of the nineteenth century, in the hands of Humphry Davy, electricity proved itself to be an immensely important ally in the great scientific project of uncovering hitherto unknown material constituents of physical substances. The direction of travel was entirely towards nature's 'lower border'—the exploration of matter, considered as existing in a sphere quite different from soul or spirit.[28] In the immediate alliance formed between current electricity, which so willingly offered itself as a tool of chemical analysis, and the scientist's wish to analyze ever further the physical composition of substances, electricity revealed another aspect of its character. It showed itself to have an affinity with analysis, with breaking things down into their constituent parts.

The discovery of current electricity has been described in this chapter as occurring as a result of the observations and experiments conducted by certain scientists. But one may also have the feeling that electricity itself had a part to play in the revelation of this further aspect of its nature. It is as if it wanted to be discovered as much as the scientists wanted to discover it. And so it put itself forward at this opportune moment, stepping into the field of vision in order to offer its services to the scientific consciousness so intent on the tasks of systematic analysis. In the chapters that follow we will see how from this point on, electricity will increasingly become an indispensable accomplice in the endeavour to forge the artificial mind that Babbage strived to create by purely mechanical means. We will see how the development of ever more refined methods of logical analysis during the nineteenth century occurs alongside ever-increasing knowledge of, and ability to control and harness, the power of electricity in the service of human ends.

Chapter Twenty-Two

TELEGRAPHY:
THE ELECTROLYSIS OF LANGUAGE

The Birth of the Electric Telegraph

Even before the invention of Volta's 'pile' and its production of an electric current, the idea that electricity could be used to transmit signals suggested itself to many people. Experiments in the mid-eighteenth century had shown that the transmission of electricity appeared to be instantaneous.[1] It was as if electricity annihilated spatial barriers, and so could be an ideal vehicle of communication over distances of space. It would clearly be a more effective means of communication than, for example, the ringing of bells whose audibility decreased over distance; and more reliable than transmitting light, because this might be obscured by mist or fog or by some physical barrier. Through the second half of the eighteenth century and the first half of the nineteenth, a fascinating range of ingenious designs was proposed for transmitting messages via electric wires. Few of them were actually tried out because their inventors tended to think in terms of transmitting different letters of the alphabet, one wire for each letter, and this involved far too many wires to be practicable.[2] In 1787 it was realized by a certain Monsieur Lomond that, by using an 'alphabet of motions' (in other words a code) employing an electrometer, one wire might be sufficient to convey a message. But this idea was not taken up at the time.[3]

With the advent of the electric current, inventors began devising ways of harnessing it for telegraphy. In 1804, the Spanish experimenter Francisco Salvá Campillo proposed a system of telegraphy utilizing electrolysis, which was followed up in 1809 by the German physician, Samuel Thomas von Sömmerring.[4] What he envisaged was a kind of 'electrolysis of language', breaking it down into a form that could be electrically transmitted and then reconstituted at the receiving end. Sömmerring's telegraph utilized 26 wires, each terminating in a separate glass tube filled with acid. When the current was sent down a selected wire, streams of hydrogen gas would be released next to associated letters at the receiving terminal. Despite the interest shown in this apparatus, which was successfully constructed and shown to work over a mile or so, it was far too cumbersome to be practicable (Fig. 22.1). Something else needed to happen in order for electric telegraphy to take off.

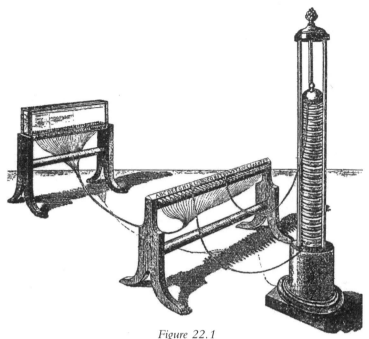

Figure 22.1
Sömmerring's electric telegraph, 1809.

The crucial event took place in 1820, when the Danish physicist Hans Christian Oersted discovered that if a current is flowing along a wire aligned north-south, and a compass is brought up to the wire, the compass needle is deflected. It is deflected in one direction if it is held above the wire, and in the opposite direction if it is placed below the wire. Likewise, if the flow of current changes direction and flows from south to north, then deflection of the compass needle will likewise be reversed (Fig. 22.2). For the first time it was shown that a magnetic force resided in the electric current, in other words *the electric current was itself magnetic.* Whereas up until this time no one had thought that there was a direct relationship between electricity and magnetism, save that of a certain similarity of behaviour between magnetized and electrified bodies, now a hitherto unsuspected aspect of electricity was unveiled. The scientific community was taken by surprise. Because the realm of electricity was essentially uncharted territory, no one really had an inkling of its nature, its extent, or what it might be capable of. And so investigators were repeatedly astonished when another facet of its character was unexpectedly revealed—often, as in the case of Oersted, quite by chance.

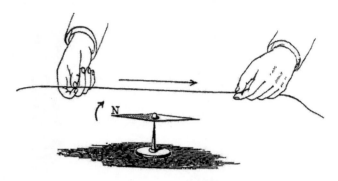

Figure 22.2

Oersted's experiment with an electric wire (above) and a compass needle (below), 1820. When the current flows from north to south, the northern end of the compass needle swings to the east. When the current flows from south to north, the northern end of the compass needle swings to the west. The reverse holds when the electric wire is placed beneath the compass.

Within months of the publication of Oersted's findings, the mathematician André-Marie Ampère conducted a series of experiments to investigate further the exact relationships between the electric current and magnetism, thereby laying the foundations of the new science of electromagnetism.[5] In just a few years, Ampère was able to set out the laws of electrodynamics in mathematical form, basing his work on the hypothesis of an 'electrodynamic molecule'. By tying the behaviour of magnetic and electric phenomena into a molecular model, Ampère enabled them to be accurately predicted and controlled.[6] Even before Ampère's work was completed, more sensitive instruments (galvanometers) were being created to detect electric currents, and much more powerful magnets (electromagnets) were being constructed that functioned by sending an electric current through wires wrapped in a spiral around a bar of iron. These electromagnets had the advantage not only of being more powerful than ordinary magnets, but also of rendering the force of magnetism controllable. When the electric current was switched off, the magnetic field immediately collapsed; the instant the electric current was switched back on the magnetic field revived. The first electromagnet was created in 1824 by William Sturgeon, and is shown in Figure 22.3.

Ampère was quick to see that, given the incredible speed at which electricity seemed to travel, if an electric current could be detected at one point on the wire conducting the current, then it would surely be detectable in exactly the same way at another point on the wire, even if the second point were potentially miles away from the first. The question

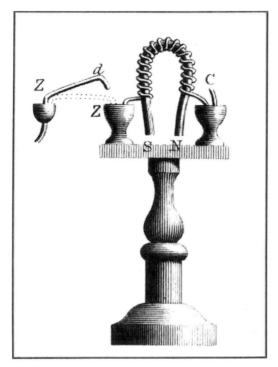

Figure 22.3

The first electromagnet, created by William Sturgeon in 1824. It consisted of a U-shaped iron core with a copper coil spiralling around it. The two ends of the coil were connected to the zinc and copper terminals of a simple battery, the electric current from which could be turned on and off by closing or opening the circuit at 'd'. The iron core became magnetized as soon as an electric current was allowed to flow through the copper coil.

as to how to employ electricity as a means of transmitting messages was then simply a technical one of how most effectively to send out a current of electricity that would maintain its strength over many miles, sufficient for a magnetized needle to be made to move, or equally an iron bar to be magnetized, at the other end of the wire.

The Electrolysis of Language

The first successful application of this principle to telegraphy was that of Baron Pavel Lvovitsch Schilling von Konstadt, a Russian diplomat, and for a while an attaché of the Russian embassy at Munich. Schilling had a great interest in electricity, and spent several years in the 1820s developing a telegraphic system. Realizing that the key to a successful system was that it should not be encumbered by too many wires, he gradually simplified it to just two, with a battery at one end and a galvanometer at the other end to detect the two possible directions of the flow of electric current. His telegraph was thus based on two signals alone, which would consist of a deflection of the needle of the galvanometer either to the left or to the right depending on the direction of the current transmitted. With each deflection of the needle, a paper disc above it, coloured white on one side and black on the other, would show either its white or its black side (Fig. 22.4).

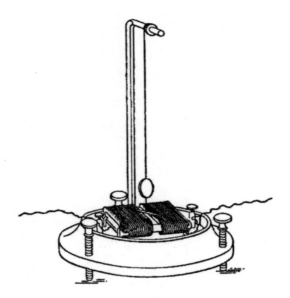

Figure 22.4

Schilling's galvanometer, circa 1825. It consisted of a coil through which an electric current was passed in one of two directions. Above it, a magnetized needle was suspended from a disc of paper, black on one side and white on the other. The deflection of the needle in response to the direction of the electric current resulted in either the white or the black side of the disc presenting itself, thus giving a choice of just two possible signals.

Schilling saw that because the electric current could be made to flow in one of two directions, a 'binary alphabet' was the obvious way of transmitting messages. Each letter of the normal alphabet could be represented by a series of just two letters, corresponding to a deflection of the galvanometer needle to the left or right, which then compelled the paper disc to show one or other of its sides. It is tempting to speculate that, being a diplomat, Schilling might have been aware of the Biliteral Alphabet of Francis Bacon, which Bacon had specifically devised for sending coded messages when working for the British government in Paris, and which he published in the Latin edition of the *Advancement of*

Learning (1623).[7] Schilling's binary alphabet was not unlike that of Francis Bacon. It differed in so far as it utilized a binary sequence of up to four letters, rather than Bacon's five-letter binary alphabet, but even within this limitation it was able to convey not only the 26 letters of the alphabet, but also ten numbers and 'four conventional signs'.[8] The 26 letters of Schilling's binary alphabet are reproduced below (Fig. 22.5).

A = b w		N = w b	
B = b b b		O = b w b	
C = b w w		P = w w b b	
D = b b w		Q = w w w b	
E = b		R = w b b	
F = b b b b		S = w w	
G = w w w w		T = w	
H = b w w w		U = w w b	
I = b b		V = w w w	
J = b b w w		W = b w b w	
K = b b b w		X = w b w b	
L = w b b b		Y = w b b w	
M = w b w		Z = w b w w	

Figure 22.5

Schilling's binary alphabet. The 'b' and 'w' stand for the 'black' and 'white' side of the disc from which the magnetized needle is suspended (see Fig.22.4 above). When the needle is deflected by the electric current, which will flow in one of two directions through the wire, the disc will turn with it, presenting either its white or black side to the observer.

By all accounts, Schilling's telegraph worked well. He even devised a more complex arrangement which could potentially use six signals simultaneously, sometimes black, sometimes white and sometimes both combined. Despite the great interest generated by this pioneering work, it did not progress beyond the experimental stage, largely due to his dying in 1837, before he was able to see any serious project through to completion.[9] His significance, however, lies less in the tangible results that he was able to produce than in his so clearly exemplifying the tendency of those working in the field to adapt their thinking towards a binary mode of thought in order to utilize the promise of electricity to convey messages.

This same tendency can be seen in others who during the 1830s dedicated their intellectual and creative energies to working out a viable telegraph system. In Germany, Gauss and Weber invented an apparatus based on the generation of electric current through electromagnetic induction, which was put to service in Göttingen over several years between 1833 and 1838. Like that of Schilling, their telegraph used the two directions of the electric current to deflect a needle either one way or the other, and a binary code—with up to four possible combinations of right and left deflections of the needle (Fig. 22.6).

r = *a*	*rlr* = *f, v*	*rrlr* = *s*	*lrlr* = 3
l = *e*	*lrr* = *g*	*rlrr* = *t*	*llrr* = 4
rr = *i*	*lll* = *h*	*lrrr* = *w*	*lllr* = 5
rl = *o*	*llr* = *l*	*rrll* = *z*	*llrl* = 6
lr = *u*	*lrl* = *m*	*rlrl* = *o*	*lrll* = 7
ll = *b*	*rll* = *n*	*rllr* = I	*rlll* = 8
rrr = *c, k*	*rrrr* = *p*	*lrrl* = 2	*llll* = 9
rrl = *d*	*rrrl* = *r*		

Figure 22.6

The binary code used in the Gauss and Weber telegraph signalling system, where 'r' stands for a deflection of the needle to the right and 'l' stands for a deflection of the needle to the left when the current is reversed.

Gauss and Weber's apparatus was taken up by Steinheil, who applied his mind to overcoming the various technical problems that telegraphy presented, especially the need to improve both the strength of the current generated and the sensitivity of the receiver at the other end. Steinheil adapted the Gauss and Weber binary code to a code of dots, printed out by little 'beaks' attached to the swinging arms of the magnetized bar at slightly different heights, so that deflections of the bar to the right produced one line of dots and deflections to the left produced a second line below it, with up to four dots being used for each letter or sign (Fig. 22.7). Using this method, Steinheil was able to print out six words per minute.[10] We see, then, how in his quest to perfect his predecessors' work, Steinheil saw the advantage of dissolving the word into a series of dots. A kind of electrolysis is performed on language itself, through which the word is broken down into essentially indistinguishable atomistic constituents. It is as if the spirit of electricity, once it offered itself as a means of communication, demanded that those who sought to utilize it adopt an atomistic mode of thought.

While the technical problem of the dwindling strength of an electric current due to poor conduction could be addressed by ever more effective means of generating the electric current as well as by improving the sensitivity of receivers, the really crucial step in producing a viable telegraphic system was the invention of the 'relay'. The principle of the

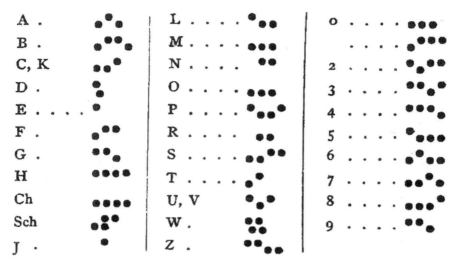

Figure 22.7

Steinheil's binary code of dots on two lines, one above the other, corresponding to deflections of a magnetized bar in response to the different directions of an electric current.

relay, first utilized by Joseph Henry in the United States in 1835, was that by dividing up the total distance to be covered by a conducting wire into a number of shorter lengths, the current could be renewed at various points by a simple electromagnetic mechanism that introduced a fresh source of electricity into the wire. This enabled signals to be sent over much greater distances. Two years later, Samuel Morse showed that, by employing the relay, a signal could be sent a distance of ten miles at least.

The transmission apparatus used by Morse, with his close colleague Vail, was essentially a switch that brought two contacts together, which then allowed an electric current to flow down a single wire, and interrupted the current as soon as the contacts were released. The idea of interrupting the current in order to send a signal was radical. Hitherto, the ability to reverse the current was regarded as the foundation of telegraphy.[11] But the Morse/Vail system, in using the ability to start and stop the flow of current as the basis of signalling, meant that the whole system could be simplified. All that was required at the transmission end was a 'key', like a piano key resting on a little spring, that the operator would push down and release to send a pulse of electricity down the line. At the other end of the line, the receiving system used an electromagnet with an armature (a length of soft iron or steel), which was pivoted horizontally with one end above the electromagnet and other beneath a roll of moving paper. As soon as the electromagnet was activated by the flow of electricity, the end of the armature nearest to it was drawn down towards it, while the other end, to which a pencil was attached, rose up towards the roll of moving paper. When the electric current was interrupted, the armature lifted away from the electromagnet, causing the pencil to fall away from the roll of paper at the opposite end (Fig. 22.8).

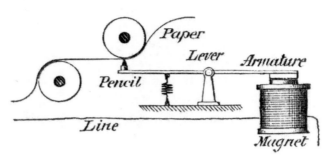

Figure 22.8

Morse's receiving apparatus, circa 1838. When an electric current passes through the electromagnet on the right, it becomes magnetized, drawing the armature down towards it. At the other end of the armature, which rests on a pivot, the pencil is raised towards a moving roll of paper. As soon as the current is interrupted, the armature rises away from the electromagnet and the pencil falls away from the paper.

The fact that the central mechanism of the apparatus was a lever reminds us of the *shaduf*, used in antiquity to draw water from a river or canal (discussed in Chapter 2). But now, instead of the lever being operated by human muscle power, the operation of the lever is powered by electricity, and the result of its operation is a series of short or long marks made on a roll of paper. The length of these marks depends on how quickly the current is broken. A current interrupted very quickly will produce a short mark; a current that lasts a little longer will produce a longer mark. The shortest signal that can be made is a dot, while a signal three times as long produces a dash. By devising a notation of dots and dashes in which, for example, the letter 'A' is represented by a dot and a dash, and the letter 'B' by a dash and three dots, the whole alphabet is readily decomposed into a binary series of dots and dashes (Fig 22.9). These correspond to audible clicks as the armature moves up and down. When a dot is sounded, the lever strikes one stop and immediately strikes again, whereas when a dash is sounded there is a longer delay between the strikes. And so this code can readily be learnt like a language: a language that is entirely mechanical, without any modulation or change of tone, and which can hammer out words at the rate of around 30 per minute.[12] In time, the use of paper rolls became more or less redundant, as

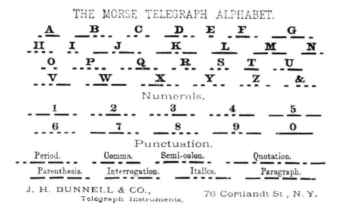

Figure 22.9
The original Morse Telegraph Alphabet, showing letters, numerals and punctuation.

experienced operators began directly to translate the audible clicks (which they came to vocalize as 'di' (or 'dit') and 'dah' for dots and dashes respectively) into recognizable words.[13]

In 1844, Morse persuaded Congress to finance an experimental line 40 miles long, between Washington and Baltimore. Its success meant that in the next two decades, the Morse/Vail telegraph spread rapidly over the whole of the United States and subsequently all over the world (with certain adaptations to the code). But as much as human beings had at last found a means of employing electricity for effectively sending communications using electric current, we might also say that through human ingenuity electricity had now found a way of entering the soul-life of humanity. At the beginning of the nineteenth century, Humphry Davy saw how current electricity could provide a new means of analytical research that would enable the researcher to *decompose substances* into their chemical elements. Now, in the mid-nineteenth century, current electricity offered itself as a medium of distance communication through a similar *decomposition of language* into abstract constituents, as if each letter were a 'compound' of more elementary binary parts. The Morse operators who learnt to think in terms of 'di' and 'dah' exemplify how a process of deconstruction of language accompanied its electrification. The price of electricity rendering service to humanity was that it placed a demand on human beings *to adapt to it*, and in adapting to it they turned away from an earlier reverential relationship to language towards the view that language is nothing more than a way of encoding and transmitting information.

The destructive effect on the human being's sense of the sacredness of language through this way of treating it seems not to have been noticed by contemporaries. Forgotten was the view of human language current in the Middle Ages that saw it as drawing forth and re-expressing the divine language embodied in nature.[14] Except for a few Romantics, like Goethe, the nineteenth century could no longer see nature as the expression of the divine—as the language of God that human beings might recapture in the articulated word. The mood of the mid-nineteenth century, a generation after Goethe, was uncritically favourable towards the new communication technology, which was received with general enthusiasm and a sense of wonder, expressed by poets such as Whitman and Taylor.[15] And so the introduction of the electric telegraph drew human beings willingly, but semi-consciously, towards a closer embrace of atomistic modes of thought and expression.

The electric telegraph especially highlighted the suitability of binary codes to communication via electrical transmission, and this was one of the main reasons for the triumph of the Morse/Vail telegraph system. It brought to a successful conclusion the efforts of researchers over several decades to find the right vehicle for a viable electrical communication system. In the elaboration of the Morse code, the profound compatibility of binary thinking with electrical transmission was made clear for all to see. It was another crucial milestone in the prehistory of the computer.

Chapter Twenty-Three

FORGING THE MIND OF THE MACHINE

George Boole: The Mathematization of Thinking

The invention of the punched card, used first in the Jacquard loom and then adopted by Babbage in his designs for an Analytical Engine, was the result of a sustained effort over many decades to find a way of creating a viable interface between human thought and machines. A similar effort can also be seen in the search for a code that could be used for communicating messages through the electric telegraph. In both cases, the solution that presented itself was a binary solution. Machines can most simply and effectively be controlled and operated when they have just one of two possible options given to them. Likewise, electricity becomes a very efficient medium of communication if it is made to speak a bivalent language of dots and dashes in varied combinations, as in the Morse Code. During the 1830s and 1840s, the eminent suitability of binary thinking both as an interface in the control and operation of complex machines and as the most effective means of communication using electricity was widely recognized. Nevertheless, the full potential of the binary number system as a medium of thought and communication was still not fully grasped or utilized. A further step needed to be taken, and it was the mathematician George Boole, who would take it.

Born in 1815, George Boole was a young man in his 20s when Babbage was working on the Analytical Engine and when the Morse Code came into existence in America. At about the same time a revolution was taking place in the purely intellectual sphere of mathematics (see Chapter 20). The main event in this revolution was that algebraic symbols were liberated from being tied to numbers. The significance of this lay in the fact that mathematicians turned their attention to how to capture the logical structures of language, and to express these structures in algebraic terms. One of the most influential of the mathematicians who belonged to this movement was Augustus de Morgan; another was George Peacock. Both were contemporaries and friends of Charles Babbage. Augustus de Morgan for a time tutored Babbage's collaborator and interpreter, Ada Lovelace. But it was George Boole, who was not connected to Babbage, and who was younger than both Peacock and de

Morgan, whose work was destined to have the most profound influence on the subsequent development of computers.[1]

What Boole set out to do was systematically to present logic as a type of algebra, in which the same laws hold whether applied to numbers, classes of things, or logical propositions. His approach can be understood if we consider the simple equation $3 + 5 = 8$. We know that this equation implies that $8 - 5 = 3$. This is basic arithmetic. Expressed in algebraic terms, if $3 = x$, $5 = y$ and $8 = z$, then we can equally well say not only that $x + y = z$ but also that $z - y = x$. This can be seen to be the case by simply substituting the appropriate numbers for the letters.

Boole then showed that if we substitute a class of *things* for numbers, so that for example $x =$ 'white sheep', $y =$ 'black sheep', and $z =$ 'all the sheep on the farm', and if we also substitute equivalent logical operations for the mathematical operations of addition and subtraction, it will be possible to cast the processes of logical inference into mathematical form. The statement 'all the sheep on the farm are either white or black' allows us to infer that all the sheep except the black sheep will be white. If we want to put our statement into algebraic terms, we need to re-express it as 'white sheep and black sheep are all the sheep on the farm'. Then it will read '$x + y = z$'. This enables us to make the logical inference that all the sheep (z) *except* the black sheep (y) will be white sheep (x), or $z - y = x$. This is a *logical inference* that we have to make, but it is now expressed in *algebraic* terms, with the operations of addition and subtraction signifying logical operations. Here the addition sign signifies what in logic is called a 'disjunction' between the two terms 'white sheep' and 'black sheep'. In ordinary language a disjunction is an 'either/or' alternative. The sheep are either black or white. The minus sign signifies what in logic is called 'exception', and in ordinary language is indicated by the word 'except' as we have seen above. We could also use the phrase 'which are not' to denote exception, so '$z - y = x$' in the example above would read, 'All the sheep *which are not* black sheep will be white sheep.'[2]

Already in the seventeenth century, Leibniz had realized that there was a resemblance between addition and the logical concept of 'disjunction', and between multiplication and the logical concept of 'conjunction', which denotes 'both/and' (as opposed to the 'either/or' of disjunction). But Leibniz was not able to formulate this in a very precise way. Boole was able to work out much more systematically the underlying logical structure of language, showing how it could be rewritten algebraically. We have seen that in the equation, $x + y = z$, the plus sign indicates that there is a 'disjunction' between terms or, to return to our example, there is *an alternative between* kinds of sheep, which are either white or black. But supposing the algebraically minded farmer wants to select fat white sheep. If he denotes fat sheep with the letter f, then he would not express the selection process as $x + f$ (white sheep and fat sheep), because the f might also include black sheep. Rather, he would express it as $x \times f$, or simply xf (fat white sheep). Here, then, there is a 'conjunction' (corresponding to the multiplication sign) between fat and white sheep, since they need to be *both* fat *and* white. In this way, Boole demonstrated that the algebraic operations of addition, subtraction and multiplication can combine and resolve not only numbers but also *conceptions* of things.[3]

Boole believed that the operations of the mind when reasoning and when doing algebra are not just analogous but are *the same*.[4] Therefore in his classic work, *An Investigation of the Laws of Thought*, he claimed: 'the ultimate laws of Logic are mathematical in their form', and set out in detail how this could be the case.[5] His thesis was unequivocal:

> The laws of thought, in all its processes of conception and of reasoning, in all those operations of which language is the expression or the instrument, are of the same kind as are the laws of Mathematics... Human thought, traced to its ultimate elements, reveals itself in mathematical forms...[6]

Boole believed he had wrested human thought away from its embodiment in ordinary spoken and written language, resolving it into a series of mathematical operations. But he was not considering the full breadth of human thought and language in their contemplative, inspired or poetic aspect, but rather was considering only that kind of thinking which was traditionally referred to as the thinking of the *ratio*, and that kind of linguistic expression which could be resolved into logical propositions. Boole was in fact following in the footsteps of Hobbes, Descartes and Leibniz, whose basic method was all along an exercise in control through constriction. Only through the constriction of thought to the purely logical thinking of the *ratio* could it be assimilated to algebra, and thereby successfully reduced to its Hobbesian status as a kind of 'reckoning' or calculation. The consequence was that ordinary language could now be by-passed as the vehicle of this reduced type of thought for, having exposed the underlying logical structure of the statements and propositions that are the product of the thinking of the *ratio*, Boole could demand that ordinary language surrender this logical structure to mathematics.

In an early essay, 'The Calculus of Logic', in which he outlined his principal ideas, Boole reveals the radical nature of his project:

The view which these enquiries present of the nature of language is a very interesting one. They exhibit it not as a mere collection of signs, but as a system of expression, the elements of which are subject to the laws of thought which they represent. That those laws are as rigorously mathematical as are the laws which govern purely quantitative conceptions of space and time, of number and magnitude, is a conclusion which I do not hesitate to submit to the exactest scrutiny.[7]

What Boole is assuming here is that all language, as a system of expression, is subject to the 'laws of thought', and because the laws of thought are mathematical, we cannot escape the conclusion that all the essential operations of thought can be expressed in a mathematical language (i.e. algebra). The fact that, for most thinking human beings, mathematics would be a totally inadequate vehicle for communicating meaningful thought does not enter the purview of Boole's argument. It is not concerned with the *content*, only with the *process*, of reasoning. And yet, against Boole's argument, we must object that if the process of reasoning is prised away from language, it is also prised away from that which transcends the narrow field of pure logic. When we think most deeply, we draw from the genius of language an ability both to access and to convey insights and truths which transcend the rules of algebra. The process of reasoning, if removed from the framework of human language, is severed from all relatedness to the inner world of soul and spirit. It is diminished to mere ratiocination: it becomes impervious to a deeper source of inspiration.

The Role of the Binary
One key reason why Boole could claim that the 'laws of thought' are the same as mathematical laws is that the binary system occupied a central place in his thinking. He saw that a reduction of logic to mathematics could only work if the arithmetical element is restricted to just two quantitative values: zero and one. When mathematics is reduced to functioning with just two numbers, or two values, only, then it corresponds to the basic demand of logical thought, originally expressed in the Law of the Excluded Middle: namely, that there is no third option between the two sides of a contradiction. A statement is either true or false, a thing either does or does not exist: there is no third alternative. Assimilating logic to algebra is an important step, but assimilating it to the binary system goes further, for the values of truth and falsehood are stripped of anything qualitative and reduced to mere number. In *An Investigation of the Laws of Thought*, Boole wrote:

Instead of determining the measure of formal agreement of the symbols of Logic with those of Number generally, it is more immediately suggested to us to compare them with symbols of quantity *admitting only of the values 0 and 1*. Let us, conceive, then, of an Algebra in which the symbols x, y, z etc. admit indifferently of the values 0 and 1, and of these values alone.[8]

We have already seen that Boole was following in the footsteps of Leibniz, who became so deeply enamoured of the binary system that he came to see it as a new theology (Chapter 16). One senses that Boole too fell under its spell, which his language betrays, for he writes of an alternative 'Universe of the Proposition' governed by a 'law of duality'.[9] This law of duality is no longer one in which the values of 'false' and 'true' could have a moral resonance. If false is reduced to *0*, and true to *1*, then these values are purely mathematical.[10] The moral component has been excised from the equation '*x = 1*' or '*x = 0*'. We are humanly disengaged from such outcomes.

In disconnecting logic from ordinary language, Boole disconnected it from the world of lived experience. This was something that was carefully guarded against in earlier times, when logic was rooted in language and directed towards knowledge of real things in the real world.[11] Now logic is given its own domain within which to function. It is a universe of logical propositions governed by the limited system of binary numbers and rigidly defined logical operations—negating, conjoining, disjoining, and so forth—which can be extended into an extraordinarily complex series of manipulations. Logic would be drawn into a sphere far removed from the pursuit of wisdom and knowledge, and concerned solely with making correct calculations. The data would be fed in, logical operations would be performed on the data, and outputs would be given in terms of ones and zeros. In this way Boole set up the conditions for the automation of logical reasoning.

Mathematics and Mechanization

Boole died in 1864, bequeathing the legacy of binary 'Boolean algebra' to future generations of mathematicians, with its promise of the automation of logic. Given the significance of this legacy it is extremely interesting to note that Boole himself acknowledged that what he was working on was limited to a narrow sphere. In this respect he was similar to Pascal. In *An Investigation into the Laws of Thought*, he wrote that mathematics should not be regarded as the sole basis of human knowledge, because there are orders of reality that can only be perceived through other human faculties: our responsiveness to beauty, our moral awareness, and our experience of love and affection. These are essential human capacities, as is 'the breadth of intellectual vision' that takes us beyond mere logic, or the 'dialectic faculty'. None of these possibilities of consciousness should be excluded from the acquisition of knowledge: they are all vitally important, because true knowledge encompasses more than the merely intellectual and secular. Boole writes all this at the end of his devastating book.[12] We can but wonder how it is that he was able both so candidly to express the limitations of his own work and yet at the same time press forward with such utter conviction in this essentially reductionist project. How is it that he could on the one hand seek to persuade us that human thought and language are ruled by rigorously mathematical laws, while at the same time expressing his love of the poetry of Homer, Dante and Milton, and the philosophy of Plato and Aristotle, referring to them with an endearing respectfulness?

Boole was a modest, open-minded and cultured man, but he placed his creative energies at the service of the binary impulse, which now became the driving force behind the automation of thinking. During the latter part of the nineteenth century, the project to resolve human thought into a machine-friendly mathematics appealed to a growing number of intellectuals, eager to dedicate themselves to advancing the cause.[13] Just five years after Boole's death, W. S. Jevons constructed a logic machine designed to function on the principles of Boolean logic. He called it a 'logical piano' because it looked similar to an upright piano. Statements could be fed into the machine in the form of equations, logically processed by it and the conclusions produced without the intervention of any human thinking at all. According to Jevons, 'Conclusions which he [Boole] could obtain only by pages of intricate calculation, are exhibited by the machine after one or two minutes of manipulation.'[14] Jevons' machine was rather crude and had many flaws, but it was significant because he saw so clearly what Boole's legacy meant: the mathematization of logic was the key to its automation.

Boole did not foresee where his work would lead, but by translating the structures of logical thinking into mathematical form, he had in fact discovered a way of automating it, and this required a new kind of language, free of words, and by implication free of human values. Having been extracted from its human container of living language, and inserted into the container of mathematics, logic was no longer constrained by moral or spiritual principles. Whereas traditional Aristotelian logic functioned within the parameters of human language, with its grammatical structures, and with its permeability to such human values as beauty of expression and such spiritual principles as the service of truth, now there was a *new kind of logic* that, by reducing human statements to mathematical equations, could function quite independently of the soul-life of human beings. This new logic was not concerned primarily with arriving at meaningful conclusions but rather with making correct calculations.

We may see how the automation of logic on the basis of Boolean algebra was developed by returning to our farmer with his flock of white sheep and black sheep. Imagine that he wants to select from his flock just white sheep, then he must make a series of judgements, 'This is a white sheep', 'This is not a white sheep' . . . and so on. Each of these judgements is a proposition that must in logical terms be either true or false. In his selection process, he will, technically speaking, employ 'logic operators' in his thinking process. The three most common logic operators are referred to as NOT (which indicates logical 'exception'), OR (which indicates logical 'disjunction') and AND (which indicates logical 'conjunction'). So, for example, if he wants to select a white sheep (x), he will NOT select a black sheep (y). If all the sheep on the farm $= z$, then this could be written as $z - y = x$, or could be tabulated in what subsequently became known as a 'Truth Table', in which 1 indicates presence, and 0 indicates absence, with the result that the farmer either *selects* (1) or *does not select* (0) the sheep:[15]

x	y	RESULT
1	0	1
0	1	0

Figure 23.1
The NOT Truth Table, in which 1 indicates presence, and 0 indicates absence.

We have already seen that if the farmer wants to select white sheep as well as black sheep this can be expressed algebraically by the plus sign, which indicates that there are two different kinds of sheep: they are either white or black. In terms of Boolean algebra, what the farmer needs to select is expressed as $x + y$, meaning x OR y. Either a white sheep or a black sheep will do, but he does not want a white goat or a brown sheep. The 'Truth Table' for this selection process is set out as follows:

x	y	RESULT
1	0	1
0	1	1
1	1	1
0	0	0

Figure 23.2
The OR Truth Table, in which 1 indicates presence, and 0 indicates absence.

We also saw that if the farmer wants to select only fat white sheep (fx), in other words sheep that are both fat AND white—not black, nor brown and certainly not goats—this could be indicated by the multiplication sign. The criteria are tighter here, and so the 'Truth Table' will be correspondingly different:

x	f	RESULT
1	0	0
0	1	0
1	1	1
0	0	0

Figure 23.3

The AND Truth Table, in which 1 indicates presence, and 0 indicates absence.

These three 'Truth Tables' express in binary form what in traditional logic are respectively known as the law of contradiction and 'sufficient' and 'necessary' conditions. In modern logic these terms correspond, as we saw, to exception (or negation), disjunction and conjunction. In the NOT Truth Table (exception), the white sheep is selected and the black excluded, in the OR Truth Table (disjunction), any sheep will satisfy the criteria—whether it be black or white, or fat or thin. In the AND Truth Table (conjunction), it is necessary that the sheep is both fat and white, or else it will not be selected.

Now suppose the farmer wants to select fat white sheep that are either over six months old or weigh over 100 pounds. Such conditions can be stated as a series of operations: (*Select fat white sheep*) = (NOT (*fat black sheep*) AND ((*sheep over six months old*) OR (*sheep over 100lbs*)). If sheep over six months old are indicated by the letters *za* and sheep over the weight of 100 pounds are indicated by the letters *zw*, then the thought 'I want to select fat white sheep, as opposed to fat black sheep, which are either over six months old or over 100 pounds in weight' can be expressed abstractly in the equation $xf = $ NOT yf AND (*za* OR *zw*). Thus a complex human thought is reduced to an algebraic equation.

In this new way of thinking, the question of the truth or falsity of propositions is not the central concern. Because thinking has become an entirely mechanical process, what is of central concern is the correct or incorrect input into this mechanical process, in order to produce outputs of one or zero. Truth is no more than a flow of energy through a system, while falsehood is a blocking of the flow. Logic is reduced to an artificial language with an artificial syntax, by means of which a type of thinking can take place that is really a selection process, which serves simply as a means to the attainment of certain specified outputs.

The Electrification of Logic

Like Charles Babbage, George Boole lived during the period when experimentation with electricity was still in its infancy. While the first half of the nineteenth century saw major breakthroughs in the ability to harness electrical energy, apart from the invention of the telegraph it was not until the second half of the nineteenth century that these experimental discoveries began to bear fruit in practical applications. The dynamo was only invented in the late 1860s, the first usable electric lamps appeared in the late 1870s, and the first public electricity supplies generated from a power station and distributed through a mains network only came on stream in the 1880s.[16] Babbage died in 1871 at the age of 80; Boole died in 1864 at the age of 49. Neither of them really belonged to the era of electricity. And yet what Boole accomplished was to lay the foundations for the electrification of thinking.

It can readily be seen that what is set out in the three Truth Tables can be replicated in electric circuits with switches, which replicate the so-called 'logic operators' NOT, OR, or AND (Figs. 23.4, 23.5 and 23.6). In Figure 23.4, The NOT Truth Table is shown as an electric circuit with just one switch. Here the input of a white sheep (x) closes the circuit, allowing the current to flow, but a black sheep will open it and then the current will be interrupted.

In figure 23.5, either a white sheep (x) or a black sheep (y) will close the

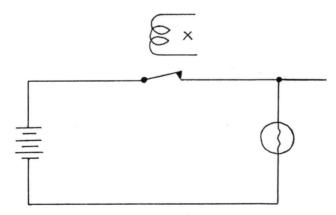

Figure 23.4
The NOT Logic Operator, shown as a switch that is either closed or open, either allowing or not allowing the electric current to pass. The symbol on the left represents a battery, the symbol on the right represents output.

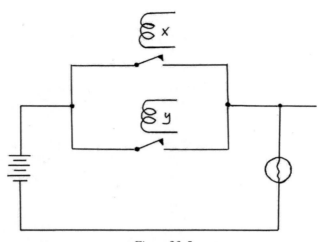

Figure 23.5
The OR Logic Operator, shown as switches in a circuit.

circuit and the current will flow. This is the kind of circuit required if the farmer wants sheep of either kind, and doesn't mind if they are white or black. Through the way the circuit is set up, the current will flow as long as at least one switch is closed, and it doesn't matter which one.

In Figure 23.6, the current will only flow if both switches are closed— white sheep (*x*) will only close one switch, and fat sheep (*f*) will only close one switch. The sheep have to be both white *and* fat in order for the current to flow, for only then will both switches be closed.

These circuits can be set up with their 'logic gates' controlling the flow of electric current, and the circuits can be made more complex simply by multiplying the number and type of gates in a given circuit, thereby

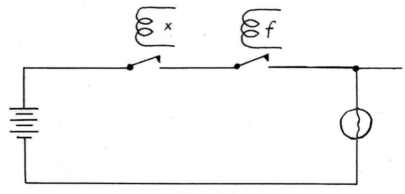

Figure 23.6
The AND Logic Operator, shown as switches in a circuit.

broadening or narrowing the criteria of analysis and selection. To return to our previous example, the farmer may want to select fat white sheep which are either over six months old or weigh over 100 pounds. Such conditions can be replicated in setting up the series of 'logic gates' expressed in the formula: (*Select fat white sheep*) = (NOT (*fat black sheep*) AND ((*sheep over six months old*) OR (*sheep over 100lbs*)). We saw that this could be expressed in the algebraic equation $xf = NOT\ yf\ AND\ (za\ OR\ zw)$. But this can also be replicated in an electric circuit with the NOT, AND and OR logic gates connected in the correct sequence.

Humans Thinking like Machines

For the sake of clarity, I have chosen only the simplest and most basic logical operations, but there are much more complex ones that can also be applied in the same way. The crucial point is that we are here considering a type of thinking that does not require a human being to be present in order for it to take place. Once the criteria of analysis are set out, this kind of thinking can take place independently of the mental activity of a human being. We have seen that shortly after Boole's death a logic machine was invented to function with the principles of Boolean logic, but it was not until 1937 that the applicability of Boolean algebra to the switching operations of electric circuits was grasped.[17] Meanwhile, late nineteenth and early twentieth century logicians, like Frege, Peano, Cantor, Peirce, Hilbert and Gödel, would develop mathematical logic to a far greater level of sophistication than Boole did.

The contribution of Gottlob Frege well exemplifies the deepening trend towards the de-humanization of thinking. Fully committed to the goal of severing human thinking from natural language and creating for it a purely formal, error-free medium, Frege published his *Begriffsschrift* ('Concept Script') in 1879.[18] Whereas George Boole had used ordinary algebra as his script representing logical relations, Frege invented a wholly new set of signs that expressed logical relationships alone. Unlike Boole, who believed mathematics is the foundation of logic, Frege believed logic to be the basis of mathematics, and so was convinced of the necessity of creating an artificial sign-language with its own precise grammatical rules that encapsulated the fundamental laws of logic. This artificial language would be the language that Leibniz had dreamed of, a *lingua characteristica*, soundless and indeed unspeakable, uncontaminated by nuances of meaning, by metaphor, paradox, or any kind of ambiguity. Frege's avowed purpose was 'to break the domination of the word over the human spirit' by devising a mathematical language in which general propositions could be reduced to unequivocal statements with their logical structure exposed.[19] Frege wrote:

> It is therefore highly important to devise a mathematical language that combines the most rigorous accuracy with the greatest possible brevity. To this end a symbolic language would be best adapted, by means of which we could directly express human thoughts in written or printed symbols *without the intervention of spoken language.*[20]

The symbols that are to replace words are not the authentic symbols of a religious outlook, but rather a language of abstract quasi-mathematical signs.[21] Frege believed that such a language, despite being so far removed from the human being that it cannot be spoken, could nevertheless 'directly express human thoughts'.

Let us briefly reflect on this. A sign language that bears no relationship to the spoken word is a language that has been stripped of its soul-content, and severed from any relationship to the *logos* in nature that the spoken word has the capacity to express. We saw in Chapter 7 that in the contemplative tradition of the Middle Ages, the world of nature was understood to mirror the divine Logos that is its creative source, which could then be drawn forth and re-expressed in human language. Hence the importance of Grammar in the Middle Ages, as the guardian of the spiritual integrity of language.[22] We also saw that poetry was regarded as the highest reach of Grammar, for through poetry there is the possibility of experiencing the Logos as creative power. Poetry and philosophy were felt to have a natural affinity with each other, as both draw on the organ of intuition, referred to as the *intellectus*, by which we grasp a spiritual truth.[23] The *intellectus* was recognized as the spiritual anchor of every act of genuine thinking, and it was contrasted with the thinking of the *ratio*, which is the instrument of logical analysis and problem-solving. From the medieval perspective, what Frege sought to do was to cut out the *intellectus* and the primordial capacity of language to express different levels of meaning, in favour of honing the analytical thinking of the *ratio* to a machine-like precision. His motivation, like that of Leibniz, was to eradicate the possibility of error. To this end, our natural language governed by grammar had to be replaced by a contrived sign-language governed by and expressive of purely logical relationships. The purpose of such a language was to peel away the calculative aspect of thinking from the wholeness of the act of reflection by providing for it its own essentially sterile medium. This medium would be a totally inhospitable environment for poetry, or indeed for any deeper spiritual intuition, because it would not allow anything to be expressed other than what has a strictly logical form. By contrast, the living language that human beings speak is regarded as no friend to logical thought:

> Language proves to be deficient ... when it comes to protecting thought from error. It does not even meet the first requirement which we must place upon it in this respect; namely being unambiguous. ... Language is not governed by logical laws in such a way that mere adherence to grammar would guarantee the formal correctness of thought processes ... We need a system of symbols from which every ambiguity is banned, which has a strict logical form from which the content cannot escape.[24]

As it turned out, the sign-language created by Frege lacked sufficient elegance to be generally adopted. Instead, a different system of notation (just as unspeakable) devised by Giuseppe Peano, and subsequently championed by Bertrand Russell, became the standard system used to this day in so-called 'symbolic logic'. The sign-language that forms the basis of modern symbolic logic serves the purpose of exposing the logical structure of sentences and propositions, and thereby making plain the validity of arguments. By transposing normal grammatical expression into a quasi-mathematical form, a new scientific exactitude is brought to the analysis of propositions, so that one can then conduct a 'logical calculus'. For example, to perform such a logical calculus on the sentence, 'All sheep are mammals', we first rephrase it as: 'For every object x, if x is a sheep, then x is a mammal.' This makes it susceptible to being cast into the purely logical form of an 'if ..., then ...' proposition (an inference). In the standard system of notation, the sign $\supset$ expresses the logical relation 'if..., then ...' and the sign $\forall$ is employed to express the universal quantifier 'all' or 'every', so the sentence, 'All sheep are mammals' (recast as 'For every object x, if x is a sheep, then x is a mammal') is now expressed:

$$(\forall x)\ (x \text{ is a sheep} \supset x \text{ is a mammal})$$

Or more briefly (substituting 's' for sheep and 'm' for mammal) as:

$$(\forall x)\ (s(x) \supset m(x))$$

Clearly, such a formula could be used to express a quite different content; x could represent all manner of different mammals—mice, horses, and cows—and the proposition would remain true. So it is not the thought *content* that is important here but the underlying logical form which the formula captures with its so-called 'logical constants', $\forall$ (All or Every) and $\supset$ (If ..., then ...), which bind the content into what Frege called a 'formula language'.[25] As we have seen, Boole had already exposed other logical constants, such as negation (NOT), disjunction (OR) and conjunction (AND), each of which is also now represented by a new sign: negation by ¬, disjunction by V, conjunction by Λ. If we return to our earlier example, the proposition 'all sheep on the farm are either black or white', we may rephrase it 'For every object x, if x is a sheep on the farm, x is either black or white', and then write it as the following formula:

$$(\forall x)\ (s(x) \supset (b(x)\ V\ w(x))$$

If we wish to propose that some sheep are fat *and* white, then we would render this as a formula by using the symbol ∃ for 'some', and the symbol Λ for 'and':

$$(\exists x)\ (s(x) \supset f(x)\ \Lambda\ w(x))$$

The defining character of a proposition is that it is either true or false, and in logic the truth or falsity of a proposition is referred to as its 'truth-value'. The proposition 'all sheep on the farm are either black or white' will be true if one or other, or both, components of what is asserted are true. Otherwise it will be false. The proposition 'some sheep are fat and white' will be true if and only if *both* its components are in some cases true, otherwise it will be false. In these propositions, there is a constant element that gives to the proposition its logical skeleton (the logical constants), and there is a variable element—the colour or size of the sheep—which will determine the proposition's truth-value.[26] The truth-values of a proposition's variable components determine the truth-value of the whole proposition, and these truth-values can be tabulated in 'Truth Tables' in exactly the way that we saw earlier (Figs. 23.1–23.3). We may put 'T' or 'F' in the columns instead of 1 and 0, but really it makes no difference: we are here engaged in the kind of analysis that can be conducted as efficiently by an electric circuit, with switches judiciously arranged, as by a human being.

The development of symbolic logic clearly belongs to the same impulse as that which was driving humanity towards the automation of thinking. By extricating logic from its home in living, spoken language, it was given the ability to function in a domain that was no longer connected to human perceptions, meanings or values. The function of logic was no longer understood as being to relate human beings to 'knowledge of things', which philosophers from antiquity through to the Middle Ages and beyond saw as its principal purpose.[27] Those human beings who immerse themselves in this world of the 'formula language' are carried away from, not towards, reality precisely because within this language the organic unity of thought and word has been torn asunder. Relationship to the *logos*, to the meaning in things that words express, is replaced by a merely mechanical process in which it is not even necessary to understand the meaning of the signs of the formula language: all that is required is that one follows the rules that the signs represent. In so doing one is no longer thinking like a human being but like a machine. One has turned oneself into an 'engine of logic'.[28] For an engine of logic, knowledge is not the aim, nor is truth, but rather the logical validity or correctness of a proposition. In conducting such a logical calculus the human being becomes a calculating machine, functioning in accordance with fixed rules, and applying its analytical skills to anything and everything susceptible of analysis. Such a form of thinking is actually *unreflective*, and this is why Martin Heidegger insisted that it is not real thinking. It is no more than a form of 'calculative organisation', ignorant of 'the essence of thinking'.[29] For Heidegger, symbolic logic does not reach the status of real thinking because real thinking always brings us into relationship with *being*. When we really think, our thoughts bring us closer to 'the truth of Being', whereas symbolic logic with its propositional calculus closes us off from it.[30]

Frege has been hailed as the founding father of modern analytic philosophy.[31] His ambitions were shared by many leading thinkers of the twentieth century, who persuaded themselves that the dehumanization of language was the key to ensuring the objectivity of knowledge. In Vienna the Logical Positivists sought to make symbolic logic the vehicle for a radically materialistic view of the world. Rudolf Carnap, one of the most influential of the Logical Positivists, was a pupil of Frege, and dedicated himself to establishing a new rigour in scientific language that would be limited on the one hand by the rules of syntax, which he held ultimately belonged to arithmetic, and on the other hand by what is expressible in terms of the science of physics. Deeply influenced by the Logical Positivists, Bertrand Russell and Wittgenstein founded the movement of Logical Atomism, which had a similar goal: to analyse language into its so-called 'atomic' constituents, stripping down human discourse to statements of 'atomic facts'. From this basis Russell believed it would then prove possible to secure the foundations of empirical knowledge, built up systematically within the framework of mathematical logic. In both Logical Positivism and Logical Atomism, the focus was on the 'purification' of language so that it could accurately reflect physical reality, purged of any intrusions of a psychological or spiritual nature. The same goal lay behind the development of Linguistic Philosophy, which was really just another expression of Logical Positivism. Each of these schools of thought, which were all different enunciations of one nihilistic passion, avoided any reference to the inner world of the psyche, and they were unremittingly hostile to any notion of transcendent realities.[32]

The nineteenth century mathematization of logic and the development of a workable 'formula language' that could function within its own independent sphere should be understood as the preliminary exploration of programming languages for computers.[33] The culmination of this process was the construction of viable working computers, operating with programming languages derived from symbolic logic, which the machine could utilize to process data according to fixed rules expressed and applied in a notation cast adrift from living human language and bound into the binary code.[34] What was brought to expression in the realm of human thought in Logical Positivism (and its various manifestations) was a mirror to what was taken forward in the realm of technology. They are the joint outcomes of the single process of the deadening of the inner life during the nineteenth century, when Bacon's dream of transforming human thinking into nothing more than an instrument for processing data, and as inherently closed to the spiritual, edged towards its final realization. The legacy of the nineteenth century was that a form of thought was achieved that was machine compatible. It therefore had an irresistible appeal to those for whom thinking like a machine had become a new standard to which all human thought, in order to be valid, should conform.

Chapter Twenty-Four

CROSSING THE THRESHOLD

Michael Faraday and the Science of Unseen Powers

Alongside the advances towards the automation of logic made during the nineteenth century, further steps were made towards a new understanding of electricity, which came to be seen no longer simply as a property of physical bodies in space, but—like light—as itself permeating all of space. As a result of this new understanding, electricity would become the medium of modern wireless communication technologies. The origins of this breakthrough go back to the work of the Danish physicist Hans Christian Oersted, who in 1820 discovered the existence of a much more intimate connection between electricity and magnetism than had hitherto been suspected. Prior to this, people had long puzzled over the relationship between electricity and magnetism, but had not been able consciously to lay hold of what this relationship was. After Volta discovered the means to produce a continuous electric current (in 1800), the approach to electricity changed radically, for electricity now revealed a new aspect of itself. It then became possible for Oersted literally to *see* that magnetism and electricity were far more closely related than had previously been realized. In the 1830s, Michael Faraday, who was an exact contemporary of Charles Babbage, followed up Oersted's initial insights, putting a huge effort into exploring further the exact nature of this relationship.[1]

It is largely due to Faraday's conscientious and systematic researches into electricity over the course of several decades, but especially during the 1830s, that the way was prepared first of all for James Clerk Maxwell to formulate mathematically the laws that govern the relationship between electricity and magnetism in the 1860s and 1870s, and then for Heinrich Hertz to experimentally propagate electromagnetic waves for the first time in the 1880s. This cleared the path for the introduction of wireless communication technologies: for example, the wireless telegraph (from 1895 onwards), radio broadcasting (from 1910 onwards) and two-way radio communication (during the 1930s and 1940s). It also opened the way for the discovery of electrons (in 1897) and the establishment of the whole field of electronics, which would prove so important for the design and functioning of computers, first using vacuum tubes and then transistors and the integrated circuit. Faraday, Maxwell and Hertz can be seen as taking humanity across a threshold, on the other side of which lay the technologies we are today familiar with: mobile electronic devices such as mobile phones, smart phones, tablets and laptops, as well as the ever more ubiquitous 'internet of things', based on the connectivity of millions upon millions of different transmitting and receiving devices placed in the outer environment, in the workplace and in the domestic environment of the home.

These later developments, so crucial to the computerization of more and more aspects of life, were only made possible because of the breakthrough in understanding the relationship between electricity and magnetism, due largely to Faraday's work.[2] Faraday was a meticulous investigator, and conducted an extremely large number of experiments. We shall consider four crucial experiments that he carried out in 1831, each concerned with the generation of an electric current involving the forces of magnetism. It was already known before Faraday began his experiments that one important characteristic of an electric current is that it has the capacity to magnetize iron, making the iron function in just the same way as an ordinary magnet, by wrapping a coil of copper wire around an iron bar and sending an electric current through the wire.[3] Faraday now wanted to see if this relationship between electricity and magnetism was reciprocal, reasoning that if an electric current could magnetize an iron bar, it should be possible for a magnetized iron bar to induce an electric current. One of the experiments that he devised involved hanging up a large iron ring, often referred to as a *torus*, around which he wrapped two separate coils of insulated copper wire, one on either side of it. The first coil he attached to a battery, so that he could send an electric current through it, thereby magnetizing the ring; and the second coil he attached to a galvanometer (a sensitive meter with a needle that moves in response to an electric current) in order to see if the magnetization of the iron ring led to the presence of electricity in the second coil (Fig. 24.1).

What Faraday found was that the needle of the galvanometer only moved at the instant of his closing the switch that sent the electric current through the first coil on the battery side of the iron ring but the needle then returned to its original position and did not move again. The electric current, having momentarily awoken in the second coil, just as suddenly died. Even though the electricity was still flowing in the first coil and the iron ring continued to be magnetized, no further sign of electric current could be detected from the second coil on the galvanometer side. But at the instant he turned the current off in the first coil by opening the switch

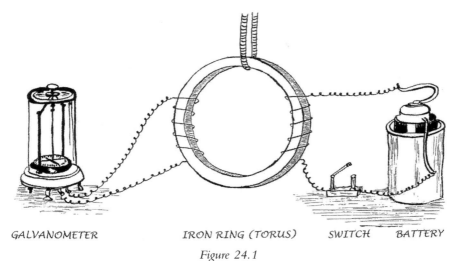

GALVANOMETER   IRON RING (TORUS)   SWITCH   BATTERY

Figure 24.1

Faraday's electrical induction experiment, using an iron ring or 'torus' around which two coils of insulated copper wire were wrapped, one attached to a battery and the other to a galvanometer.

Faraday noticed the galvanometer needle move again. It was therefore only at the moment when the electric current was *changing* in the first coil, from off to on or from on to off, that an electric current was induced in the second coil.[4] Furthermore, in the first instance, when the iron ring became magnetized, the galvanometer needle was deflected in one direction; in the second instance, when the iron ring lost its magnetism, the needle was deflected in the other direction. The direction of the electric current therefore changed, depending on whether the iron ring was being magnetized or demagnetized.[5]

Faraday proposed that the induction of the electric current was due to the creation of a state of tension, in which the particles of matter in both the iron ring and the copper coil were put into a 'constrained condition', a 'regular but forced arrangement', which sought release through being discharged. It was a 'forced state', a 'forced electrical arrangement' out of which the electric current was induced.[6] In this language of tension, constraint and forcing, we see Faraday reaching for a vocabulary that would somehow express what lay behind the phenomena involved in electromagnetic induction. It was not so much that he was trying to formulate a theory, but rather that he was trying to articulate a deeper level of the phenomena he was witnessing.

Faraday followed up this experiment with a second experiment, in which he held in his hand a bar magnet, which he passed in and out of a coil (or 'helix') of copper wire. The wire coil was not attached to a battery

Figure 24.2
Michael Faraday's experiment with a bar
magnet and a coil of copper wire.

(and so no electricity was flowing through it), but it was connected to a galvanometer, so if any electric current was induced in the wire it would be detected (Fig. 24.2).

Sure enough, he noticed the needle of the galvanometer deflect as he pushed the magnet into the coil and deflect once again (in the opposite direction) as he pulled it out, but while the magnet was stationary, either within the coil or outside it, no electric current was generated.[7] The significance of this experiment was enormous, for Faraday was showing that it was possible to generate an electric current mechanically, simply through moving a magnet in and out of a coil of wire. Because the direction of the current (that is, its electric polarity) alternated as the magnet entered and withdrew from the coil, the current was termed 'alternating current'. The implication of what Faraday discovered was that *mechanical motion*, powered for example by a steam engine, could then be made to produce an alternating electric current, potentially on an industrial scale.[8]

But what exactly was occurring? In Figure 24.3, we see that when a magnet is plunged (now vertically rather than horizontally) into a coil of wire, an electric current is induced in the coil, and this also creates a

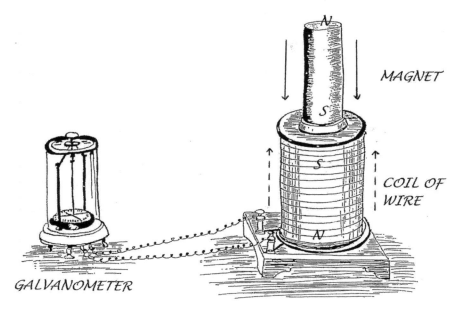

Figure 24.3
Faraday's experiment with a bar magnet and a coil of copper wire, showing how the bar magnet overpowers the resistance of the south pole of the magnetized coil of wire to the south pole of the magnet as the bar magnet is forced down inside the coil.

magnetic field around the coil. However, the coil's magnetic field is contrary to the magnetic field of the magnet. As the magnet enters the coil, the south pole of the magnet therefore meets resistance from the south pole of the newly magnetized coil, but the magnet overpowers this resistance as it is forced down inside the coil.

Once the magnet is within the coil, simultaneously both the electric current and the coil's magnetic field die away. Only when the magnet is pulled out again is the current once more induced in the coil, which again becomes magnetized. But the magnetic field of the coil is now reversed, so that the south pole of the magnet which is being withdrawn has to pull against the attraction exerted by the north pole of the coil, as can be seen in Figure 24.4.[9]

In this experiment, we see how electricity is born from the strife of opposing magnetic forces. On the one hand there is the overcoming of the weaker by the stronger as the resistance offered by the magnetized coil to the bar magnet is overpowered as the latter is pushed into it. On the other hand, there is a violent wrenching asunder of the stronger from the weaker as the bar magnet is then pulled away from the magnetized coil of wire, despite their mutual attraction. A year after the publication of these

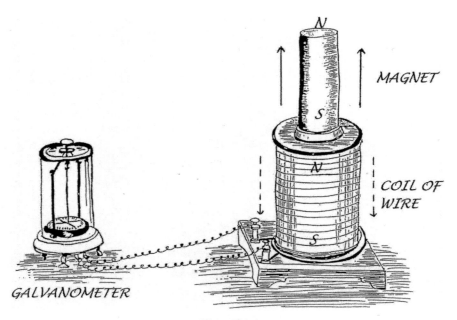

Figure 24.4

Faraday's experiment with a bar magnet and a coil of copper wire, showing how attraction between the south pole of the bar magnet and the north pole of the magnetized coil of wire is overcome as the bar magnet is withdrawn from the coil.

experiments, Faraday made the following observation concerning the nature of electricity:

> It has never been resolved into simpler or elementary influences, and may perhaps best be conceived as *an axis of power having contrary forces, exactly equal in amount, in contrary directions.*[10]

The words 'power' and 'force' or 'forces' were frequently used by Faraday, but rarely defined with precision. Often he seems to use them interchangeably. When he uses such terms, it is as if he wants to point us towards something at work in the phenomena greater than mere material particles, or atoms, in motion (Faraday was no atomist). One has the impression that he has reached a certain boundary—a boundary of knowledge—and he is not able to bring his consciousness to bear on what lies upon the other side of it. In this case one wonders what kind of power the 'axis of power' is. Is it solely material or does it have a spiritual aspect as well? And what is the nature of the 'contrary forces' that he believes are implicated in it?[11]

In a third experiment, Faraday took a length of copper wire, both ends of which were connected to a galvanometer, and drew the wire into and then out of a magnetic field created by two powerful magnets placed nine inches apart. As the wire was pulled into the magnetic field an electric current was induced, causing the galvanometer needle to deflect in one direction. As the wire was pulled out, an electric current was once again induced, causing the galvanometer needle to deflect in the opposite direction. The same effect occurred using strips of metal.[12] In this way, Faraday showed that whether a magnet is moved in relation to a stationary coil of copper wire or whether a copper wire (and also various other metals) is moved through a magnetic field created by stationary magnets, the same result would be obtained: namely, an electric current is generated. But notice how Faraday expresses this. The phrase he uses when describing this specific result is: 'a power is called into action which tends to urge an electric current through it'.[13] Throughout his career as a researcher, Faraday was aware not only that he was dealing with 'powers', but also that through his experiments these powers were being 'called into action' as if from a state of somnolence.

In a fourth experiment, he constructed a simple machine composed of a copper disk mounted on an axle so that it could be continuously rotated between two magnets.[14] On one side was a north pole, on the other a south pole, so a field of magnetic lines of force was set up between the two magnets, and the copper disk was set spinning within this field. In Figure 24.5 below, the two magnets are replaced by a large horseshoe magnet. Against the rotating copper disk, Faraday set two conducting brushes made of an amalgam of lead and copper, which were placed in contact with the rim of the disk and the axle respectively. Copper wires

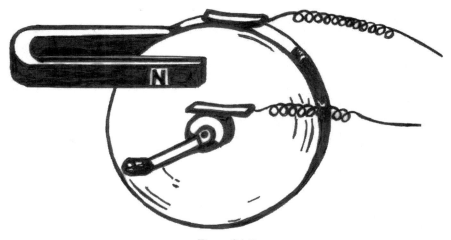

Figure 24.5
Faraday's disk dynamo, or electric generator, producing a continuous direct current.

led from them to a galvanometer, which enabled the electric current to be registered. The spinning copper disk became in effect an uninterrupted succession of wires, conceived as the radii of the disk, rapidly moving through the magnetic field, 'cutting' its 'lines of magnetic force'.[15] It was this action of cutting through the lines of magnetic force that generated the direct current. If the disk was rotated in the reverse direction the electric current would then also be reversed.

The 'disk dynamo', so simple in conception, demonstrates the principle that underlies the generation of electricity in power stations around the world, which is that when the lines of force of a magnetic field are cut by a moving wire, a rotating copper disk or an armature, then an electric current will be generated. Upon this principle is based the supply of electricity to households and businesses up and down the land, in every country, the world over. Without the application of this principle in the turbines that generate electricity in our power stations, which then connect to national grids of electricity cables, sub-stations and transformers, every computer that runs off (or needs to be recharged by) a mains electricity supply would be inoperable.

Clearly, the phenomena Faraday was observing—the magnetic field artificially set up between two opposite magnetic poles, the rotation of the copper disk and the subsequent generation of the electric current—were dependent on his active involvement. He had to produce them in order to observe them. Faraday's experimental apparatus was in this respect already *technology* that was compelling nature to render certain effects, which would not otherwise become manifest without human intervention. When Martin Heidegger observed that modern science is less the foundation of modern technology than technology the foundation of modern science, he could well have had Faraday in mind.[16] In Faraday's case, the records of his experiments illustrate how experimental results were so entangled with knowledge of how to produce certain effects, that the question of the nature of the 'powers' that were being 'called into action' was usually overlooked. The experimentalist's activity ran ahead of his capacity to know what it was he was really engaging with, and from what domain of nature these phenomena were being raised to manifestation.

One might think that, being a religious man (Faraday was an evangelical Christian belonging to an obscure sect called the Sandemanians), a spiritual perspective might have illumined his understanding of the forces he was awakening. But Faraday always sought to keep his scientific researches free of religious conceptions. He once wrote in a letter to Ada Lovelace:

I do not think it at all necessary to tie the study of the natural sciences and religion together, and, in my intercourse with my fellow creatures, that which is religious and that which is philosophical [i.e. scientific] have ever been two distinct things.[17]

This was a consistently held view of Faraday's. His science went alongside a sincerely held religious faith, but his science and his religion existed in different compartments of his life. To the extent that his religious faith did inform his scientific work through what was, *outside* the laboratory, a fundamentally religious orientation to the natural world, it did not equip him with conceptions that would help to shed light on the phenomena he was investigating *within* the laboratory. Rather the opposite: for Faraday, it was an unquestioned presupposition that electricity and magnetism were infused into creation by God, and he regarded electricity as one of the most powerful of God's agents, indeed the highest of God's powers known to humanity.[18] The exalted position of electricity in his world-view seemed to him repeatedly confirmed in the religious awe that he felt in the face of thunderstorms, which he experienced as sacred theophanies of God's power.[19] While this may have provided the driving force behind his investigations, it was not a realistic view of the nature of electricity. The very naivity of his religious faith closed the door on a more nuanced metaphysical perspective on the spiritual provenance of the powers and forces that he was evoking through his researches. In the laboratory context, metaphysics had no place.

Faraday intuitively felt that there was a relationship between gravity, magnetism and electricity, and devoted much of his later research to trying to demonstrate the commonality between them. Although for the most part he showed little interest in light, his quest for a unified force involved his demonstrating that rays of light could be brought under the dominion of magnetic forces.[20] In order to accomplish this highly symbolic act of subordinating light to magnetism, the light had first to be passed through heavy glass (with a high concentration of lead), something that would never occur in nature. In this respect his work can be seen as continuing the Gilbertian/Galilean tradition, with its affirmation of the pre-eminence of nature's 'lower border' and its emphasis on compelling the world of qualities to conform to laws that could be expressed with mathematical exactitude. At the same time, however, Faraday was unable to accept the atomistic view of nature so central to Galileo's thought. His phenomenological approach to research meant that he sought for explanations that viewed the parts from the perspective of a greater whole, and to the extent that it was permissible to speak of atoms and particles they were to be understood as 'centres of forces or powers' of varying intensity, which extend throughout space. This was a radically new perspective. For Faraday, matter comes into existence where lines of force are concentrated, and so the particles of matter are to be understood as 'centres of force' extending throughout the solar system.[21] Furthermore, along with the notion of 'lines of force' stretching through space, Faraday introduced the concept of a greater 'field of force' within which the lines of force existed.[22] Once again, in this concept, as with his language of 'powers' and 'forces', we may feel that Faraday is trying to find words for a realm of existence which humanity was ill-equipped to comprehend.

Maxwell Crosses the Threshold

Faraday's holistic approach, including his 'field' concept, especially appealed to James Clerk Maxwell, who was born in 1831 when Faraday, aged 40, was most intensely involved in his induction experiments. As a young man, Maxwell steeped himself in Faraday's *Experimental Researches on Electricity*, which became the starting point for his own work.[23] Like Faraday, Maxwell turned his gaze away from the magnetized or electrified physical bodies towards the invisible medium in which they exist, attributing to the 'aetherial' medium surrounding and pervading these bodies the source of their magnetized or electrified state.[24] He could not accept that one electrified or magnetized body acts directly on another at a distance, but believed rather that it sets up a disturbance in the surrounding, pre-existent field in the form of waves, which cause certain effects then to take place in other bodies. For Maxwell, when it comes to electromagnetic phenomena, it makes less sense to speak of electric and magnetic fields affecting one another (as Faraday had assumed), than to speak of a single field that is *both* electric *and* magnetic. Because changes to a magnetic field will induce changes in an electric field, which will in turn induce changes in the magnetic field as a wave ripples through it, Maxwell argued that we are dealing with a single *electromagnetic* field, with both magnetic and electric properties. In his seminal paper, 'A Dynamical Theory of the Electromagnetic Field' (1864), Maxwell wrote:

> The energy in electromagnetic phenomena is mechanical energy. The only question is, Where does it reside? On the old theories, it resides in the electrified bodies, conducting circuits, and magnets, in the form of an unknown quality called potential energy, or the power of producing certain effects at a distance. On our theory it resides in the electromagnetic field, in the space surrounding the electrified and magnetized bodies, as well as in those bodies themselves, and is in two different

forms, which may be described without hypothesis as magnetic polarization and electric polarization, or, according to a very probable hypothesis, as the motion and strain of one and the same medium.[25]

Maxwell envisaged the electromagnetic field on the analogy of a fluid medium, subject to pressures and strains, through which energy flows in wave-like patterns. While this was meant as an analogy rather than as a physical hypothesis, Maxwell was in no doubt that the electromagnetic field was a real entity. The existence of a new realm of nature was being proposed, a realm invisible to the eye, whose 'dynamical' attributes Maxwell believed could be revealed only through mathematics.[26] Here Maxwell differed from Faraday, the experimentalist, who felt at sea when it came to mathematical reasoning.[27] For Maxwell, just because Faraday's methods were restricted to experiment, they were too 'rough'. Only by subjecting the phenomena to rigorous mathematical treatment could one bring this new realm of existence within the sphere of human knowledge and control. In a lecture given at Cambridge in 1871, Maxwell declared:

> In every experiment we have first to make our senses familiar with the phenomenon; but we must not stop here,—we must find out which of its features are capable of measurement, and what measurements are required in order to make a complete specification of the phenomenon.'[28]

We might feel inclined to ask whether a 'complete specification of the phenomenon' can be achieved without our bringing to it concepts which would illumine it from a spiritual perspective. But Maxwell, who from his youth was nurtured on the thought of Descartes, Leibniz and Hobbes, clearly felt no such inclination. In the same lecture he went on to say that by improving the accuracy of numerical measurement, science:

> is preparing the materials for the subjugation of new regions, which would have remained unknown if she had been contented with the rough methods of her early pioneers.[29]

Here we see the depth of Maxwell's commitment to mathematics, as the main instrument both of scientific knowledge and of the 'subjugation' of these lately discovered 'new regions'.

In the shift in the locus of electromagnetic phenomena from electrified and magnetized bodies to an invisible and all-pervasive electromagnetic field, which exists in a state of potential until actualized by motion in, or disturbance of, some part of it, a metaphysical threshold is crossed. Primacy is taken away from the phenomenal world of natural qualities and perceptible forms, and is given to a realm of forces and potentials that are essentially sub-phenomenal. Maxwell the mathematician believes that we must turn to mathematics as the only means by which this realm can be brought within the sphere of exact scientific knowledge. And yet the characteristic of this realm is precisely that it is a realm of potentials.[30] What, then, is its metaphysical status? Where does it sit within the greater order of nature, whose upper and lower borders were once so clearly recognized? Despite being a deeply religious man, Maxwell was not willing to ask, let alone answer, such questions. For him, all scientific knowledge is necessarily restricted to the material sphere, and there is no truth beyond this sphere accessible to science.[31]

In his early essay 'On Faraday's Lines of Force', a manifesto for his later work on the electromagnetic field, Maxwell wrote: 'the aim of exact science is to reduce the problems of nature to the determination of quantities by operations with numbers'.[32] Thereby the greatest possible mathematical exactitude is sought, but the deeper metaphysical question as to the nature of what is being investigated is avoided. It is simply assumed that what is being investigated is material, and any moral or spiritual factors are left out of the enquiry.[33] By setting strict quantitative parameters around research, a crucial threshold of knowledge is then crossed *semi-consciously*, for the scientist's attention is resolutely directed towards what is measurable, as if this amounted to an exhaustive treatment of what is real.

One casualty of this approach was light, whose time-honoured place in theology and metaphysics Maxwell must surely have been aware of.[34] Maxwell proposed that the 'aetherial medium' that is the source of electromagnetic phenomena is no different from the medium that transmits light. We have already seen that Faraday had succeeded in 'magnetizing' light. Now Maxwell went one step further. He argued that light should be regarded as no more than an 'electromagnetic disturbance' taking the form of waves or 'undulations' of the electromagnetic field.[35] Light, which in traditional metaphysics and theology had been understood as the purest natural phenomenon expressive of spirit, was now to be regarded as a by-product or epiphenomenon of electromagnetism. This became the explanation for why electromagnetic waves and light waves travel at exactly the same speed—i.e. the speed of light. By implication, what the human eye is able to see is only one small fraction of a much greater spectrum of electromagnetic waves invisible to the eye. If all these waves travel at the same speed of light, then some will have longer and some shorter wavelengths. And if this is the case, then it should be possible for human beings to artificially propagate electromagnetic waves of different lengths and frequencies. In one move, light is divested of its spiritual sovereignty within nature, and human beings are charged with the task of subjugating the 'new regions' which have so alluringly appeared on the horizon.

Heinrich Hertz: Letting the Genie Out

Maxwell died in November 1879 when Heinrich Hertz, a young student in Berlin, had just begun to conduct his initial researches into electricity.[36] Hertz was destined to become the first person to experimentally demonstrate the existence of electromagnetic waves. In a series of experiments pursued between 1886 and 1888, Hertz was able to build up a picture of how these electromagnetic waves emanated from the transmitting apparatus he constructed. This consisted of a high voltage electric circuit, which he set up in such a way that sparks would be made to fly across a 'spark gap' in the circuit, causing an oscillation along two horizontal copper wires either side of the spark gap, thereby generating electromagnetic waves that radiated out into the laboratory. In order for the energy of the electric discharge to radiate out into space, the voltage had to be intensified, and Hertz did this by including a special kind of induction coil in the circuit, that had the effect of greatly increasing the voltage.[37] As the sparks flew across the gap in the circuit, the waves they created could be detected by a simple detector made of a copper ring, positioned on a stand. In this ring there was also a small spark gap across which sparks would be induced in response to the wave (Fig. 24.6). The spark gap in the transmitting apparatus and that of the detector could be mutually calibrated, through minute adjustments made to each by a micrometer screw. In order for the detector to 'catch' a spark, it had also to be exactly placed, and this enabled both the frequency and the length of the wave to be determined.[38]

Hertz was able to adjust the length and frequency of the waves by changing the number of turns of copper wire in the induction coil. With fewer turns, higher frequency waves resulted; with more turns, lower frequency waves were generated.[39] With the transmitting apparatus charged, Hertz would move back and forth with his detector in the large lecture hall in which he conducted his experiments, catching the sparks in it and measuring the differences in their size, depending on the position of the detector. In this way he was able to chart the wave-patterns produced by the transmitter. Figure 24.7 depicts the progress of a wave as it emanates from the transmitting apparatus, each successive map showing, around the oscillator in the centre of the picture, the rapid enlargement of a circle which the waves fill.[40]

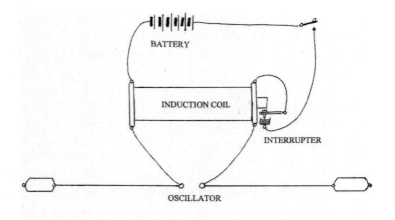

Figure 24.6

A simplified diagram of Hertz's apparatus. It shows an induction coil in the centre, which greatly amplifies the voltage coming from the six-cell battery at the top. An 'interrupter' is connected to the induction coil, which repeatedly interrupts the Direct Current from the batteries, thereby collapsing the magnetic field and at once re-establishing it, thus enabling induction to take place in the coil. Sparks flying across the spark gap of the oscillator setting up an oscillation along its two horizontal wires, thereby generating electromagnetic waves that radiate into the atmosphere, to be detected by a simple detector made of a copper ring with a spark gap, placed on a stand.

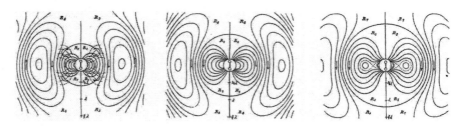

Figure 24.7

Three charts of rapidly expanding electromagnetic waves, radiating out from an oscillator. From left to right, the space filled by the waves is indicated by circles R1, R2 and R3 in each successive map.

According to Hertz, who like Maxwell and Faraday before him, believed in the primacy of the electromagnetic field as an entity in its own right, the conditions of the surrounding space were crucial to understanding what was occurring; for the field, rather than the transmitting apparatus, was the 'seat' of the action:

> Fundamentally the waves which are being developed do not owe their formation solely to processes at the origin, but arise out of the conditions of the whole surrounding space, which latter, according to our theory, is the true seat of the energy.[41]

For Hertz, this is not a matter of conjecture but of his research results, which were confirming over and over again that what he was dealing with was precisely the 'new regions' of reality that Maxwell referred to in his Cambridge lecture of 1871. Electricity could no longer be regarded simply as a property of certain physical substances or processes: its true mode of being is to be spread out in space. It is vast. It is everywhere. It is as omnipresent as light. But it is far greater than light, for light should be regarded simply as one amongst many electrical phenomena.[42] Light is just:

> a small appendage to the great domain of electricity. We see that this latter has become a mighty kingdom. We perceive electricity in a thousand places where we had no proof of its existence before. In every flame, in every luminous particle we see an electric process. Even if a body is not luminous, provided it radiates heat, it is a centre of electric disturbances. *Thus the domain of electricity extends over the whole of nature.*[43]

And yet, while its domain had so dramatically increased in size and importance, Hertz was not able satisfactorily to answer the question of what electricity actually *is*. Certainly the mathematical approach taken by Maxwell contributed greatly to the ability to predict and control electromagnetic phenomena, and it underpinned all of Hertz's experimental work.[44] It had the effect of opening up a vast new world. However, this approach—to which Hertz was committed—was not equipped to address the deeper question of the existential status of *what* had been opened up. If the mathematical approach could successfully apply equations to it, then it seemed to Hertz that the equations 'were wiser than ourselves, indeed wiser than their discoverer'.[45] As Hertz candidly admitted:

> To the question 'What is Maxwell's theory?' I know of no shorter or more definite answer than the following:—Maxwell's theory is Maxwell's system of equations.[46]

A system of equations, while it may bring order and coherence to the phenomena, falls short of human insight into their spiritual provenance. Humanity was insufficiently prepared to recognize the true nature of the genie that, after so many millennia of being spellbound, had at last escaped confinement to claim for itself a 'mighty kingdom' to rival the whole of nature.

Chapter Twenty-Five

TOWARDS A HUMAN FUTURE

The Electrification of the Atom

When Heinrich Hertz propagated the first electromagnetic waves in the 1880s, the scientific consensus was that matter was made up of tiny, indivisible atoms. By the end of the following decade, this consensus had changed, and the notion that atoms were themselves composed of much smaller component parts became generally accepted. What was remarkable in this development was that the component parts of the atom had been shown to be *electrically charged*. Electricity was present at the very heart of matter.

This shift in the conception of the structure of matter came about initially as a result of experiments conducted within vacuum tubes. We have already seen how electricity has a certain affinity with the vacuum, as if it flourishes best in an environment where life cannot be sustained. In the experiments, high voltages of electricity were passed through the vacuum tube from one end of the tube to the other, producing a ghoulish green glow. This effect, already well known before the 1880s, was widely understood as produced by 'rays' of electricity emanating from the (positive) cathode at one end of the tube, and travelling towards the (negative) anode at the other end. But then in 1897, J. J. Thomson was able to demonstrate conclusively that the so-called 'cathode rays' were in fact tiny particles much smaller than atoms. These particles carried a negative electrical charge, and they subsequently became known as 'electrons'.

Thomson's work was carried forward by Ernest Rutherford, who in 1912 proposed a new model of the atom, consisting of a central nucleus surrounded by a 'cloud' of electrons. Because electrons were understood to be negatively charged particles, the search began for positively charged particles within the nucleus to balance them, for without positively charged particles, the atom would not be electrically neutral. In due course the positively charged 'proton' was discovered, so by the 1920s the electrical picture of the atom was beginning to be filled out. Atoms were now thought to consist of a central nucleus harbouring positively charged protons, surrounded by a cloud of orbiting negatively charged electrons. Since the number of protons and electrons needed to be the same in order for the atom to be electrically neutral, scientists could determine the number of protons in an atom by determining the number of its electrons. What this new conception of the atom meant was that electricity had now taken its seat at the centre of the structure of matter.

To many scientists it seemed that a definitive explanation of electricity could at last be given. The static electricity produced when we rub glass or plastic with a cloth should be explained as due to an interaction at the subatomic level between positively and negatively charged particles present in the materials involved. Similarly, an electric current carried along a copper wire should be explained as occurring because of a movement of electrons from one atom to another within the copper wire. Conduction takes place more easily in substances like copper because their electrons are more loosely bound to the atomic nucleus, whereas in other substances (like glass) they are more tightly bound. This way of thinking, entirely consonant with that established by Galileo in the seventeenth century, accounts for what occurs in the lived world of human experience in terms of what occurs at atomic and subatomic levels hidden from our experience. Electricity at one level of nature (as we experience it when we rub glass or plastic, or send a current through a copper wire) is explained by appealing to electricity at another level of nature or, more accurately, a level *beneath* nature, which is out of the range of our direct experience of electricity in the world we actually live in. However, the appeal to the electrical structure of matter does not leave us any the wiser as to what electricity actually *is*. The conception of the electrified atom does not so much provide an *explanation* of electricity, as a description of electrical events occurring at this subnatural level. And this description crucially misses out the human role in the production of electrical effects through actions performed on the macro-level, such as rubbing glass, constructing a battery or firing up an electric generator.

In Rutherford's atomic model, which became the basis of subsequent experimentation within particle physics, the typical size of an atom containing the nucleus and the 'electron cloud' surrounding the nucleus is 100,000,000th (one hundred millionth) of a centimetre across. The central nucleus in which protons are embedded is considerably smaller: 10,000,000,000,000th (ten trillionth) of a centimetre across, in other words one hundred thousandth of the diameter of the atom taken as a whole.[1] One would therefore need one hundred thousand nuclei to fill the space occupied by the whole atom. Such relationships within a scale so tiny are hard, if not impossible, to picture accurately, and are anyway virtually meaningless in terms of our experience of the physical world that we inhabit humanly, as embodied beings. The structure of the atom conceived and pictured in the mind of the physicist can only be a 'model', because the picture of the atom is necessarily an abstract mental picture projected onto phenomena of which we can have no direct experience. As Neils Bohr candidly admitted in 1934:

> Isolated material particles are abstractions, their properties being definable and observable only through their interactions with other systems.[2]

The newly discovered electrons, which were seized upon as the ultimate cause of electricity, turned out to be intrinsically problematic. They confounded physicists who attempted to answer the most basic questions about them, such as: What is their size? Can they be measured? What are they actually made of? Is it possible to accurately locate them? Are they even entities? Electrons could not be pinned down. They were found to exist in a curious no-man's land between being and non-being. As J. R. Oppenheimer said:

> If we ask, for instance, whether the position of the electron remains the same, we must say 'no'; if we ask whether the electron's position changes with time, we must say 'no'; if we ask whether the electron is at rest, we must say 'no'; if we ask whether it is in motion, we must say 'no'.[3]

At best, one could speak of electrons only in terms of probabilities or tendencies to exist, which is the way medieval metaphysicians characterized *materia prima*.[4] The electrification of the atom, far from providing a final explanation of electricity, rather indicated that an endpoint had been reached in a process begun many hundreds of years previously—a process that involved a transference of human allegiance away from the upper border of spiritual archetypes to the lower border of *materia prima*, from the supernatural to the subnatural. In the electrification of the atom the lower border received its ultimate validation as the locus on which human understanding of nature would rest. Upon this uncertain ground, this ground of the indeterminate and the potential, at greatest remove from the luminous world of spiritual archetypes, a new culture was erected, in which electricity would play a key role.

Towards a Metaphysical Perspective on Electricity

Because of the deeply entrenched Galilean habit of directing attention to occurrences at the atomic and subatomic levels in order to account for what happens in the macrocosmic world, the true character of electricity has remained largely unrecognized. A different *kind* of knowing is required, concerned with electricity's non-measurable features, which cannot be expressed in quantitative terms. This kind of knowing would be based on a qualitative approach to electricity, which builds up a picture of electricity's character by contemplating its many and diverse modes of manifestation in the sphere of our actual experience. Faraday was inclined to refer to electricity as a 'power' which human beings have called forth. Although he drew back from following up the implications of this characterization of electricity, Faraday intuited that electricity is not simply a physical force, and he strongly resisted any attempt to account for it in terms of the movement of atoms or particles: he saw that it is something more, but was unable to say exactly what.[5] A purely materialistic approach, dominated by quantitative thinking and mediated by technological interventions, no matter how sophisticated, will not lead us to the answer to that question. A different approach is needed in order to come to the right assessment of the inner nature of electricity.

In earlier chapters we saw that, in the ancient world, it was the inward aspect of electricity that was the principal focus of knowledge. This inward aspect was associated with certain gods and was directly encountered by those initiated in their Mystery rites. In the ancient Mystery teaching, the provenance of electricity was identified as the Underworld, a region of existence inimical to life, which lies 'below' the natural world. This may account for why so many people today, despite the collective approbation of electricity, nevertheless feel an instinctive wariness towards it. What is needed today is a renewal of the kind of knowledge of electricity that was acquired in the Mysteries, namely a knowledge oriented towards understanding electricity as a spiritual power with distinct moral qualities. In other words we need to move beyond the usual objectifying mindset with its accompanying utilitarian agenda in order to apprehend electricity with an awakened moral and spiritual awareness, alert to its inner nature. This kind of consciousness of electricity would provide a much needed counterbalance to the scientific and technological approach that treats it as a merely physical force resulting from the interactions of particles, to be manipulated and controlled by us for human benefit. It would be a different kind of knowing, based on a very different philosophical perspective, which would be able to encompass the moral and spiritual dimension of the natural order that, since the seventeenth century, it has been the central project of mainstream science to exclude.

Herein lies a great challenge, living in a culture which has not begun to reckon electricity in moral or spiritual terms. We have collectively thrown in our lot with a power that we utilize, and that we ourselves bear responsibility for bringing into manifestation, but it is a power we do not humanly understand. Few of us question our growing dependency upon it, hardly noticing its background presence in almost every aspect of our lives, and the pervasive influence it exerts on us. We are asleep to the real nature of this power that we have accepted as our most intimate companion. Instead of treating it with the circumspection and respect due to something so lethal, we have allowed ourselves to be seduced by what it can do for us, for it imbues everything it touches with its allure. The electricity-mediated virtual world that glows invitingly from so many screens, from televisions to iPads and smart phones, can seem more appealing than the world of nature to which we really belong, and we witness today a mass migration of people's desires, affections, and loyalties from the natural world to the virtual. But the virtual world is parasitic upon the real world, and because it presents to us a simulation of reality, there is a danger that we embrace an illusion while turning our backs on reality. Our peril today is that we become spellbound by a relationship that gives us a false sense of power, comfort and reassurance, but does not truly nourish the soul. The danger is that the soul is hollowed out, our relationship to nature devitalized, and our physical wellbeing undermined. And all the while we far too readily fall back on a merely utilitarian attitude, naively thinking that because the technology is so useful to us it is entirely benevolent, despite innumerable studies pointing to the contrary.[6]

A more meaningful knowledge of electricity requires that we take into account how we *feel* towards it, for our felt response can reveal to us objective qualities in things that physical instruments may not detect or measure. It is precisely such felt qualities that the Baconian/Galilean approach to knowledge deliberately sought to exclude, with the consequence that scientific knowledge became desensitized to the inner nature of what it investigated. Consider our experience of light. We saw in earlier chapters that natural light was regarded, both in antiquity and in the Middle Ages, as a manifestation within the world of nature of an inner, spiritual light, whose provenance was the upper border of spiritual archetypes, and ultimately proceeded from the divine source of existence itself. For Plato, Plotinus, Augustine and Aquinas, there was no question that the light bore within it a kind of reminiscence of the divine goodness. It had a distinctive moral quality, conveying to the natural world something intrinsically pure, innocent and good. Feelings of reverence for the light came spontaneously to people in earlier times, and can still be felt today by anyone who steps out into nature with their inner and outer senses alert. Electricity, by contrast, does not evoke these feelings, for it lacks this moral quality. Few people feel towards electricity anything approaching the pious devotion that filled the soul of the medieval monk or nun when they contemplated the light that illumines the natural world, and yet modern science tells us that light is a form of electricity. As Heinrich Hertz said: 'Light of every kind is *an electrical phenomenon* . . . Take away from the world electricity, and light disappears . . .'[7] If the relationship between electricity and light is really as Hertz describes it, why is it that we do not feel the same way towards electricity as we feel towards light? We should take these feelings seriously: they direct us towards a qualitative aspect that a scientist such as Hertz would dismiss as irrelevant.

The discovery of electricity took place when, following the Reformation, the relationship of human beings to the upper border changed, and the new mechanistic science began to breach the lower border of nature. While there may be a close physical relationship between electricity and light, their spiritual provenance is quite different. This is something that we must learn to recognize: we must once again learn to recognize the qualities of the two borders, and to understand that just as light emerges from the upper border, so electricity emerges from the lower border. Electricity is a subnatural power conjured into manifestation in the world of nature from out of the realm of *materia prima*. From a spiritual point of view electricity belongs to a completely different sphere from that of light. Far from light being a phenomenon of electricity, it would be more true to regard electricity as a degraded form of light, which in certain respects it mimics, but without the life-giving and morally uplifting qualities of light.[8]

In the early days of the discovery of electricity, people were distracted from questioning its spiritual provenance by all the magical and entertaining effects produced by the first 'electricians'. Today there is a similar difficulty in getting past what electricity offers to us, in order to make a more objective judgement as to its real character. Electricity can seem a most willing, obedient and in certain respects 'intelligent' servant, bringing our machines to 'life' and doing our bidding simply through our flicking a switch or pressing a button. But we should be concerned less with how it can serve us, and more with what effects it has on us, and on all of nature. Our world has changed. It now has incorporated into its invisible fabric a power of subnature, which permeates the very atmosphere we breathe. Nature is no longer what it used to be. It has been infiltrated by subnature. Through twenty-five chapters, we have traced how this has happened, but the transformed situation in which we now live means that we cannot simply carry on as we have been doing, as if nothing fundamental has changed.

Electricity and the Analytical Mind

In 1925, Rudolf Steiner made the following pertinent remark:

> [Electricity] must be recognized in its true character—in its peculiar power of leading down from nature to subnature. Only human beings must themselves become aware lest they slide downwards with it.[9]

Steiner was calling our attention to a danger that electricity poses to the human soul: its 'peculiar power' of leading us down towards subnature, towards a realm beneath that to which human beings naturally belong, for this is its home and its place of origin. Should we 'slide down' towards that realm, then we would be in danger of losing our inner bearings. Precisely because its provenance is the lower border, the tendency of electricity is to divert us away from awareness of both nature and spirit, and to lead us to regard the electrically mediated phenomena that we encounter in radio, television, computer, smart phone, and so on, as the locus of reality. Thereby we lose our connection with the world to which we truly belong as human beings.

In previous chapters, we saw how shortly after the discovery of the electric current, Humphry Davy was able to use it as an aid in the analysis of substances into their chemical constituents. He called this 'electro-chemical analysis'. Electro-chemical analysis showed the affinity between the impulse of the analytical mind towards reducing things to their component parts and the disintegrating effect of electricity on various substances. We saw how subsequently in the mid-nineteenth century, current electricity was exploited as a medium of long-distance communication through a comparable decomposition of language into abstract constituents, resulting in the Morse Code, in which each letter of the alphabet was replaced by a series of dots and dashes. And we also saw how electricity was finally able to become the vehicle for a fully automated logical analysis of data, through the clever arrangement of switches in electric circuits.

In each of these stages towards the automation of logic, which paved the way towards the first computers, it became more and more evident how well matched was the impulse towards analysis with what electricity was able to offer. If the sages of medieval Europe recognized the affinity between the inner nature of light and the contemplative thinking of the *intellectus*, then the modern rationalists discovered a similar affinity between electricity and the logical analysis of the *ratio*. It was not by chance that the period in history when electricity began to be stirred up and awakened from dormancy was the same historical period when a reduced conception of human reason as no more than a faculty of calculation was put forward by thinkers such as Bacon, Hobbes and Leibniz. The nature of electricity cannot be understood unless we take into account such changes in human consciousness. But then it is precisely our consciousness that is challenged today by the advent of electricity into our lives.

So long as human thinking was guided by the contemplative intelligence of the *intellectus*, its orientation was towards the transcendent realm of spiritual archetypes. Because of this orientation of pre-modern consciousness towards the upper border, its back was turned on electricity, which remained outside the purview of a science that was conducted within a metaphysical and religious outlook that was, as it were, facing the other way. Only at the dawn of the scientific revolution, when the *intellectus* was jettisoned in favour of the *ratio*, did the lower border command the attention of human thinking, which became irresistibly drawn towards the penetration of subnature, and with it the discovery of electricity. Without this change in human consciousness, electricity would not have been unveiled and released into nature.

Human consciousness must be understood as interwoven with the world, for what we become conscious of is not simply 'out there', having nothing to do with us. The discovery of electricity reflects back to us something that is a part of our own being. We need, therefore, rightly to assess what aspect of our nature is being reflected back to us in electricity, and to bring this into a healthy relationship with the whole human being. Just because the emergence of electricity was inextricably linked to certain changes in human consciousness, we need both to understand what it is in ourselves that resonates so strongly with it, and at the same time to find the counterpoint within the human being that would save us from 'sliding downwards with it'.

Logic and the History of Consciousness

Human consciousness is not static. If the emergence of the computer has occurred at a certain historical moment, then the computer should—like the discovery of electricity—be understood as closely related to an evolutionary change in human consciousness. This means that we cannot simply view the computer, which today takes the form of so many different 'smart' devices, as just another useful physical tool, an object among other objects that exist outside us. Of course it does exist outside us, it is a physical object, but it also concentrates within itself the fruits of a historical development of human consciousness without which it would never have come into being. This historical development has occurred in relation to the changing role of logic in human thought-processes.

In the great civilizations of ancient Egypt and Mesopotamia, logic was at first confined entirely to the practical, craft-focused activities of making and constructing things. It did not exist as a self-conscious science with formal laws. Only during the third millennium BC in Mesopotamia did its ambit expand, by its being applied to the interpretation of the will of the gods. In ancient Mesopotamia, the interpretation of omens and the elaboration of divinatory techniques called upon powers of rational analysis that, even if deficient by later standards, nevertheless had the effect of liberating logical thinking from the restricted sphere of the crafts. This was a significant development in the evolution of human consciousness, as it extended the application of logical thinking beyond the accomplishment of concrete practical tasks into a sphere in which, up until that time, clairvoyant vision or spiritual intuition had predominated. Now reasoned interpretation of the will of gods, manifested in various physical events and signs, assumed a paramount role in the exercise of priestly functions in ancient Mesopotamia and thereafter also in Greece. Although this expansion of the use of logic may seem a long way from the invention of the computer, it was a necessary step towards it.

Divination and omen reading continued for many hundreds of years to be an important part of both Greek and Roman religion. But with the birth of philosophy in ancient Greece, the scope of logical reasoning was further broadened to include the weighing of the truth of arguments conducted between human beings—not just between philosophers but also between disputants in the law courts and political assemblies. For the first time rules of logic were explicitly formulated. From classical Greece to medieval Europe, an understanding of these rules and skill in applying them formed the basis of Dialectic. Dialectic was cultivated as one of the seven Liberal Arts and regarded as one of the foundations of the development of the whole person. The well-rounded human being was someone able to think logically, and able to accept the rules of logic as the basis of correct reasoning, and hence as a guide to the truth. During this period logic was commissioned to serve a specifically human developmental goal, which was that individuals through their own powers of thought and reflection should understand the purpose of human existence in the greater cosmic order, and assume full moral responsibility for their actions.

Greater acuity in logical thinking had practical consequences too. It made it possible for mechanical inventions like the camshaft and the mechanical clock to come into existence. It also enabled unscrupulous people to deceive others by false arguments and to further their own ends by working out cunning schemes. Logic could be used both morally and immorally, and its double-edged nature was the cause of its being regarded by many with distrust and suspicion. Such was the threat posed by the ascendancy of logic at this time that within the philosophical and theological mainstream, from Plato and Aristotle in ancient Greece through to Thomas Aquinas in medieval Europe, logic was consistently positioned as the junior partner to the higher contemplative thinking, which steered it towards the pursuit of wisdom, truth and goodness. But while logic was held in subservience to contemplation, and technological advances were seen as of less value than spiritual advances to the wellbeing of humanity, its assiduous cultivation during this period was nevertheless an important precondition for the later development of the computer.

With the disintegration of the medieval consensus, which began in the fourteenth century, logic more and more acquired its own independent domain. We see this first of all in Ockham's conception of logic as a science of reasoning isolated from knowledge of reality. For more and more people, knowledge of reality was felt to derive from our perception of individual things, which were experienced as having no *intrinsic* intelligibility. The human mind increasingly experienced itself as outside of nature, and this served to undermine the previous view that contemplative thinking could gain access to a spiritual meaning intrinsic to creation. In subsequent centuries, those who sought to re-found knowledge on the basis of empirical observation and experiment regarded the contemplative aspect of thinking with more and more distrust. In a nominalist age, which had lost awareness of the world of spiritual archetypes, only the systematic and strictly logical analysis of accumulated data could be relied upon. For Francis Bacon, the careful sifting of research results through this kind of analysis amounted to a much needed 'new instrument' (a *novum organum*), for acquiring reliable knowledge. Made entirely of human minds working according to strict, rule-based procedures, this 'new instrument' was a prototype computer. Bacon was the first person to conceive the mechanization of reasoning as a realizable goal. His contribution was to give a powerful new impulse to the project to mechanize the mind.

In Bacon's successors, Hobbes, Descartes and Leibniz, reasoning became more and more identified with calculation, and logic increasingly came to be seen as having a greater affinity with mathematics than with spoken language. The invention of the first calculating machines during the seventeenth century provided an important stimulus to the shift towards the mathematization of logic. The notion that if logic could be captured in a mathematical form, it might be able to function independently of the human mind, was first clearly conceived by Leibniz towards the end of the century. However, the quest by Leibniz and others for an artificial language based on mathematics only began to be realized in the nineteenth century. This was crucial to the attempt to free logic from the human domain and introduce it into machines, but the introduction of logic into machines was not easily done, as Charles Babbage's failed attempt to build an Analytical Engine demonstrated. Babbage's attempt to build such an engine nevertheless showed that by the mid-nineteenth century logic was no longer harnessed to the pursuit of truth, for its role in the Analytical Engine was merely to ensure the validity of certain mechanical operations performed on collections of data. The achievement of Boole, Frege and others was to strip down logic to its barest machine-compatible components. When in the twentieth century electricity and logic were able to join forces, purely mechanical operations, tied into the binary code, could become electrically mediated. This then allowed a certain kind of machine 'intelligence' to come into existence based on the deployment of the stripped down logical operations that were now electrically replicated, enabling vast amounts of data to be processed very fast. And so the computer was born, concentrating within itself the results of a long historical struggle in the shadow of the machine, which had been cast into the human soul. While the fruits of this struggle now exist outside us in the innumerable devices that fill our world, the shadow has become established within the human soul, as a condition of human existence.

The prehistory of the computer is *our history*: it is the history of human consciousness. It is the history of the project to mechanize the mind, advancing slowly over many centuries, and often against determined and effective resistance from those who saw that it would entail the eclipse of higher cognitive faculties within the human being, receptive to the world of spirit. This project to mechanize the mind has now been partially accomplished, as we know every time we use a computer, for the computer is a 'mechanical mind'. But we must not make the mistake of regarding the computer as the locus of this project: the real locus is the *human* mind. While the exponential growth of processing power in computer technologies may seem to be the most recent stage of this long and complex history, this history is really about the changing relationship of human beings to their own inner life of thought.

Today, we have entered a period of profound uncertainty as to how human thinking differs from machine 'thinking'. We are repeatedly told that the human mind is just a biological computer, and that thinking is a physical activity of the brain, the product of neural networks whose main activity is a kind of data-processing.[10] We are told that a computer does what the brain does, and does it rather better than the human brain. But it is not the computer that is responsible for this assault on what it means to be human, it is the assault on what it means to be human that stands behind the development of the computer.

This assault began in earnest when Bacon, Hobbes, Descartes, Leibniz and others identified reason solely with logical thought, so that the higher reaches of the mind were disowned, disparaged and in time became increasingly forgotten. As a result the very notion of what it means to be human underwent a drastic contraction from how this was conceived in earlier times. If in antiquity and the Middle Ages, philosophers sought to articulate how human beings differ from animals, then we have arrived at a position today when it is imperative to articulate how we differ from machines. In pre-modern times, it was the danger of descending to the level of the beast, driven by untamed desires and wild passions, that seemed to pose the greatest threat to our humanity. Today this seems a lesser danger to the human essence than redefining (and then recreating) ourselves in the image of the machine. To avoid this danger, the compact we have forged with computers demands from us a complementary awareness of what it is within the human being that distinguishes us from them.

The Source of Human Freedom

While the impulse towards mechanization could be seen as a rigidifying impulse—a creeping sclerosis of the soul—we must also acknowledge that it had a powerfully liberating effect, for it freed human consciousness from the dominance of the gods. The gift of logic, which lay behind the mechanistic impulse, was that it transferred the fulcrum of consciousness from the dreamlike, feeling-imbued relatedness to the archetypal world, to an ever more wide-awake centre of awareness largely cut off from the archetypal realm.[11] Logic wakes us up. It is an essential component in our being free, autonomous agents. The ability to reason things out using logical thought-processes is crucial to the modern sense of freedom. And yet it is not the real source of freedom. For we are only truly free when we give to ourselves ideals, which then become the guiding principles of our actions. Thought must have a *content* if it is to make us free, and logic does not provide thought with any content.

If we regard the highest reach of the mind as extending no further than the ability to analyze and calculate, then we do not so much approach the source of our freedom as identify with an essentially empty and meaningless activity, for analysis and calculation alone do not give us access to meaning. Unless nourished and guided by another kind of thinking that introduces value and meaning to the content of thought, the human being is not fully present: the analysis and calculation takes place in a vacuum. And so we become La Mettrie's *L'homme machine*, and fall under the spell of an extraneous archetype. It is the archetype of the Inhuman— inhuman because the 'machine-man' has lost the capacity to think contemplatively, has lost contact with the real source of thought and has indeed lost contact with what is essentially human. A deeper, wiser, contemplative intelligence must be present alongside logical thought-processes and must act as their guide, for it alone recalls us to ourselves, and to our freedom.

Without this inner star to guide the discursive reason, we would be obliged to conceive any future enhancement of consciousness in terms merely of an increase of the speed and accuracy of computation, a conception that can be traced back to Hobbes and Leibniz. And since we have today out-sourced computational power to our digital devices, it is no surprise that there are many who see our human future as requiring a merger with machines that can compute far more rapidly and efficiently than we can. Neural implants have been on the agenda for many years, and there is every likelihood that the pressure to accept them will only grow stronger as this reduced view of what it means to be human becomes more and more prevalent.[12] Should this come about, the possibility of true freedom would be seriously jeopardized. Human consciousness would become ensnared in artificial neural networks, and a debased creature would arise, in which augmented computational efficiency would be gained at the cost of the ability to meaningfully reflect on the moral and spiritual dimensions of life. The challenge that confronts us, therefore, is to know from direct experience the root of our freedom, through our engaging in a wholly unconditioned activity, whose origin is within us and whose purpose is set by us.

It is this quality of unconstrained purposive activity, which we ourselves initiate, that is above all else the characteristic of an activity that is free. At its kernel is a contemplative thinking which is practised for its own sake, rather than as a means to some other end. Contemplative thinking calls us back to ourselves in an experience that is qualitatively different from when we are engaged in logical analysis or problem-solving. In contemplative thinking, we enter our own interior sacred space, which is characterized by stillness and receptivity towards an inspiration transcendent of the analysing, problem-solving mind. This is the inner ground on which the human being must stand in order to come into contact with what lives beyond the upper border. In the Bible it is referred to as a 'pondering in the heart', which it is said the Virgin Mary practised.[13] Pondering in the heart, the pondering mind is *in* the heart, and it is only when it takes its place there that we regain our connection with the spiritual world. When we centre ourselves in that deeper intelligence of the heart, we are then also centred in our essential freedom.

It is above all the inner quality of freedom that should guide us as we go towards a future in which it seems we shall be increasingly hemmed in on all sides by an ever more pervasive computer 'intelligence' which, while purporting to serve us, ties us into an ever greater conformity. For the quality of freedom belongs to every truly human act, and its value will be inestimable in a world under the dominion of the machine. Consider the following activities whose very meaning stems from the fact that they are rooted in human freedom: acts of compassion that spring from the selfless impulse to do good to another; creative visualization, through which we freely enter into imaginative contact with the archetypal realm; imaginative empathy through which we are able to participate in the inner life of another person or creature; and that profoundest of all human activities, prayer, in which the mind, grounded in the heart, turns towards its transcendent source. The meaning and purpose of each of these activities belongs entirely to the activities themselves, and does not reside in some external end product. But such activities nevertheless have a spiritual effect. This effect is to fecundate the world with impulses from the 'heavenly' sphere beyond the upper border that can balance and counteract humanity's descent towards the machine. Through engaging in such activities we attain our full stature as human beings, because each activity has its origin in the deeper contemplative intelligence, and in each we know with the certainty of direct experience that our action is unconditioned and free. To live from this deeper intelligence in ourselves, tending such capacities that so clearly raise us above what is merely mechanical, is the lesson that the prehistory of the computer teaches, and is one of the surest ways to secure a truly human future.

NOTES

Chapter One

1. Berthold Laufer, *The Prehistory of Aviation* (Chicago: Field Museum of Natural History, 1928), p. 10.
2. For a discussion of the rules of association in Mesopotamian liver divination, see Ullah Jeyes, 'The Act of Extispicy in Ancient Mesopotamia: An Outline' in *Assyriological Miscellanies*, vol. 1, ed. Bendt Alster (Copenhagen: University of Copenhagen Institute of Assyriology, 1980), p. 23f; and Ivan Starr, *The Rituals of the Diviner* (Malibu: Undena Publications, 1983), pp. 8–12.
3. Few people have understood this better than Owen Barfield, *Saving the Appearances: A Study in Idolatry* (London: Faber and Faber, 1957).
4. Plutarch, *De Isis et Osiride*, 70, in J. Gwyn Griffiths, *Plutarch's De Isis et Osiride* (Swansea: University of Wales Press, 1970), p. 228.
5. Henri Frankfort, *Kingship and the Gods* (Chicago: University of Chicago Press, 1978), p. 187.
6. Plutarch, *De Isis et Osiride*, 39. See also Frankfort, op. cit., p. 391.
7. Herodotus, *Histories*, II.60–63.
8. Heliodorus, *Aethiopica*, IX.ix. Author's translation.
9. See, for example, Samuel A. B. Mercer, *The Religion of Egypt* (London: Luzac, 1949), pp. 363–365; Frankfort, op.cit., pp. 185–194.
10. Frankfort, op. cit., p. 314.
11. Frankfort, op. cit., p. 316.
12. Hammurabi's law code, dating from the eighteenth century BC, was by no means the first as there are similar documents from the reign of Ur-Nammu (c. 2100 BC) and his successors, for which see George Roux, *Ancient Iraq* (Harmondsworth: Penguin, 1980), p. 190ff.
13. For Mesopotamian diviners as 'scholar priests', see Ulla Jeyes, *Old Babylonian Extispicy: Omen Texts in the British Museum* (Istanbul: Nederlands Historisch-Archaeologisch Instituut, 1989), pp. 14–16 and p. 20. See also Leo Oppenheim, *Ancient Mesopotamia: Portrait of a Dead Civilization* (Chicago: University of Chicago Press, 1977), p. 214, who observes:

    > divination in Mesopotamian civilization [was] not only ... an essential means of orientation in life but also ... an area for the display of intellectual endeavours and aspirations.

14. It is interesting to note that the pulley was never adopted in ancient Egypt, despite its use by the Assyrians from at least the eighth century BC. See Brian Cotterell and Johan Kamminga, *Mechanics of Pre-industrial Technology* (Cambridge: Cambridge University Press, 1990), p. 89.
15. Richard Wilkinson, *Symbol and Magic in Egyptian Art* (London: Thames and Hudson, 1994). As well as the symbolism of materials, Wilkinson also discusses the symbolism of form, size, location, colour and so on, all of which strongly influenced the products of ancient Egyptian craftsmanship.

16. Eugen Strouhal, *Life in Ancient Egypt* (London: Opus Publishing Ltd., 1992), Chapter 11. The bow-drill was not employed until the New Kingdom.

Chapter Two

1. It is a later version of Coffin Text 335. See T. G. Allen, *Occurrences of Pyramid Texts* (Chicago: University of Chicago Press, 1951), p. 138.
2. R. O. Faulkner, *The Ancient Egyptian Book of the Dead* (London: British Museum Publications, 1985), p. 44. These statements of the sun god belonged to a tradition going back roughly 700 years before the Book of the Dead papyri, for which see R. T. Rundle Clark, *Myth and Symbol in Ancient Egypt* (London: Thames and Hudson, 1978), p. 78f.
3. According to Rundle Clark, op. cit., p. 78, the underlying theme of Chapter 17 of the Book of the Dead is that of inner transformation. It is:

   a compendium of the stages through which the soul had to pass from the time it was liberated during the lustrations at the funeral to the time when it emerged from the Underworld to join the sun.

   See also James Wasserman, ed., *The Egyptian Book of the Dead*, translated by R. O. Faulkner with an introduction and commentary by Ogden Goelet (San Francisco: Chronicle Books, 1994), p. 159, for commentary on this passage.
4. H. and H. A. Frankfort, 'Myth and Reality' in Henri Frankfort et al., *Before Philosophy* (Harmondsworth: Penguin, 1949), p. 19, make the following remark concerning the thinking of the ancients:

   They could reason logically; but they did not often care to do it. For the detachment which a purely intellectual attitude implies is hardly compatible with their most significant experience of reality.

5. See H. Te Velde, *Seth, God of Confusion* (Lieden: E. J. Brill, 1967), pp. 5–7. The main hieroglyphs associated with Seth all speak of division: the curious animal hieroglyph, with its forked tail, is shown here as a determinitive for the word for 'storm', above the sky hieroglyph from which rain falls:

Fig. 2.11

Another common hieroglyph, discussed in Te Velde, op. cit., p. 31, is probably some kind of cutting implement, which is divided at the bottom and carries the meaning of separation or severance:

Fig. 2.12

A third hieroglyph associated with Seth is the *was* scepter, which R. A. Schwaller de Lubicz, *The Temple of Man* (Rochester, Vt.: Inner Traditions International, 1998), vol. 2, p. 1001, has shown is based on the natural division of plant growth reversed. It is as if the Sethian force is the very antithesis of the generative forces of life. In Figure 2.13, the naturally branching stem on the left is turned upside down to produce three different forms of *was* scepter, with the head of Seth clearly defined on the one on the far right:

Fig. 2.13

For the hieroglyphs, see Alan Gardiner, *Egyptian Grammar* (Oxford: Oxford University Press, 1973), pp. 442–542 (E20, Aa21 and S40).

6. Thus Dumuzi's dreams were interpreted by Geshtinana, those of Gilgamesh by his mother, and the famous dream of Gudea by the priestess of Nanshe. For 'The Dream of Dumuzi', see Diane Wolkstein and Samuel Noah Kramer, *Inanna: Queen of Heaven and Earth* (London: Hutchinson, 1984), pp. 74–84; for Gilgamesh's dreams, see Alexander Heidel, *The Gilgamesh Epic and Old Testament Parallels* (Chicago: The University of Chicago Press, 1949), pp. 23–27: Tablet 1, column 5; Tablet 2, column 1. For Gudea and the priestess of the goddess Nanshe, see Samuel Noah Kramer, *The Sumerians: Their History, Culture and Character* (Chicago: University of Chicago Press, 1963), p. 138f.

7. For the *sha-ilu*, see A. Leo Oppenheim, *The Interpretation of Dreams in the Ancient Near East*, Transactions of the American Philosophical Society, vol. 46 (Philadelphia: The American Philosophical Society, 1956), p. 221. For the *barû*, see Ulla Jeyes, *Old Babylonian Extispicy*. pp. 14–20.

8. O. R. Gurney 'The Babylonians and Hittites' in Carmen Blacker and Michael Loewe, eds., *Divination and Oracles* (London: George Allen and Unwin, 1981), p. 148f.

9. Morris Jastrow, *Religious Belief in Babylonia and Assyria* (New York: Benjamin Blom, 1911), p. 183.

10. For the earlier Sumerian priestess being a medium, see O. R. Gurney, op. cit., p. 145. For the significance of this whole development, see also Jeremy Naydler, *The Future of the Ancient World: Essays on the History of Consciousness* (Rochester Vt: Inner Traditions, 2009), Chapter 9.

11. Charles Singer, E. J. Holmyard and A. R. Hall, eds., *A History of Technology*, vol. 1 (Oxford: Oxford University Press, 1954), p. 524.

12. Martin Gardner, *Logic Machines and Diagrams* (Brighton: Harvester Press, 1982), p. 116.

13. Eugen Strouhal, *Life in Ancient Egypt*, p. 97. The earliest evidence of the use of a *shaduf* in ancient Egypt comes from the eighteenth dynasty tombs of Neferhotep and Merire II at Akhetaten, and from a number of nineteenth dynasty Theban tombs, for which, see Strouhal, op. cit. p. 97.

14. Mark Lehner, *The Complete Pyramids* (London: Thames and Hudson, 1997), p. 206.

15. In the famous Middle Kingdom 'Hymn to Hapy', translated in Miriam Lichtheim, *Ancient Egyptian Literature*, vol. 1 (Berkeley: University of California Press, 1975), p. 208, we read:

> Songs of the harp are made for you,
> One sings to you with clapping hands . . .
> When you overflow, O Hapy
> sacrifice is made for you;
> oxen are slaughtered for you,
> a great oblation is made for you.

16. Singer, Holmyard and Hall, op. cit., p. 211.
17. Ibid., p. 209.
18. For the early history of the wheel, see Singer, Holmyard and Hall, op. cit., pp. 200–212.
19. For the wheel as a form of lever, see Sigvard Strandh, *A History of the Machine* (London: Hutchinson, 1984), p. 34.
20. Henri Frankfort, *Kingship and the Gods*, p. 3, referring to the conception of kingship in the ancient Near East, states:

> The purely secular—in so far as it could be granted to exist at all—was purely trivial. Whatever was significant was imbedded in the life of the cosmos, and it was precisely the king's function to maintain the harmony of that integration.

For Mesopotamia, see Frankfort, *Kingship and the Gods*, pp. 251–258; for Egypt, p. 51f. Also Erik Hornung, *Idea into Image: Essays on Ancient Egyptian Thought* (New York: Timken Publishers, 1992), Chapter 7.

21. Owen Barfield, *Saving the Appearances*, p. 109f.
22. Thorkild Jacobsen, *The Treasures of Darkness* (New Haven and London: Yale University Press, 1976), p. 73.
23. Wolkstein and Kramer, *Inanna*, p. 71.
24. Frankfort, *Kingship and the Gods*, p. 326:

> They placed a garment in their midst
> And said to Marduk, their firstborn:
> 'O Lord, thy lot is truly highest among gods.
> Command annihilation and existence, and may both come true . . .'

25. See Jacobsen, op. cit. p. 129, where we learn how Ninurta captured Imdugud in a battle in the mountains and hitched the lion-bird to his war chariot and drove it across the skies. For Inanna's control and mastery of Imdugud, see Jacobsen, op. cit., p. 136.
26. Rudolf Steiner, *The Problem of Faust*, Lecture 1, 30 September, 1916, Dornach (unpublished typescript, p. 27f), refers to the way in which knowledge of electricity and other such forces was guarded in antiquity as follows:

> The knowledge connected with these Mysteries, these secrets of nature, did not consist merely of concepts, ideas and feelings, nor merely of dogmatic imaginations. Whoever wished to acquire it had first to show himself wholly fitted to receive it; he had to be free from any wish to employ the knowledge selfishly; he was to use both knowledge and the ability derived from it solely in the service of the social order.

27. Thus, in the Old Babylonian (early second millennium) *Epic of Anzu*, Tablet 1.3, lightning is described as Adad's 'weapon' and as an 'arrow' of the gods (Tablet 2), for which see Stephanie Dalley, *Myths from Mesopotamia* (Oxford: Oxford University Press, 1991), p. 207 and p. 213. In the *Epic of Creation*, Tablet 4.39, Marduk advances against Tiamat on his storm chariot, with 'lightning in front of him'. See Dalley, op. cit., p. 251. In numerous portrayals, both on cylinder seals and in temple reliefs, lightning is imagined as a deadly thunderbolt that the raging storm god or goddess hurls at his or her opponents.

28. See p. 12 above, with n. 5. For Seth as god of destruction, illness, strife and confusion, see H. Te Velde, *Seth, God of Confusion*, p. 25.

29. See Jeremy Naydler, *Temple of the Cosmos: the Ancient Egyptian Experience of the Sacred* (Rochester, Vt.: Inner Traditions International, 1996), pp. 73–78.

30. In *The Book of the Dead*, Chapter 39, Seth proclaims:

> I am Seth, who causes confusion and thunders in the horizon of the sky; I am one in whose heart is destruction. [Author's translation]

See Te Velde, op. cit., p. 85 for a discussion of this passage. See also n. 5 above.

31. As in the Sed Festival rite of 'Giving the Head', for which see Eric Uphill, 'The Egyptian Sed-Festival Rites', *Journal of Near Eastern Studies*, 24 (1965), pp. 378–380. See also Jeremy Naydler, *Shamanic Wisdom in the Pyramid Texts: The Mystical Tradition of Ancient Egypt* (Rochester, Vt: Inner Traditions, 2005) pp. 76–78.

Chapter Three

1. Rosalind Thomas, *Literacy and Orality in Ancient Greece* (Cambridge: Cambridge University Press, 1992), p. 3. See also W. V. Harris, *Ancient Literacy* (Cambridge, Mas: Harvard University Press, 1989), p. 328, who estimates the literacy rate in classical Greece to have been no more than 5 per cent of the population.

2. John Herington, *Poetry into Drama: Early Tragedy and the Greek Poetic Tradition* (Berkeley: University of California Press, 1985), Chapter 1.

3. Herington, op. cit., p. 3f. See also Bruno Gentili, *Poetry and its Public in Ancient Greece* (Baltimore: The Johns Hopkins University Press, 1988), p. 3ff.

4. Herington, op. cit., p. 96.

5. Plato, *Ion*, 535d.

6. Gentili, op. cit., p. 10f. See also Bennett Simon, *Mind and Madness in Ancient Greece* (Cornell University Press, 1978), p. 83, who, commenting on the immersion of both the rhapsode and audience in the poem, writes:

> In a sense, the experience of audience and poet can be considered an artistically controlled blurring of boundaries of the self or as a series of transient experiences of merging with the poem and its characters.

7. Penelope Murray, *Plato on Poetry* (Cambridge: Cambridge University Press, 1996), p. 8 and p. 110.

8. F. M. Cornford, *Principium Sapientiae: A Study of the Origins of Greek Philosophical Thought* (Gloucester Mass: Peter Smith, 1971), p. 104:

> The poets who composed the theogonies and cosmogonies of the post-heroic age—those 'theologians' in whom Aristotle recognized the precursors of the natural philosophers—would all have claimed, with Hesiod, that they were uttering revealed truth.

See also J-P. Vernant, *Myth and Thought Among the Greeks* (London: Routledge and Kegan Paul, 1983), p. 354, according to whom Mnemosyne (the Muse of memory) was a goddess who 'gives the poet—like the diviner—the privilege of beholding unchangeable and permanent reality; she brings him into contact with the primal being, a mere infinitesimal fraction of which is revealed to human beings in the march of time, and then only to be veiled once again.'

9. Cornford, op. cit., p. 111. The 'thinking soul' translates *to logon echon*.

10. The origin of the universe was seen by Pythagoras as due to the interaction of a limiting principle with the principle of the Unlimited, both implicit within the One. In a movement of contraction away from the One, the Unlimited is drawn in or 'breathed in' by the limiting principle, and in so doing the heavens are formed. With this initial contraction and separation, the Dyad emerges from the Monad as a principle of differentiation and as the origin of duality. Through the Dyad the opposites of Odd/Even, Right/Left, Male/Female and so on enter into manifestation. The reconstruction of Pythagoras' cosmogony rests largely on the surviving fragments of Philolaus (second half of the fifth century BC), and the doxographical tradition that goes back to Aristotle (e.g. *Physics*, IV.6, 213b:22–23) and Theophrastus. For Philolaus, see Charles H. Kahn, *Pythagoras and the Pythagoreans: A Brief History* (Indianapolis: Hackett Publishing Company, 2001), Chapter 3. For the doxographical tradition, see G. S. Kirk and J. E. Raven, *The Presocratic Philosophers* (Cambridge: Cambridge University Press, 1957), p. 5. From a relatively early date, it seems that the Dyad was regarded as a principle of evil, as this is how Aetius refers to it in his summary of Theophrastus' now lost history of early Greek philosophy. In this characterization of the Dyad, we may detect echoes of an earlier Mystery teaching. For Aetius, see W. K. C. Guthrie, *History of Greek Philosophy*, vol. 1 (Cambridge: Cambridge University Press, 1962), p. 248.

11. Fr. 2, quoted in Proclus *Commentary on the Timaeus I*, 345, translated in G. S. Kirk and J. E. Raven, *The Presocratic Philosophers* (Cambridge: Cambridge University Press, 1957), p. 269. See also Jonathan Barnes, *Early Greek Philosophy* (London: Penguin, 1987), p. 132.

12. Oswyn Murray, *Early Greece* (London: Fontana, 1980), p. 211 comments that craftsmanship never became socially respectable in early Greece.

13. H. G. Liddell and R. Scott, *Greek-English Lexicon*, revised and augmented throughout by H. S. Jones (Oxford: Oxford University Press, 1968), p. 1785a.

14. Homer, *Odyssey*, 1.205. Translation by A. T. Murray (London: Harvard University Press, 1995), p. 27.

15. Odysseus has *mêtis*, 'skill' or 'cunning' (e.g. *Odyssey* 2.279), and is referred to as *poikilomêtês*, 'full of various wiles', as in *Odyssey* 3.163.

16. Ibid., 9.177–566.

17. Ibid., 9.408. The Greek word for 'guile' here is *dolos*, which also means a trick or treachery. 'Force' translates *bia*, which could also be rendered 'violence' or 'brute strength.'

18. Ibid., 9.414. 'Flawless scheme' translates *mêtis amumôn*.

19. Steve Talbott, *Devices of the Soul* (Sebastopol, CA: O'Reilly Media, 2007), pp. 3–8 has interpreted it in this way, and to his penetrating insights this section is deeply indebted.

20. See Jane Harrison, *Myths of the Odyssey in Art and Literature* (London:Rivingtons, 1882), pp. 24–30, for a comparison of the positive light in which the Cyclopes were

regarded by Hesiod and others, as compared with the Homeric and post-Homeric view of them as monsters and demons.

21. The Greek verb *odussasthai* means both 'to suffer pain' and 'to cause pain'. In *Odyssey*, 1.62 it is implied that it is Odysseus' fate to suffer or receive pain from the gods. Elsewhere, the meaning of his name could equally be interpreted as one who causes pain to others, as in *Odyssey*, 19.405–409.

Chapter Four

1. Liddell and Scott, *Greek-English Lexicon*, p. 1785a.
2. Homer, *Odyssey*, 1.205, p. 27.
3. Ibid., p. 445f. For example, *mêchanaomai* meant both 'to make by art, to construct,' and equally 'to contrive, to scheme', The adjective *mêchanikos* meant both 'inventive' and 'clever'.
4. Marchant and Charles, *Cassell's Latin Dictionary* (London: Cassell, 1915), p. 328.
5. Owen Barfield, *History in English Words* (Edinburgh: Floris, 1985), p. 185. The noun 'machination' and the verb 'to machinate' are still used today.
6. Hesiod, *Works and Days*, 926–928.
7. Robert Flaceliere, *Greek Oracles* (London: Elek Books, 1965), p. 8f.
8. This had become all too apparent by the mid-fifth century BC. The new, anti-divinatory consciousness is well exemplified by Anaxagoras, for whom physical phenomena should be explained by physical causes alone. See Diogenes Laertius, *Lives of the Philosophers*, 2.6, quoted in J. Barnes, *Early Greek Philosophy* (London: Penguin, 1987), p. 237. See also Plutarch, 'Life of Pericles' §6 in *The Rise and Fall of Athens*, trans. Ian Scott-Kilvert (Harmondsworth: Penguin, 1960), p. 170, who gives a dramatic account of Anaxagoras' challenge to the practice of divination. Anaxagoras denied that the malformation of a ram's horn could have any bearing on the outcome of the political contest between Pericles and Thucydides (as had been claimed), but was due entirely to natural causes.
9. For Xenophanes, see Kirk and Raven, *The Presocratic Philosophers*, p. 168f.
10. This view has also been attributed to Euripides. See Walter Burkert, *Greek Religion* (Oxford: Basil Blackwell, 1987), pp. 314–315.
11. See Martin Nilsson, *Greek Folk Religion* (New York: Harper, 1961), p. 135.
12. Rudolf Steiner, *The Riddles of Philosophy* (New York: The Anthroposophic Press, 1973), p. 18:

> Connected with this circumstance is the fact that the birth of thought life brought with it a shattering of the foundations of the inner feelings of the soul. This inner experience should not be overlooked in a consideration of the time when the intellectual world conception began.

See also Steiner, op. cit., p. 30.
13. F. M. Cornford, *Principium Sapientiae: A Study of the Origins of Greek Philosophical Thought* (Gloucester Mass: Peter Smith, 1971), p. 135ff. See also Eduard Zeller, *Outlines of the History of Greek Philosophy* (New York: Dover, 1980), p. 105.
14. Werner Jaeger, *Paideia*, vol. 1 (Oxford: Oxford University Press, 1965), p. 298ff.
15. Plato, *Euthydemus*, 284A–D and 298D–E. The arguments, briefly stated, are (1) 'No-one can speak about that which is not; to lie is to speak about that which is not; therefore no one can lie,' and (2) 'That dog is a father; that dog is yours; therefore that dog is your father.' While to us such reasoning may seem puerile, at the time

when Plato lived no one had yet had to contend with arguments such as these, and it was not immediately apparent where the fallacy in the sequence of thoughts lay. It was only with Aristotle's articulation of the rules of syllogistic reasoning that this became clear.

16. As Plato says in the *Republic* (London: William Heinemann, 1942), 505A:

> And if we do not know it [the Good] then, even though we should know all other things perfectly, you must understand that such knowledge would be of no benefit to us, just as no possession is of any benefit without the possession of the Good.

To put across this single point, Plato devotes a whole dialogue—the *Lesser Hippias*. See also Robert Earl Cushman, *Therapeia: Plato's Conception of Philosophy* (New Brunswick, NJ: Transaction Publishers, 2002), p. 69.

17. Aristotle, *Nicomachean Ethics* 1140a11.
18. Aristotle, *Metaphysics*, 981a–982a and 1003a.
19. Fr. 2, quoted in Proclus *Commentary on the Timaeus I*, 345, translated in Kirk and Raven, *The Presocratic Philosophers*, p. 269. See also Jonathan Barnes, *Early Greek Philosophy* (London: Penguin, 1987), p. 132.
20. William and Martha Kneale, *The Development of Logic* (Oxford: Oxford University Press, 1962), p. 14f, make the following comment:

> It is significant that in the Peripatetic tradition, which throughout its history bears traces of its Platonic origin, logic never became a part of philosophy, a subject in its own right, but was treated as a capacity (*dunamis*) which might be acquired or as an art (*technê*) to be learnt.

21. For the Law of Identity, see Aristotle, *Metaphysics*, IV.iv.10–11 (1006b, 9–11):

> ... for not to have one meaning is to have no meaning; and if words have no meaning there is an end to discourse with others; because it is impossible to think of anything if we do not think of one thing; and even if this were possible, one name might be assigned to that of which we think. Now let this name, as we said at the beginning, have a meaning; and let it have *one* meaning.

For the Law of Contradiction, see Aristotle, *Metaphysics*, IV.iii.9f; 1005b.19–21, where he states that it is 'the most certain of all principles' and defined it as follows:

> It is impossible for the same attribute at once to belong and not to belong to the same thing and in the same relation ...

For the Law of the Excluded Middle, see *Metaphysics*, IV.vii.1f; 1011b.23–25, where Aristotle states:

> Nor indeed can there be any intermediate between contrary statements, but of one thing we must either assert or deny one thing, whatever it may be.

22. 'The third that is not given' is what falls between, and actually transcends, the two parts of a contradiction, as Jung never tired of pointing out. For the ancient consciousness of the divine in nature, see Jeremy Naydler, *The Future of the Ancient World* (Rochester, Vt: Inner Traditions, 2009), pp. 190–196.
23. Aristotle identified three *figures* of the syllogism: in the first figure, the major premise is always universal and the minor premise is always affirmative (as for example in 'All men are mortal; Socrates is a man ...' which leads to the conclusion that Socrates is

mortal). In the second figure, the major premise is always universal but the minor premise contradicts it, leading to a negative conclusion (as for example in 'Whoever has measles has spots; this child has no spots; therefore this child does not have measles'). In the third figure the minor premise is always affirmative and the conclusion is always particular (as for example in 'Some Athenians are wise; all Athenians are Greeks; therefore some Greeks are wise.') The different *moods* of the syllogism are determined by the type of statement made in each of the premises and the conclusion: whether it is universal affirmative ('All...'), universal negative ('None/ No...'); particular affirmative ('This/Some...') or particular negative ('This is not/ Some are not...').

24. Aristotle, *Prior Analytics*, 25b, 32–35. In the Middle Ages this came to be known as the *dictum de omni et nullo* (the maxim concerning everything and nothing). The first part of the *dictum* (*de omni*) is the principle that what is universally affirmed of a whole is also affirmed for every part of that whole. In our example, if all horses have four legs, and this animal is a horse, then this animal has four legs. The second part of the *dictum* (*de nullo*) is the principle that whatever is universally denied of a whole is also denied of any part of that whole. In our example, if all horses have four legs, and this animal does not have four legs, then this animal is not a horse.

25. Plato, *Republic*, 577D. See also *Republic*, 611C–612A.
26. Aristotle, *Nicomachean Ethics*, 1113a17.
27. Plato, *Timaeus*, 90B–C.
28. Plato, *Timaeus*, 90C.
29. Plato, *Republic*, 511D.
30. Aristotle, *Metaphysics* XI.3.8, 1061b.
31. Aristotle, *Nicomachean Ethics*, 10.7.8.
32. Plato, *Republic*, 507D–509B.
33. Plato, *Republic* 517C,

> [The Idea of the Good] is indeed the cause for all things and of all that is right and beautiful, giving birth in the visible world to light and the author of light [the sun], and itself in the intelligible world being the authentic source of truth and reason (*nous*).

34. Aristotle, *On the Soul*, 418b, 14–15.
35. Aristotle, *On the Soul*, 430a, 14–25. See also *Metaphysics*, XII.7.8, 1072b.
36. For 'first activity' see Plotinus, *Enneads* V.3.12; for the One as the Good, see *Enneads*, VI.9.3; for 'generative radiance', *Enneads* VI.7.36; for 'formative principles and archetypes', *Enneads* I.6.3.
37. Plotinus, *Enneads* V.3.17. For Plotinus' 'metaphysics of light' see F. M. Schroeder, *Form and Transformation: A Study in the Philosophy of Plotinus* (Montreal/Kingston, Ontario: McGill-Queen's University Press, 1992), pp. 24–39.

Chapter Five

1. Archimedes, *On the Equilibrium of Planes*, Postulate 1, in Morris R. Cohen and I. E. Drabkin, *A Sourcebook in Greek Science* (Cambridge, Massachusetts, 1948), p. 186.
2. Plutarch, *Life of Marcellus*, 17:

> He [Archimedes] would not deign to leave behind him any commentary or writing on such subjects; but repudiating as sordid and ignoble the whole trade of engineering, and every sort of art that lends itself to mere use and profit, he placed

his whole affection and ambition in those purer speculations where there can be no reference to the vulgar needs of life . . .

3. Vernant, *Myth and Thought Among the Greeks* (London: Routledge and Kegan Paul, 1983), p. 282ff. As Vernant explains (p. 285), from the point of view of the ancient Greek scientist,

> reasoning remained rigorous so long as it did not venture beyond the realm of pure theory. A concern for effectiveness and for taking purely technical details into account belongs to a different type of thought operating on a different level. . . . As one gets closer to physical reality theory loses its rigour and ceases indeed to be theory. It is not applied to but rather becomes debased in concrete facts.

See also Charles Singer et al., *A History of Technology*, vol. 2 (Oxford: Oxford University Press, 1956), p. 604:

> Through the study of nature they sought the wisdom and intellectual satisfaction which each free citizen should endeavour to acquire in his leisure (*otium*). Application to the mechanical arts (*neg-otium*) was definitely inferior; by exercising their trade the *banausoi* (craftsmen) killed the spirit.

4. Plutarch, *Life of Marcellus*, 16.
5. Vitruvius' description of the watermill in *De Architectura* 10.5 is extremely brief and one has the impression that it was a rather rare phenomenon. See Terry S. Reynolds, 'The Medieval Roots of the Industrial Revolution', *Scientific American*, 251 (July 1984), p. 109, col. 2. For the evidence for watermills in the Roman period, see Örjan Wikander, 'The Water-Mill' in Örjan Wikander, ed., *Handbook of Ancient Water Technology* (Leiden: Brill, 2000), pp. 371–400.
6. For the number of authenticated watermill sites, see Örjan Wikander, 'The Water-mill' in op. cit., p. 372. For industrial applications, see Örjan Wikander, 'Industrial Applications of Water-Power' in op. cit. p. 401–410.
7. Örjan Wikander, 'Industrial Applications of Water-Power' in op. cit. p. 401–410.
8. Jan Koster, 'From Carillon to IBM: The Musical Roots of Information Technology' published online, 2003, at <http://www. let.rug.nl/koster>
9. A useful overview is given in Adam Robert Lucas, 'Industrial Milling in the Ancient and Medieval Worlds' in *Technology and Culture*, 46 (January 2005), pp. 7–11.
10. Strandh, *A History of the Machine*, p. 37.
11. Henry Hodges, *Technology in the Ancient World* (Harmondsworth: Penguin, 1971), p. 183.
12. Ibid.
13. Commenting on the relative slowness of the spread of watermills, R. J. Forbes and E. J. Dijksterhuis, *A History of Science and Technology*, vol. 1 (Harmondsworth: Penguin, 1963), p. 76, write:

> Many ancients believed the forces of nature to be the domain of supernatural power and saw in their harnessing an act of blasphemy.

A similar explanation is given by R. J. Forbes, 'Power', in Singer, et al., *A History of Technology*, vol. 2, p. 606, where the conversion of the Roman Empire to Christianity is seen as a decisive factor in bringing a more open attitude towards the more widespread adoption of technologies:

The ancient world did not dream of man's harnessing these supernatural powers until Christianity, by its opposition to animism, opened the door to a rational use of the forces of nature.

14. 'Cease from grinding, ye women who toil at the mill; sleep late, even if the crowing cocks announce the dawn. For Demeter has ordered the nymphs to perform the work of your hands, and they, leaping down on the top of the wheel, turn its axle which, with its revolving spokes, turns the heavy concave Nisyrian mill-stones.' Quoted in R. J. Forbes, 'Power', in Singer, et al., *A History of Technology*, vol. 2, p. 593.

15. Vernant, op. cit., p. 298f., n. 28, makes the following observation concerning Hero's *Baroulkos*:

> Now, in his *Baroulkos*, Hero emphasizes that, at the base of all the difficulties in mechanical problems and of the obscurity that surrounds the investigation of causes in this science, lies the fact that one cannot actually see the forces at work in heavy bodies nor how they are distributed. Since these forces belong to the sphere of the invisible, it is inevitable that *logismos*, reasoning, is the dominating force in mechanics.

16. R. J. Forbes and E. J. Dijksterhuis, *A History of Science and Technology*, vol. 1, p. 58.
17. E. J. Dijksterhuis, *The Mechanization of the World Picture* (Oxford: Oxford University Press, 1961), p. 74:

> The slight development of technology directed at a useful economic end is undoubtedly connected with the institution of slavery, which prevailed all through antiquity and was considered perfectly normal even by thinkers such as Plato and Aristotle ... The fact that living machines could be used at will no doubt diminished the need for inanimate implements, the more so because no humanitarian considerations formed a motive for calling in the aid of such implements. If machines, however, were deemed superfluous because there were slaves ... a vicious circle must have been created: for want of machines the slaves could not be dispensed with.'

The author continues:

> It may be left an open question whether manual labour was held in contempt because it was performed by slaves, or was left to the slaves because it was considered inferior, for here again a fatal interaction probably took place. It is, however, a fact that whatever called for purely mental work (the later *artes liberales*) was regarded by society as greatly superior to handicraft, whether aided by machines or not, to the trades, and to mechanical engineering, and to which the plastic arts may also be added. The *artes mechanicae* do not befit the free Hellene: any activity bringing man into too close a contact with matter has a degrading effect.

18. Cicero, *De Republica*, 1.22.
19. Cicero claimed (*De Republica*, 1.22) that the tradition went back as far as Thales, but this seems unlikely.
20. T. Freeth et al., 'Decoding the ancient Greek astronomical calculator known as the AntiKythera Mechanism' in *Nature*, vol. 444, Issue 7119 (2006), pp. 587–591.
21. R. J. Forbes, 'Power', in Singer, et al., *A History of Technology*, vol. 2, p. 606.
22. E. J. Dijksterhuis, *The Mechanization of the World Picture*, pp. 54–68.

Chapter Six

1. Rudolf Steiner, *Faust and the Mothers* (lecture given on 2 November, 1917, Dornach. Unpublished typescript).
2. W. K. C. Guthrie, *The Greeks and Their Gods* (Boston: Beacon Press, 1955), p. 50. For the chthonic Zeus as 'subterranean counterpart to the sky father', a role which continued into later times, see Walter Burkert, *Greek Religion* (Oxford: Blackwell, 1985), p. 200f.
3. Porphyry is the source of this story, which is discussed along with the thunder-rites of Zeus in Jane Harrison, *Themis: A Study of the Social Origins of Greek Religion* (London: Merlin Press, 1977), pp. 56–60.
4. Harrison, op. cit., p. 61. According to Harrison, the vision was simply of a 'thunder stone', but there must surely also have been some simulation of the lightning flash.
5. Probably having passed first through a Cretan phase. See F. M. Cornford, *Principium Sapientiae* (New York: Harper and Row, 1965), Chapt. 15.
6. Hesiod, *Theogony*, 140. The names of the three Cyclopes were 'Thunder' (Brontes), 'Lightning' (Steropes) and 'Flash' (Arges).
7. Hesiod, *Theogony*, 501–506. In bestowing their powers on Zeus, the Cyclopes effectively put themselves under his command (for they were identical to their powers). This is comparable to Ninurta, Inanna and Marduk's mastery of the lion-bird Imdugud, with the obvious difference that the Cyclopes gave their allegiance freely.
8. *Theogony*, 119.
9. Hesiod, *Theogony*, 727–728. For Hesiod's distinction between Gaia and *chthôn*, see H. J. Rose, *A Handbook of Greek Mythology* (London: Methuen, 1964), p. 19. It is interesting to note that according to Rudolf Steiner, *Faust and the Mothers*, 'the being of electricity ... definitely does not belong to the earth'.
10. According to Hesiod, *Theogony*, 721–728:

    A bronze anvil falling down from the sky
    Would fall nine days and nights and on the tenth would reach earth (*gaia*).
    It is just as far from earth down to murky Tartarus.
    A bronze anvil falling down from earth
    Would fall nine days and nights and on the tenth reach Tartarus.
    There is a bronze wall beaten round it, and Night
    In three circles flows round its neck, while above it grow
    The roots of earth (*ge*) and the barren sea.

11. Carl Kerényi, *Eleusis: Archetypal Image of Mother and Daughter* (Princeton: Princeton University Press, 1967), pp. 83–94.
12. Sophocles has Oedipus explain that what he is experiencing is 'a holy mystery that no tongue may name' (*Oedipus at Colonus*, line 1525). Hermes (as psychopomp) then leads him down to the Underworld (*Oedipus at Colonus* lines 1547ff). See Sophocles, *The Theban Plays* translated by E. F. Watling (Harmondsworth: Penguin, 1947), pp. 115–118. See also Kerényi, op. cit., p. 85.
13. Mircea Eliade, *The Forge and the Crucible*, p. 102f. Eliade connects these rites with both the earlier Cretan Mysteries already discussed.
14. Walter Burkert, *Greek Religion*, p. 283f.
15. Park Benjamin, *A History of Electricity (The Intellectual Rise in Electricity): From Antiquity to the Days of Benjamin Franklin* (New York: John Wiley, 1898), p. 35.

16. Sources of the myth are given in Robert Graves, *The Greek Myths* (Harmondsworth: Penguin, 1955), vol. 1, 42.d, p. 156.
17. Guthrie, *The Greeks and Their Gods*, pp. 40–51, reminds us that originally not only was Zeus addressed as 'Greatest of the Kuroi' just as was Dionysus, but was also known by the epithet 'Bacchus', and his worshippers were called 'Bacchoi'. Like Dionysus, those who identified with Zeus originally underwent a ritual death and rebirth as part of his mystery cult, and devoured his sacred animal.
18. Kirk and Raven, *The Presocratic Philosophers*, pp. 104–117.
19. Quoted in Kirk and Raven, op. cit., p. 138.
20. Anaximenes, for instance, is said to have adopted the same explanation of thunder and lightning as Anaximander. See Kirk and Raven, op. cit., p. 158.
21. The key principles that Aristotle expounds are that electricity arises through the tension between polarized moistness and warmth on the one hand, and of the forces of levity being overcome by those of gravity on the other. On the former, see Aristotle, *Meteorologica*, I.4 (341b: 6–22) and II.4 (359b: 28–360a: 27). See also D. Ross, *Aristotle* (London: Methuen, 1971), p. 109f. On the latter, see *Meteorologica*, II.9 (369a: 22–25). Briefly, Aristotle understood the origin of lightning in terms of a 'dry exhalation' from the earth rising up and expanding into the sky but being caught in the moist condensing clouds, which trap it and overwhelm it, ejecting it as lightning as it is forced hurtling back towards the ground. In the interaction between the two qualities of dryness and moistness, and the forces of expansion (levity) and contraction (gravity), one can catch an echo of Zeus the sky god and Zeus Chthonios.
22. Hans Daiber, 'The Meteorology of Theophrastus in Syriac and Arabic Translation' in William Wall Fortenbaugh and Dimitri Gutas, *Theophrastus: His Psychological, Doxographical and Scientific Writings* (New Brunswick, NJ: Translation Publishers, 1992), p. 237f.
23. The Stoics, for example, understood thunder to be the clashing of clouds together, which also produces the fiery flash of lightning. See Aetius, *Placita Philosophorum*, in Pseudo-Plutarch, *Sentiments Concerning Nature With Which Philosophers Were Delighted* in *The Complete Works of Plutarch: Essays and Miscellanies* (New York: Crowell, 1909), vol. 3, Book 3, Chapter 3.
24. Cicero, *De Divinatione*, 2.19.44.
25. As our source is Diogenes Laertius, writing in the third century AD, it is somewhat unreliable, although Diogenes does himself draw on the much earlier account of Hippias, a contemporary of Socrates and Plato. It is known that Hippias was one of the earliest systematic doxographers, making a collection of passages from Homer and Hesiod, as well as Orphic and other early philosophical writings. See Kirk and Raven, *The Presocratic Philosophers*, p. 94. See also Theophrastus, *On Stones*, Greek text, translation and commentary by John F. C. Richards and Earle R. Caley (Columbus, Ohio: Ohio State University, 1956), p. 117.
26. Aristotle, *De Anima*, 405a19:

> Thales, too, as is related, seems to regard the soul as somehow producing motion, for he said that the (Magnesian) stone has a soul because it moves iron.

27. According to Park Benjamin, *A History of Electricity*, p. 35, Thales' doctrine of the soul inherent in the magnet 'was intended to be in direct contrast with the prevailing theories fostered by the priests of the Cabiric mysteries, namely, that the stone was supernaturally influenced'.

28. Theophrastus, *On Stones*, 28, p. 51.

29. Whether in the ancient world any non-initiatory knowledge of electricity existed beyond the propensity of rubbed amber to produce electro-static effects seems unlikely. The so-called 'Baghdad batteries', apparently dating from the Parthian period (third century BC to third century AD), consist of unglazed ceramic vessels with copper cylinders inserted inside them, discovered near Baghdad in 1936. One of these vessels had an iron spike inside the copper cylinder, and it has been speculated that this and the others could have been used as galvanic cells, had the vessels been filled with an electrolyte such as lemon juice, vinegar or grape juice. It is impossible to arrive at any certain judgement, but the fact that no other comparable finds have yet come to light from the ancient world does suggest that even if there was some knowledge of how to produce electrical phenomena beyond rubbing amber, this was not widespread, neither has any written record survived. See D. E. von Handorf, 'The Baghdad Battery: Myth or Reality?' in *Plating and Surface Finishing*, 89 (2002), pp. 84–87.

30. The most famous of these treatises was by Cicero, and was written in 45 BC. But some years before him Posidonius of Apameia wrote a treatise of the same title, as did Chrysippus of Tarsus in the third century BC. Both are now lost, but Cicero's treatise *De Natura Deorum* has survived. For an English translation, see Cicero, *On the Nature of the Gods*, translated by Horace C. P. McGregor (Harmondsworth: Penguin, 1972).

Chapter Seven

1. While St Augustine did not adhere to a rigid terminology, he consistently held to the distinction between these two levels of thinking. In his *Confessions*, for example, he distinguished the lower 'reasoning power' (*ratiocinans potentia*) from the higher 'intelligence' (*intelligentia*) in a way that strongly echoed Plato's distinction between *dianoia* and *nous*. As one of the most influential and revered authorities during the medieval period, the importance of Augustine's views on the nature of thinking at this time should not be underestimated. For him, the lower reason (*ratio* or *ratiocinatio*) is restricted by our sensory experience and so has as its object temporal things, whereas the intelligence (*intellectus* or *intelligentia*), which is seated in the heart, is directed towards that which is immutable and eternal. See *Confessions*, 7.17, in *The Confessions of St Augustine*, translated by F. J. Sheed (London: Sheed and Ward, 1948), p. 117f. Augustine used terms such as *ratio*, *ratiocinatio* and *cogitio* to denote the 'lower reason' (*ratio inferior*), and *intellectus* and *intelligentia* with the verb *intellegere* to denote the 'higher reason' (*ratio superior*) and its activity. See also Allan D. Fitzgerald, ed., *Augustine Through the Ages* (Michigan, Grand Rapids: Wm. R. Eerdmans Publishing Co., 1999), p. 452–4. A similar understanding is to be found in Boethius, who likened the relationship of reason (*ratiocinatio*) to intellect (*intellectus*) to that of time to eternity. Boethius also described it as being like that of the circumference of a circle to its central point, because the circle is conceived as being ever in motion, while the central point is still. If discursive reasoning functions through moving from one thought to another in order to arrive at a conclusion, then the thinking of the *intellectus* is characterized by arrival at the still point of insight. See Boethius, *The Consolation of Philosophy*, 4.6.78–83, translated by S. J. Tester (London: Harvard University Press, 1973), pp. 362–3.

2. Thus St Augustine could write in *De Magistro*, xii.40 (quoted in Mary T. Clark, *Augustine* (London: Continuuum, 1994), p. 18:

But when things are spoken of which we perceive through the mind, that is, through intellect and reason, we are talking about things which, being present, we see in that inner light by which he himself who is called the inner man is illuminated, and in which he delights.

Augustine's teaching on the *intellectus* and on divine illumination is expounded in *De Trinitate*, 12.14–15. See also n. 1 above.

3. For Thomas Aquinas, *Summa Theologiae*, 1a, 83, 1, 'human beings have free will to the extent that they are rational'. In his *Commentary on Aristotle's Nicomachean Ethics*, translated by C. I. Litzinger (Chicago: Henry Regnery Company, 1964), vol 1, Bk 3, Lecture 5, §457, Aquinas affirms:

Choice itself must be accompanied by an act of reason (*ratio*) and intellect (*intellectus*).

4. An explanation of syllogistic reasoning is given in Chapter 4, pp. 37–38.

5. Thus George Zarnecki, 'The Monastic World' in Joan Evans, ed., *The Flowering of the Middle Ages* (London: Thames and Hudson, 1966) could write:

It can be said, without much exaggeration, that until the development of the universities, the intellectual life of Europe was based on the monasteries.

6. Christopher Brooke, *The Monastic World: 1000–1300* (London: Paul Elek, 1974), p. 235f.

7. The tradition goes back to St Augustine, *De Trinitate*, 9–13. See especially Bk. 12:22–24. Augustine's doctrine of divine illumination should be understood as admitting of different degrees of knowledge, in which the soul is more or less consciously absorbed within the divine. See the discussion in Mary T. Clark, *Augustine*, pp. 19–25. See also Allan D. Fitzgerald, ed., *Augustine through the Ages*, pp. 5–6.

8. Thomas Aquinas, *Commentary on Boethius, De Trinitate*, 2, Q.5, Art.1, ad 2, in Thomas Aquinas, *Selected Philosophical Writings*, translated by Timothy McDermott (Oxford: Oxford University Press, 1993), p. 9:

So logic is not classified as a major branch of theoretical science, but as something that serves philosophy, providing it with tools of speculation: the syllogisms and definitions we need in the theoretical sciences.

9. Ibid., Q.6, Art.1, ad 3, pp. 36–38:

It is also for this reason that divine science provides the starting points for every other science, rational considerations having their beginning in intellectual ones...

10. Martin Irvine, *The Making of Textual Culture: 'Grammatica' and Literary Theory, 350–1100* (Cambridge: Cambridge University Press, 1994), p. 1 quotes a much repeated definition of grammar as 'the art [or science] of interpreting the poets and other writers and the principles for speaking and writing correctly', which turns up in the writings of Sergius, Marius Victorinus, Maximus Victorinus, Alcuin and others. The definition shows that rhetoric was basically covered by grammar. For the rules of composition of the formal letter and the sermon, the most common kinds of text composed in monasteries, see Jean Leclercq, *The Love of Learning and the Desire for God* (London: SPCK, 1978), Chapter 8.

11. For the breadth of classical learning in cathedral schools, see Paul Abelson, *The Seven Liberal Arts: A Study in Medieval Culture* (New York: Columbia University Press, 1906), pp. 26–29. For the monastic schools, see Leclercq, op. cit., pp. 141–143. See also Ernst Robert Curtius, *European Literature and the Latin Middle Ages* (London: Routledge and Kegan Paul, 1979), pp. 48–54 for lists of curriculum authors.

12. Hugh of St Victor, *De Tribus Diebus*, 814, quoted in Peter Ellard, *The Sacred Cosmos* (Scranton: University of Scranton Press, 2007), p. 57. See also the discussion in Friedrich Ohly, 'On the Spiritual Sense of the Word in the Middle Ages' in Friedrich Ohly, *Sensus Spiritualis: Studies in Medieval Significs and the Philology of Culture* (Chicago and London: University of Chicago Press, 2005), p. 5.

13. John of Salisbury, *Metalogicon*, 1.14, translated by Daniel D. McGarry (Philadelphia: Paul Dry Books, 2009), pp. 39–41.

14. John of Salisbury, *Metalogicon*, 1.22, p. 72, where John quotes the Roman grammarian Quintilian:

> Let no one despise the principles of grammar as of small account. Not that it is a great thing to distinguish between consonants and vowels, and subdivide the latter into semivowels and mutes. But, as one penetrates farther into this, so to speak, sanctuary (*sacrum*), he becomes conscious of the great intricacy (*subtilitas*) of grammatical questions.

For the view of Grammar as sacred, see Irvine, *The Making of Textual Culture: 'Grammatica' and Literary Theory, 350–1100*, pp. 13–14.

15. According to John of Salisbury, writing in the mid twelfth century (*Metalogicon*, 1.13), 'Grammar is the cradle of philosophy' and is 'the first of the arts to assist those who are aspiring to increase in wisdom'. For the importance of Grammar for the study of philosophy, see also *Metalogicon*, 1.21. The primacy of Grammar was repeatedly affirmed by teachers associated with Chartres during the twelfth century, for example Thierry of Chartres, Bernard of Chartres as well as John of Salisbury, for which see Édouard Jeauneau, *Rethinking the School of Chartres* (Toronto: University of Toronto Press, 2009), p. 42. This assessment of the importance of grammar was widespread in subsequent centuries too. For example, Dante refers to grammar as 'la prima arte' in *Paradiso*, 12.138. According to Curtius, *European Literature and the Latin Middle Ages*, p. 42f:

> Of the seven *artes*, those of the *trivium* were far more thoroughly cultivated than those of the *quadrivium*, and grammar the most exhaustively of all. It was, indeed, the foundation for everything else.

16. Logic did this by equipping its students with the means of discerning correct and incorrect forms of reasoning that so often are obscured within ordinary language. As St Anselm, writing in the eleventh century, put it:

> We ought not to be held back by the way in which the inaccuracies of speech hide the truth, but should rather aspire to the precision of the truth which lies hidden under the multiplicity of ways of speaking. [*Non tantum debemus inhaerere improprietati verborum veritatem tegenti quantum inhiare proprietati veritatis sub multimodo genere locutionum latenti.*]

The quotation is from St Anselm, *De Causa Diaboli*, quoted in D. P. Henry, *Medieval Logic and Metaphysics* (London: Hutchinson, 1972), p. v (translation adapted). Unlike in our own times, in the Middle Ages there was no non-linguistic logical notation, employing signs in place of words: this developed at a much later period.

17. For example, in the sentence 'The two brown dogs barked and then were fed' the dogs are the Subject of the sentence and the word 'dogs' is a noun, 'brown' is both a Quality and an adjective, 'two' is a Quantity and an adjective, 'bark' falls under the category of Action and is a verb, while 'were fed' falls under the category of Passivity and is a verb.

18. Taki Suto, *Boethius on Mind, Grammar and Logic* (Leiden: Brill, 2012), p. 9f.

19. John A. Trentman, 'Logic' in *Contemporary Philosophy. A New Survey*, Vol.6: *Philosophy and Science in the Middle Ages*, Part 2 (Dordrecht: Kluwer Academic Publishers, 1990), pp. 807.

20. Thomas Aquinas, *Commentary on Aristotle's 'De Interpretatione'* 1.2,n.3, quoted in Robert W. Schmidt, *The Domain of Logic According to St. Thomas Aquinas* (The Hague: Martinus Nijhoff, 1966), p. 318

21. As Schmidt, op. cit., p. 318, has stated, according to Aquinas:

> Its whole purpose and reason for existing is to guide human knowledge to truth in its quest of the real. Truth, and truth about real being, is the end and final cause of logic.

22. Jean Leclercq, *The Love of Learning and the Desire for God*, pp. 249–269. See also C. H. Lawrence, *Medieval Monasticism* (London: Longman, 1989), pp. 142–144. In the conflict that arose between Abelard and St Bernard, the abbot of Clairvaux, during the first part of the twelfth century, we see a dramatic manifestation of an intrinsic antagonism between the logical and the contemplative approaches to theological understanding.

23. C. H. Haskins, *The Renaissance of the Twelfth Century* (New York: Meridian Books, 1957), p. 98. See also Louis John Paetow, Introduction to Henri d'Andeli, *The Battle of the Seven Arts* (Berkeley: University of California Press, 1914), p. 16.

24. C. H. Haskins, *The Renaissance of the Twelfth Century*, p. 99. For the early twelfth century schoolman, John of Salisbury, the study and writing of poetry was the culmination of grammatical studies. In his *Metalogicon*, 1.22, p. 63, John quotes Seneca's statement with approval:

> The subject of the grammarian is language, and if he goes further, history, and if he proceeds still further, poetry.

25. John of Salisbury, *Metalogicon*, 1.22, p. 63. In *Metalogicon*, 1.24, p. 67, John writes in the same vein:

> Carefully examine the works of Virgil or Lucan, and no matter what your philosophy, you will find therein its seed (*conditura*).

The word *conditura* means both its 'seed' or 'germ' and also its 'seasoning'.

26. Such as Bernardus Silvestris and Alan of Lille. It also found favour in the monasteries where the love of poetry led many monks also to writing in verse, for which see Jean Leclercq, *The Love of Learning and the Desire for God*, pp. 170–172. It is interesting that even the logician, Abelard, so renowned for the sharpness of his intellect, bowed to the underlying spirit of the times in choosing to distil his ethical teaching in the form of a poem (written to his son), known as the *Carmen ad Astralabium*. See David Luscombe. 'Peter Abelard and the Poets' in John Marenbon, ed., *Poetry and Philosophy in the Middle Ages: A Festschrift for Peter Dronke* (Leiden: Brill, 2001), p. 156.

27. Haskins, *The Renaissance of the Twelfth Century*, p. 98f.
28. A more detailed description is given in Adolf Katzenellenbogen, 'The Representation of the Seven Liberal Arts' in Marshall Clagett et al, eds,. *Twelfth Century Europe and the Foundations of Modern Society* (Madison: University of Wisconsin Press, 1961), p. 46.
29. The image is discussed in Margaret Gibson, 'A Picture of *Sapientia* from S. Sulpice, Bourges' in *Transactions of the Cambridge Bibliographical Society*, vol. 6, no. 7 (1973), p. 126. According to Gibson, the five logical texts are a selection of those available at the time (comprising the *logica vetus*), by Boethius, Porphyry and Aristotle, for which see n. 4 above.
30. *De Nuptiis Philologiae et Mercurii*, translated in W. H. Stahl and R. Johnson with E. L. Burge in *Martianus Capella and the Seven Liberal Arts*, vol. 2 (New York: Columbia University Press, 1977).
31. Ibid., p. 107.
32. For the serpent-dragon iconography of Satan, see Jeffrey Burton Russell, *The Devil: Perceptions of Evil from Antiquity to Primitive Christianity* (New York: Cornell University Press), pp. 244–246.
33. The word 'dialectic' derives from the Greek suffix *dia*, meaning 'twice', 'double' or 'divided down the middle' and *legein*, meaning 'to reckon' or 'to speak'. See Ernest Klein, *A Comprehensive Etymological Dictionary of the English Language* (Bingley: Emerald, 2008), p. 209. See also William Harris Stahl, *Martianus Capella and the Seven Liberal Arts*, vol. 1 (New York: Columbia University Press, 1971), p. 105, n. 63.
34. At Rheims cathedral, for instance, a huge sculpture of Eve was commissioned in the thirteenth century to be placed above the north door to the right of the rose window relating the story of Adam and Eve (Fig. 7.5). Eve is shown tenderly holding a dragon in her hands, just as Dialectic does at Chartres (Fig. 7.6), the main difference between the two creatures being that at Rheims the fruit of the tree of the knowledge of good and evil is revealed inside the dragon's mouth.

Figure 7.5

35. Haskins, *The Renaissance of the Twelfth Century*, p. 136f.
36. Henri d'Andeli, *The Battle of the Seven Liberal Arts*, lines 450–451.

Chapter Eight
1. William and Martha Kneale, *The Development of Logic*, p.186, trace the first logic diagrams to Alexander of Aphrodisias (who lived during the late second and early third centuries AD) 'or possibly to some earlier scholar'.
2. John of Salisbury, *Metalogicon*, 2.10, translated by Daniel D. McGarry (Philadelphia: Paul Dry Books, 2009), p. 98.
3. The Square of Opposition was by no means the only logic diagram used in the Middle Ages. Another more complex diagram, known as the *pons asinorum* or 'bridge of asses', also had its origins in antiquity and was designed to make it easy to construct syllogistic arguments in a more or less mechanical way. See William and Martha Kneale, op. cit, pp. 185–187. See also C. L. Hamblin, 'An Improved *Pons Asinorum*?' in *Journal of the History of Philosophy*, 14 (1976), pp. 131–136.
4. For this reason, the diagrammatic rotating circles devised by Ramon Lull in the thirteenth century as an aid to theological reflection deviated from the medieval norm. Lull's devices served to combine ready-made concepts and terms, for example expressing the formal attributes of God ('goodness', 'greatness', 'eternity' and so on), in order to demonstrate the major truths of Christianity, but thereby his method bypassed any living experience of thinking as a spiritual activity. The prevalent attitude was that the application of purely logical thinking to the deeper truths of theology, in the desire to make what is divine intelligible to reason, leads to a profanation of God's mysteries. For this attitude, see Leclercq, *The Love of Learning and the Desire for God*, pp. 245–269. For a description of Lull's *ars magna*, see Martin Gardner, *Logic Machines and Diagrams* (Brighton: The Harvester Press, 1982), Chapter One. Gardner credits Lull as the inventor of a primitive logic machine (Gardner, op. cit., p. 1), but this claim is seriously called into question in W. Marciszewski and R. Murawski, *Mechanization of Reasoning in a Historical Perspective* (Amsterdam: Editions Rodopi, 1995), Chapter Two. These authors argue that Lull's *ars magna* consisted less in a systematic attempt to mechanize logic than in a somewhat extravagant and unverifiable combinatorial mixing of terms.
5. The term 'complex machine' is used here to distinguish them from the so-called five 'simple machines' of the inclined plane, the wedge, the screw, the lever and the wheel.
6. Adam Robert Lucas, 'Industrial Milling in the Ancient and Medieval Worlds' in *Technology and Culture*, 46 (January 2005), p. 4, n. 6.
7. Jean Gimpel, *The Medieval Machine: The Industrial Revolution of the Middle Ages* (London: Futura, 1979), p. 29. As Gimpel remarks, 'Such a concentration of mills formed a real industrial complex in the centre of Paris.' But note that the scale of development was not necessarily new. We know that floating mills were built under the Grand Pont in Paris during the reign of Louis VII (1137–80), and were destroyed with this bridge in 1296. See R. J. Forbes, 'Power' in Charles Singer et al., eds., *A History of Technology*, vol. 2 (Oxford: Oxford University Press, 1956), p. 608.
8. Gimpel, *The Medieval Machine: The Industrial Revolution of the Middle Ages*, p. 27. For ore and other industrial uses, see Terry S. Reynolds, 'Medieval Roots of the Industrial Revolution' in *Scientific American*, 251 (July 1984), p. 111.
9. Reynolds, op. cit., p. 112.
10. In Baghdad, they were developed by the Musa brothers at the court of Abu Jafar

al Ma'mun ibn Harun, the son of the famous Harun al-Rashid. The Musa brothers drew on the pioneering designs of both Hero of Alexandria and Philo of Byzantium.

11. See T. Koetsier, 'On the prehistory of Programmable Machines: musical automata, looms, calculators' in *Mechanism and Machine Theory*, 36 (2001), p. 590–591.

12. Eduard Farré-Olivé, 'A Medieval Catalan Clepsydra and Carillon', in *Antiquarian Horology*, 18.4 (1989), pp. 371–380. Interestingly, the author discounts any Arabic influence (p. 380).

13. Jan Koster, 'From Carillon to IBM: The Musical Roots of Information Technology', §3, published online, 2003 at <http://www.let.rug.nl/koster>. According to Koster,

> The repinnable drums of our carillons since the Middle Ages are automata programmable in the true sense, with the information representation (of music in this case), as in punched cards, depending on the distribution of interpretable points on a 'map' of columns and rows.

14. Gerhard Dohrn-van Rossum, *History of the Hour: Clocks and Modern Temporal Orders* (Chicago and London: University of Chicago Press, 1996), p. 107.

Chapter Nine

1. The first public mechanical clock to strike the sequence of 24 equal hours was installed in the tower of the church of San Gottardo in Milan in 1336. It was followed shortly thereafter by Jacopo de' Dondi's clock installed in Padua in 1344. See Gerhard Dohrn-van Rossum, *History of the Hour: Clocks and Modern Temporal Orders* (Chicago and London: University of Chicago Press, 1996), p. 108–109.

2. Silvio A. Bedini and Francis R. Maddison, *Mechanical Universe: The Astrarium of Giovanni de' Dondi* (Philadelphia: American Philosophical Society, 1966), p. 14f.

3. For example Johannes de Sacrobosco, following the Roman poet Lucretius, *De Rerum Natura*, uses the phrase 'the machine of the universe' (*machina mundi*) in his treatise *De Sphera Mundi* (1230), I, 2. Robertus Anglicus, in his 1271 commentary on Sacrobosco's *De Sphera*, implies that the *machina* is a clockwork mechanism, in arguing for the advantages of a mechanical representation of the cosmos. See Lynn Thorndike, *The Sphere of Sacrobosco and Its Commentators* (Chicago: University of Chicago Press, 1949), pp. 10–12. See also Rossum, *History of the Hour*, pp. 89–90.

4. Jean Buridan, *Questiones Octavi Libri Physicorum*, 8, 12, quoted in A. C. Crombie, *From Augustine to Galileo* (London: Heinemann, 1957), p. 253:

> One does not find in the Bible that there are Intelligences charged to communicate to the celestial spheres their proper motions; it is permissible then to show that it is not necessary to suppose the existence of such Intelligences. One could say, in fact, that God, when he created the Universe, set each of the celestial spheres in motion as it pleased him, impressing on each of them an *impetus* which has moved it ever since.

For a discussion of Buridan's theory of impetus, see Dijksterhuis, *The Mechanization of the World Picture*, pp. 181–185.

5. Thomas Bradwardine, *De Causa Dei* (1344), for which see George Molland, 'Addressing Ancient Authority: Thomas Bradwardine and *Prisca Sapientia*' in *Annals of Science*, 53 (1996), pp. 213–233.

6. Nicole Oresme, *Le Livre du Ciel et du Monde* (1377), 2.2, where, referring to the

relationship of God to the universe, he writes: '. . . it is similar to the way in which a man has made a clock, and lets it go and be moved by itself.' See Crombie, *From Augustine to Galileo*, p. 255.

7. Dante, *Paradiso*, 26: 64–66:

> Le fronde onde s'infronda tutto l'orto
> de l'ortolano eterno, amo io cotanto
> quanto da lui a lor di bene è porto.

[The leaves with which all the garden of the eternal gardener are enleaved, I love so much; as much as he has bestowed on each his blessing.]

8. Aristotle, *Physics*, IV.1, 208b19–23:

> 'Up' or 'above' always indicates the 'whither' to which things buoyant tend; and so too 'down' or 'below' always indicates the 'whither' to which weighty and earthy matters tend and this shows that 'above' and 'below' not only indicate definite and distinct localities, directions and positions, but also produce distinct effects.

Aristotle's *Physics*, IV, is a most important source for understanding the consciousness that existed prior to our modern gravity-dominated consciousness. Dante's ascent to heaven in the *Paradiso* was only possible on the basis of Aristotelian physics, for 'the moving force in mortal hearts', once freed of the weight of sin, draws us upwards to God (*Paradiso*, 1.109–141).

9. This scheme was not immune to change. For example, None gradually moved from mid-afternoon to midday, becoming our English 'noon' during the twelfth century, although on the Continent it continued to remain mid-afternoon until the fourteenth century. See Alfred W. Crosby, *The Measure of Reality: Quantification and Western Society* (Cambridge: Cambridge University Press, 1997), p. 33.

10. The episodes in the life of the Virgin celebrated in her Hours were:

> *Vigils/Matins*: The Annunciation;
> *Lauds*: The Visitation;
> *Prime*: The Nativity;
> *Terce*: Angels Announce Christ's Birth to the Shepherds;
> *Sext*: The Adoration of the Magi;
> *None*: The Presentation in the Temple;
> *Vespers*: The Flight Into Egypt;
> *Compline*: The Coronation of the Virgin.

11. John Harthan, *Books of Hours* (London: Thames and Hudson, 1977), p. 13–14:

> [The Virgin Mary] became, through the mystery of the Incarnation, the central figure in an unprecedented devotion in which many of the deepest emotions of men and women were involved. The multiplicity of traces left in cathedral, chapel and shrine bear withness to this. Through their Books of Hours, with their personal prayers and private images, layfolk were to identify themselves with *Dei genetrix*, the Mother of God.

12. Rossum, *History of the Hour*, p. 42.

13. In Paris, King Charles V ordered all the churches throughout the city to ring their bells on the hour (i.e. at 60 minute intervals) when the royal clock struck the hour. See Gimpel, *The Medieval Machine*, p. 155, who comments:

By making the churches ring their bells at regular sixty-minute intervals, Charles V was taking a decisive step toward breaking the dominance of the liturgical practices of the Church. The Church would bow to the materialistic interests of the bourgeois and turn its back on eternity.

14. Jacques Le Goff, *Time, Work and Culture in the Middle Ages* (Chicago: University of Chicago Press, 1980), pp. 29–42.
15. Rossum, *History of the Hour*, p. 144. According to Gimpel, *The Medieval Machine*, p. 155f, only the Greek Orthodox Church refused to compromise with the secularization of time. No mechanical clock was allowed to be installed in an Orthodox church until the twentieth century. Gimpel (p. 156) explains:

> For them it would have been blasphemy; for them the mathematical division of time into hours, minutes, and seconds had no relationship with the eternity of time.

16. Rossum, op. cit., p. 138ff and p. 271.
17. The earliest surviving spring-driven clock dates to 1450. See Eric Bruton, *The History of Clocks and Watches* (London: Orbis Publishing, 1979), p. 48.
18. Bruton, op. cit., p. 109. Queen Elizabeth 1 had a complete section of her inventory of 1572 devoted to watches, categorized as jewels, for which see Bruton, op. cit., p. 56.
19. Lewis Mumford, *Technics and Civilization* (Chicago and London: University of Chicago Press, 2010), pp. 12–18.
20. Ibid., p. 14.

Chapter Ten
1. For this reason the conception of a mechanical universe entailed re-conceiving God's relationship to it as being that of an engineer or mechanic who exists outside what he has designed.
2. William of Conches, *Commentary on Macrobius*, quoted in Peter Ellard, *The Sacred Cosmos* (Chicago, University of Scranton Press, 2007), p. 92. See Peter Ellard, op. cit., Chapter 5 for an illuminating discussion of exemplarism and the realist tradition at Chartres.
3. Thus for St Augustine, *De Diversis Quaestionibus 83*, the *ideae principales* 'are themselves not formed . . . but are contained in the divine understanding'. Quoted in C. G. Jung, *The Archetypes and the Collective Unconscious* (London: Routledge, 1980), p. 4.
4. The process has been well documented by Carolyn Merchant, *The Death of Nature* (London: Wildwood House, 1980).
5. Ockham, *Commentary on the Sentences (of Peter Lombard)*, I, 2, q. 28, quoted in Meyrick H. Carré, *Realists and Nominalists* (Oxford: Oxford University Press, 1946), p. 113. See also Etienne Gilson, *History of Christian Philosophy in the Middle Ages* (London: Sheed and Ward, 1980), p. 492.
6. Gordon Leff, *The Dissolution of the Medieval Outlook* (New York: Harper and Row, 1976), p. 10. See also Carré, *Realists and Nominalists*, p. 109.
7. Thomas Aquinas, *Summa Theologiae*, 1a.34.1 in Timothy McDermott, ed., *Summa Theologiae: A Concise Translation* (London: Eyre and Spottiswoode, 1989), p. 73:

> Knowing implies a relatedness of knower to known . . . it implies taking in the form of what we know so as to make the mind actively knowing . . .

8. Both *scientia rationalis* (science of reason) and *scientia realis* (science of reality) deal in propositions, but the propositions of logic are composed of terms that signify other terms, while the propositions of *scientia realis* are composed of terms that signify things outside the mind. Since for Ockham all human knowledge is expressed in propositions, which alone can be true or false, the emphasis is thrown onto *scientia rationalis* as the guarantor of knowledge of a world of things lacking any inherent meaning. See Ernest A. Moody in Ross, op. cit., pp. 290–291.

9. St Thomas Aquinas, Commentary on Aristotle's *De Interpretatione* 1.2, n. 3, quoted in Robert W. Schmidt, S. J., *The Domain of Logic According to St. Thomas Aquinas* (The Hague: Martinus Nijhoff, 1966), p. 318. See also Chapter 4, p. 37f, above.

10. See Leff, *The Dissolution of the Medieval Outlook*, p. 12, who comments:

> From attempting to explain the place of individuals in a world of universal natures or essences, the problem was henceforth to account for universals in an exclusively individual world. It led to the substitution of a logical for a metaphysical order, with universals treated solely as concepts whose meaning depended upon the grammatical form and logical signification of the terms standing for them.

For incisive comments on the significance of Ockham's terminist logic in relation to the natural world, see William J. Courtenay, '*Antiqui* and *Moderni* in Late Medieval Thought' in *Journal of the History of Ideas*, 48, 1 (Jan–March,1987), p. 7. For a good summary of Ockham's terminist logic and its relation to the physical sciences, see Frederick Copleston, *A History of Medieval Philosophy* (London: Methuen, 1972), pp. 244–247.

11. *Summulae*, 1, 14, quoted in Carré, *Realists and Nominalists*, p. 119. For the subjectivism of Ockham's epistemology, and for a fuller discussion of his nominalist philosophy, see Jeremy Naydler, 'The Regeneration of Realism and the Recovery of a Science of Qualities' in *The International Philosophical Quarterly*, XXIII, 2 (June, 1983), pp. 162–164.

12. Strictly speaking, Ockham recognized only the categories of substance and quality as applying to individual things, but the tenor of his thought was nevertheless towards a quantitative view of the world.

13. Referring to this deeply significant moment of the loss of universals in nature and the new materialistic outlook that opens up from it, Carré, *Realists and Nominalists*, p. 109, observes:

> We are here at a crisis in the history of European thought. It is here that there occurs most markedly the breach with the magnanimous tradition of *philosophia perennis*.

14. It is important to understand that we are here concerned not with the perception of any new content of the world, but with what Henri Bortoft called *a way of seeing*, which reconstitutes the world in the form of 'parts external to one another'. For Bortoft, as for Aristotle, the quantitative way of seeing is manifest wherever there are parts external to one another, regardless of whether number is explicitly included in the content or not. See Henri Bortoft, *The Wholeness of Nature: Goethe's Way of Science* (Edinburgh: Floris Books, 1996), p. 173.

15. Aristotle, *Metaphysics*, V.13, 1020a.7–9, p. 257.

16. For Nicholas of Autrecourt and other medieval atomists, see Crombie, *From Augustine to Galileo*, pp. 235–239. Also Gilson, *History of Christian Philosophy in the*

Middle Ages, p. 509. Atomism is, of course, the natural container of 'the quantitative way of seeing', for which see Bortoft, *The Wholeness of Nature*, p. 174–175, and note 14 above.

17. Crombie, *From Augustine to Galileo*, pp. 250–253. See also, Edward Grant, *Physical Science in the Middle Ages* (Cambridge: Cambridge University Press, 1977), pp. 50–53.

18. For example, a right-angled triangle for uniformly accelerated motion. See David C. Lindberg, *The Beginnings of Western Science* (Chicago: University of Chicago Press, 2007), pp. 301–306, who points out that Oresme's attempt at geometrical representation was a forerunner of modern graphing techniques. By resolving the intensity of a quality into a geometrical figure, its qualitative characteristics are mathematized and so become measurable. See also Crombie, *From Augustine to Galileo*, pp. 258–261.

19. William J. Courtenay, '*Antiqui* and *Moderni* in Late Medieval Thought' in *Journal of the History of Ideas*, 48, 1 (Jan–March,1987), pp. 3–5.

20. The process is well documented in Leff, *The Dissolution of the Medieval Outlook*.

21. William Cobbett, *A History of the Protestant Reformation* (Dublin: James Duffy, 1826), II.37.

22. See Peter King, *Western Monasticism* (Kalamazoo, Michigan: Cistercian Publications, 1999), pp. 273–74.

23. Leclercq, *The Love of Learning and the desire for God*, p.93, comments:

> The spiritual men of those days counsel the renunciation of carnal images; but this is in order to substitute for them a holy imagination. The sanctification of the imagination results in their attachment to the slightest particulars of the text, and not merely to the ideas it contains.

For the monks, it was this 'holy imagination' that allowed them 'to picture, to "make present", to see beings with all the details provided by the texts'.

24. A favourite text for this kind of commentary was *The Song of Songs*, ostensibly about carnal love but regarded by the monastic tradition as a treatise on the love of God and, as such, an inexhaustible spiritual treasure, capable of rendering new insights to all who meditated upon it. In the prologue to his commentary on *The Song of Songs*, Origen remarks that 'the whole body of it consists of mystical utterances'. See Origen, *The Song of Songs: Commentary and Homilies*, Prologue, translated by R. P. Lawson (Mahway, N.J.: Paulist Press, 1957), p. 22. For the commentary tradition on *The Song of Songs* see Leclercq, *The Love of Learning and the desire for God*, pp. 106–109, and p. 120.

25. Origen, *On First Principles*, 4.1.11, explains that the first is the literal sense (corresponding to its 'body'), the second is its moral sense, relating to how one should live one's life (corresponding to its 'soul'), and the third is its allegorical sense, understood as the eternal truths that it conveys (corresponding to its 'spirit'). Origen's teaching is discussed in Peter Harrison, *The Bible, Protestantism and the Rise of Natural Science* (Cambridge: Cambridge University Press, 2001), p. 18f.

26. Thus Origen's three senses were frequently expanded to four: the literal, the allegorical, the moral (or 'tropological') and the mystical (or 'anagogical'), and different meanings on all four levels would be brought out in Biblical commentaries. As Dante explained in his Letter to Can Grande, as well as its literal sense, the Exodus of the Israelites from Egypt *allegorically* signifies the redemption of humanity by Christ, *morally* the conversion of the soul from a state of sin to a state of grace, and *mystically*

the return of the soul to God. See William Anderson, *Dante the Maker* (London: Hutchinson, 1983), pp. 329–335.

27. For example, Hugh of St Victor wrote four books commenting on the story of Noah's Ark. His approach was to utilize the Ark as a symbolic image of an interior 'ark' of spiritual wisdom, which, when reconstructed in the mind of the contemplative, would function as a container of Christian doctrine. Through such imaginative inner work, the soul could fortify itself and thereby avoid being swept away by the flood-waters of worldliness. See Paul Rorem, *Hugh of St Victor* (Oxford: Oxford University Press, 2009), p. 129f.

28. St Augustine, *Contra Faustum Manichaeum*, 32.20, discussed in Peter Harrison, 'The Bible and the Emergence of Modern Science', in *Science and Christian Belief*, vol. 18, No. 2, (October 2006), p. 118. See also Friedrich Ohly, *Sensus Spiritualis: Studies in Medieval Significs and the Philosophy of Culture* (Chicago: University of Chicago Press, 2005), pp. 7–14.

29. For example, the pelican was well known as a symbol of Christ's sacrificial love, just as the lily was a symbol of the Virgin Mary's purity. For Hugh of St Victor, *De Tribus Diebus*, 4, 'the whole sensible world is like a kind of book written by the finger of God—that is, created by divine power—and each particular creature is somewhat like a figure ... instituted by the divine will to manifest the invisible things of God's wisdom.' For Bonaventure, *Breviloquium*, 2.12, 'The creature of the world is like a book in which the creative Trinity is reflected, represented and written.' Both quoted in Harrison, 'The Bible and the Emergence of Modern Science', p. 119.

30. Harrison, 'The Bible and the Emergence of Modern Science', p. 122:

> The interpretation of the book of nature and of the book of Scripture was part of a single hermeneutical enterprise. It follows that denial of the validity of allegorical interpretation would have far reaching consequences not just for the reading of Scripture, but for a whole religious, symbolic conception of the natural order.

31. Galileo, *The Assayer*, in *Discoveries and Opinions*, translated by Stillman Drake (New York: Anchor Books, 1957), p. 237f.

32. See note 12 above.

33. According to E. A. Burtt, *The Metaphysical Foundations of Modern Physical Science* (London: Routledge and Kegan Paul, 1932), p. 56, these thinkers included Vives, Sanchez, Montaigne and Campanella. One can, however, already observe that the tendency towards the prioritization of the quantitative over the qualitative is occurring from the fourteenth century onwards, to compensate the collapse of the realist viewpoint brought about by the rise of nominalism. See Crombie, *From Augustine to Galileo*, pp. 274–288.

34. Burtt, *The Metaphysical Foundations of Modern Physical Science*, p. 57 comments:

> Again, Kepler's position led to an important doctrine of knowledge. Not only is it true that we can discover mathematical relations in all objects presented to the senses; *all certain knowledge must be knowledge of their quantitative characteristics, perfect knowledge is always mathematical* ... Therefore quantity is the fundamental feature of things, the *primarium accidens substantiae*, 'prior to the other categories'. Quantitative features are the sole features of things as far as the world of our knowledge is concerned. Thus we have in Kepler the position clearly stated that the real world is

the mathematical harmony in things. The changeable, surface qualities which do not fit into this underlying harmony are on a lower level of reality; they do not so truly exist.

35. Crombie, *From Augustine to Galileo*, pp. 290–293.
36. Bortoft, *The Wholeness of Nature*, p. 174f, comments:

> The quantitative way of seeing discloses a world fragmented into separate and independent units, and it is therefore not surprising to find that the philosophy of atomism was readily incorporated into the new quantitative science of physics. Atomism fits the form of quantity like a hand fits a glove. The two are congruent with each other, so that atomism functions as a picture of the quantitative way of seeing. It pictures the mode of conception ... Atomism is really a container that carries the quantitative way of seeing.

37. Quoted in Burtt *The Metaphysical Foundations of Modern Physical Science*, p. 78.
38. Bortoft, *The Wholeness of Nature*, p. 167.
39. For Descartes, just as for Galileo, only that which can be put into the form of mathematics gives us reliable knowledge. Sensible qualities, in so far as they cannot be quantified, do not exist outside the human mind. Only those qualities that can be quantified objectively belong to things: these are extension (length, breadth, depth), weight, figure and motion. Descartes is discussed further in Chapter 15.
40. Dijksterhuis, *The Mechanization of the World Picture*, p. 431.
41. Descartes, *Principles of Philosophy*, I.69–70 and IV.197. As Burtt, *The Metaphysical Foundations of Modern Physical Science*, p. 115, comments:

> The universe of mind, including all experienced qualities that are not mathematically reducible, comes to be pictured as locked up behind the confused and deceitful media of the senses, away from this independent extended realm, in a petty and insignificant series of locations inside of human bodies.

42. René Descartes, *Discourse on the Method*, in *The Philosophical Works of Descartes*, vol. 1, ed. Elizabeth S. Haldane and G. R. T. Ross (Cambridge University Press, 1980), p. 119.

Chapter Eleven

1. Aristotle, *Physics*, 4.8, translated by Philip H. Wicksteed and Francis M. Cornford (London: Heinemann, 1970), p. 357:

> But as to differences that depend on the moving bodies themselves, we see that of two bodies of similar formation the one that has the stronger trend (*rhopē*) downward by weight (*barous*) or upward by buoyancy (*kouphotētos*), as the case may be, will be carried more quickly than the other through a given space in proportion to the greater strength of this trend (*rhopē*).

2. For a modern discussion of the 'two borders' of nature, see Ernst Lehrs, *Man or Matter* (London: Rudolf Steiner Press, 1985), p. 247ff. The distinction is of particular importance for a spiritual approach to the understanding of electricity, as both Lehrs, op. cit., and Hedley Gange, 'A Comprehensive Approach to Electricity', *Science Forum*, 1 (1979), pp. 12–28, make clear.

3. Thomas Aquinas, *Commentary on the Gospel of John: Chapters 6–12*, translated by

Fabian R. Larcher and James A. Weisheipl (Washington DC: Catholic University of America Press, 2010), §1142, p. 107:

> Furthermore it [the supreme light] makes all things to be actually intelligible inasmuch as all forms are derived from it . . .

Such a view is consistent with the teaching of Dionysius the Areopagite and John Scotus Erigena, for whom all created beings are conceived as lights, for their very essence consists in their being so many reflections of the divine light emanating from the 'Father of Lights'. See Etienne Gilson, *History of Christian Philosophy in the Middle Ages* (London: Sheed and Ward, 1980), p. 120. See also Dionysius the Areopagite, *The Celestial Hierarchy* I.1–3 in Pseudo-Dionysius, *The Complete Works* translated by Colm Luibheid and Paul Rorem (New York: Paulist Press, 1987), pp. 145–147.

4. Thomas Aquinas, *Summa Theologiae*, I.104.1, edited by Timothy McDermott (London: Eyre and Spottiswoode, 1989), p. 154:

> Now the whole of creation relates to God as [illuminated] air to the sun: God alone exists by nature—existing is what he is—whereas creatures only share in existence—existing is not what they are. Hence, as Augustine says, 'If, at any time, the ruling power of God were to desert what he created, his creation would immediately lose its form, and all nature would collapse.'

For the 'Father of Lights' see Dionysius the Areopagite, *The Celestial Hierarchy* I.1 in Pseudo-Dionysius, *The Complete Works*, pp. 145.

5. Gospel of John, 8.12. Genesis 1.3 records that on the first day God said 'Let there be light' whereas the sun, moon and stars were not created until the fourth day.

6. For the 'Father of lights', see notes 3 and 4 above. For the light as an image of the divine goodness, see Dionysius the Areopagite, *The Divine Names*, IV.4 in Pseudo-Dionysius, *The Complete Works*, p. 74.

7. Dionysius the Areopagite, *The Divine Names*, IV.4 in Pseudo-Dionysius, *The Complete Works*, pp. 74–75. Just as the Good is the agent of cohesion in all things, maintaining and preserving them in unity, so too does the light—as visible image of the Good—gather and hold together all perceptible things.

8. Dionysius the Areopagite, *The Divine Names*, IV.4 in Pseudo-Dionysius, *The Complete Works*, p. 74. See also *The Celestial Hierarchy*, I.1 in Pseudo-Dionysius, *The Complete Works*, p. 145: 'It returns us back to the oneness and deifying simplicity of the Father who gathers us in.'

9. Gilson, *History of Christian Philosophy in the Middle Ages*, p. 120:

> All created beings are lights: *Omnia quae sunt, lumina sunt*, and their very essence consists in being so many reflections of the divine light. Made up of that multitude of tiny lamps that things are, creation is only an illumination intended to show God.

10. Robert Grosseteste, *On Light (De Luce)*, translated by Clare C. Reidl (Milwaukee: Marquette University Press, 1942), p. 10. See also James McEvoy, *Robert Grosseteste* (Oxford: Oxford University Press, 2000), p. 87f.

11. McEvoy, *Robert Grosseteste* p. 91.

12. See note 3 above. Thomas Aquinas is somewhat ambivalent regarding the relationship of natural light to spiritual light, at times arguing that the relationship is one of analogy, but at other times seeing natural light as an image or exemplar of the

divine light, and therefore as participating in its essence. In his commentary on Aristotle's *De Anima*, II.14, 415f, he sharply differentiates natural from spiritual light, whereas in his commentary on Dionysius' *Divine Names*, IV.1 he argues that natural light participates in the divine goodness, which is its exemplar. See the discussion in David L. Whidden, *Christ the Light: The Theology of Light and Illumination in Thomas Aquinas* (Minneapolis, MN: Fortress Press, 2014), pp. 74–80, who concludes that Aquinas sees God 'as the exemplar cause of sensible light and the means by which we can understand light's nature' (p. 80).

13. Isidore of Seville (early seventh century) makes the distinction between *lux* and *lumen* in *Etymologies* XIII.x.14. For the relationship of *lux* to *lumen* as one of archetypal source to image, see Yael Raizman-Kedar, 'Plotinus's Conception of Unity and Multiplicity as the Root to the Medieval Distinction between *Lux* and *Lumen*' in *Studies in History and Philosophy of Science*, 37.3 (2006), pp. 379–397. See also Whidden, *Christ the Light*, pp. 69–70.

14. Galileo's views on the nature of light are recorded in his letter to Piero Dini, 23 March, 1615, in which he elaborates an entirely physical conception of light, ironically quoting in support of this view none other than Dionysius the Areopagite! In the same letter, he reduces the Holy Spirit to radiant heat. For the text of the letter, see Maurice A. Finocchiaro, *The Galileo Affair* (Berkeley: University of California Press, 1989), pp. 60–67.

15. For Galileo's conception of gravity, see Burtt, *The Metaphysical Foundations of Modern Physical Science*, pp. 91–94. It is worth noting that not all thinkers in the Middle Ages regarded gravity as an inherent tendency of heavy things to fall. Many prominent thinkers (not least Thomas Aquinas) regarded it as an external force to which heavy things were subject, just as light things were subject to levity. See Crombie, *Augustine to Galileo*, p. 242f.

16. Galileo, *Il Saggiatore*, quoted in Burtt, *The Metaphysical Foundations of Modern Physical Science*, p. 78.

17. Galileo, *Dialogue Concerning the Two Chief World Systems* (Berkeley: University of California Press, 1962), p. 207f.

18. Aristotle, *Metaphysics*, XI.iii.7 (1061a29–32), translated by Hugh Tredennick in Aristotle, *Metaphysics* (London: William Heinemann, 1977), pp. 67–69.

19. Aristotle, *Metaphysics*, VI.i.5–7 (1025b25–1026a6), in op. cit., p. 295, where he admonishes us not to fail 'to observe how the essence and the formula (*logos*) exist, since without this our enquiry is ineffectual' and argues that physics 'should look for and define the *essence in physical things*'.

20. As Burtt, *The Metaphysical Foundations of Modern Physical Science*, p. 80, observed:

> The features of the world now classed as secondary, unreal, ignoble and regarded as dependent on the deceitfulness of sense, are just those features which are most intense to man in all but his purely theoretic activity, and even in that, except where he confines himself strictly to mathematical method. ... Man begins to appear for the first time in the history of thought as an irrelevant spectator and insignificant effect of the great mathematical system which is the substance of reality.

21. Isidore of Seville, for example, gives both options, in *Etymologies*, XIII.9. See *The Etymologies of Isidore of Seville*, translated by Stephen A. Barney et al. (Cambridge: Cambridge University Press, 2006), p. 273f.

22. Albertus Magnus, *De Mineralibus*, II.1.1 and II.1.4, translated by Dorothy Wyckoff as *The Book of Minerals* (Oxford: Oxford University Press, 1967), pp. 56–58 and p. 64f. See also note 23 below.

23. Isidore mentions the attractive properties of amber in *Etymologies*, XVI.8.7, in *The Etymologies of Isidore of Seville*, p. 324. He points out that when amber is rubbed it receives the 'spirit of heat' and becomes attractive to 'leaves and chaff and the fringes of clothing, just as a magnet attracts iron'. Writing in the late twelfth century, Alexander of Neckam, *De Naturis Rerum*, 98, argues that if the ability of lodestone to attract iron is due to a strong innate virtue in the lodestone overcoming the weaker virtue of the iron and dominating it, then the ability of amber and jet to attract chaff is due to a virtue acquired by them when rubbed, which enables them to overcome the weaker virtue of the chaff and dominate it. Writing in the following century, Albertus Magnus, *The Book of Minerals (De Mineralibus)*, II.2.7, p. 93 mentions the attractive power of both amber and jet. A better observer than Isidore, Albertus also remarked on the attractive properties of many other precious stones when rubbed, for which see *The Book of Minerals (De Mineralibus)*, II.2.9, p. 102. Typical of his times, Albertus, op. cit., II.1.4, ascribed this power of attraction to the specific substantial form of the stones, and the influence of heavenly forces.

24. The concept of the weight-driven mechanical clock was referred to in 1271 by Robertus Anglicus, in his commentary on the astronomical treatise of Johannes de Sacrobosco, *Tractatus De Sphera Mundi*. Anglicus indicated that at the time efforts had for some years been underway to construct such a clock. See Derek J. de Solla Price, 'On the Origin of Clockwork, Perpetual Motion devices and the Compass' in *Contributions from the Museum of History and Technology* (Washington DC: Smithsonian Institution, 1959), p. 106.

25. In his treatise, Petrus Peregrinus clearly differentiated the poles of the magnet, revealed that unlike magnetic poles mutually attract, and demonstrated that when a magnet is dissected the two unlike poles will continue to exist. He also showed that iron needles placed on a lodestone will align themselves on a north-south axis and he invented the first mariner's compass with a pivoted needle. Prior to this, mariners' compasses were magnetized needles inserted into a reed or a short piece of wood and placed in a bowl of water. This is described by Alexander of Neckam in his treatise *De Naturis Rerum* ('On the Natures of Things'), Chapter 98, and again in his *De Utensilibus*, both composed at the beginning of the thirteenth century. See Alexander Neckam, *De Naturis Rerum et De Laudibus Divinae Sapientiae*, edited by Thomas Wright (London: Longman, Roberts and Green, 1863), pp. xxxiv–xxxviii.

26. *Epistola de Magnete*, I.4 in *The Letter of Petrus Peregrinus on the Magnet (1269)*, translated by Brother Arnold (New York: McGraw Publishing Co., 1904), p. 6f:

> I wish to inform you that this stone bears in itself the likeness of the heavens, as I will now clearly demonstrate. There are in the heavens two points more important than all others, because on them, as pivots, the celestial sphere revolves: these points are called, one the arctic or north pole, the other the antarctic or south pole. Similarly you must fully realize that in this stone there are two points styled respectively the north pole and the south pole.

27. According to Petrus Peregrinus, it is 'by natural instinct' that the strangely ensouled lodestone aligns itself to the heavenly poles, just as it is 'according to its natural desire' that it attracts iron, and alters the innate 'virtue' of the iron, which is forced to

become obedient to the stronger virtue of the lodestone. Thus 'The virtue of the stronger becomes active , whilst that of the weaker becomes obedient or passive' (*Epistola*, IX, p. 14). For orientation by 'natural instinct' see *Epistola*, V, p. 9 and IX, p. 16; for 'natural desire' *Epistola*, VIII, p. 13; for the inherent 'virtue' of the iron, see *Epistola*, VIII, p. 13.

28. In this important respect, Petrus Peregrinus differs little in his outlook from William the Clerk who flourished 100 years prior to him, and Cardinal de Vitry who flourished in the early thirteenth century, for both of whom the lodestone receives its virtue from the Pole Star. Only Petrus Peregrinus, with his sharper observation, notices that it is not actually the Pole Star towards which the magnetized needle points but the starless celestial poles.

29. The full title is *De Magnete, Magnetisque Corporibus, et de Magno Magnete Tellure* (London: Peter Short, 1600), which translates as 'On the Magnet and Magnetic Bodies, and on the Great Magnet of the Earth'. English translation by S. P. Thompson, *On the Magnet* (London: Chiswick Press, 1900), p. 48.

30. These researchers included most notably the English instrument maker Robert Norman (fl.1580s), the Neapolitan natural philosopher John Baptista Porta (1540–1615), and the Venetian historian, theologian and statesman Fra Paolo Sarpi (1552–1623).

31. William Gilbert, *De Magnete*, 1,3.

32. According to Park Benjamin, *A History of Electricity: From Antiquity to the Days of Benjamin Franklin* (New York: John Wiley and Sons, 1898), p. 344f, Galileo learned of Gilbert's work through Sagredo, and repeated his experiments very soon after the publication of *De Magnete*.

33. William Gilbert, *De Magnete*, 2,2.

34. William Gilbert, *De Magnete*, 2,2. This was not entirely new, as Albertus Magnus had already observed the attractive properties of diverse substances, in *The Book of Minerals (De Mineralibus)*, II.2.9.

35. William Gilbert, *De Magnete*, 2,2 (*On the Magnet*, p. 52). The Latin term derives from the Greek word for amber, *elektron*, meaning 'that which shines'. Gilbert observed that 'electrics' all share the characteristic of having been a liquid that has undergone solidification. Amber is a good example of this, as it is a fossilized resin. Their 'humour' is therefore intrinsically moist, and it is this hidden moistness that stands behind the idea of the effluvium that they exhale. In contrast to magnetism, the body drawn to an electrified body is simply 'allured' but does not otherwise undergo any change. In the case of something magnetized, a force seems to be implanted into it by the magnet.

36. In *De Magnete*, 2,2 (*On the Magnet*, p. 60).

37. *De Magnete*, 2,2 (*On the Magnet*, p. 60).

38. See note 7 above.

39. Thomas Aquinas, *Questiones Disputatae de Potentia Dei* ('On the Power of God') translated by the English Dominican Fathers (Westminster, Maryland: The Newman Press, 1952), 5, a1:

> Without any doubt whatever it must be admitted that things are preserved in existence by God, and that they would instantly be reduced to nothing were God to abandon them.

See also notes 3 and 4 above.

40. This can already be seen in relation to magnetism in Athanasius Kircher, *Magnes sive de arte magnetica* (1643), for whom God is re-envisaged in the image of the magnet as a kind of magnetic, all-pervasive power, which holds the world together. In the eighteenth century, as knowledge of electricity became more widespread, Swabian Pietists such as Friedrich Christoph Oetinger, Prokop Divisch and Johann Ludwig Fricker embraced the view of electricity as the primeval creative spirit of the world. See Nicholas Goodrick-Clarke, 'The Esoteric Uses of Electricity: Theologies of Electricity from Swabian Pietism to Ariosophy' in *Aries*, 4, 1 (2004), pp. 69–73.

Chapter Twelve

1. Martha Ornstein Bronfenbrenner, *The Role of Scientific Societies in the Seventeenth Century* (New York: Arno Press, 1975), p. 101. Just four years after the founding of the Royal Society, the Académie des Sciences was founded in France in 1666. The Society of Sciences of the Elector of Brandenberg, under the patronage of the Prussian King Frederick I, followed in 1700.

2. *Saggi di Naturali Esperienze*, CCLXV, translated in W. E. Knowles Middleton, *The Experimenters: A Study of the Accademia del Cimento* (Baltimore and London: Johns Hopkins Press, 1971), p. 249. The experiment followed the same method as Galileo's unsuccessful attempt to measure light, recorded in his *Dialogue*, First Day.

3. The experiments against levity are recorded in *Saggi di Naturali Esperienze*, CCVII–CCXVI, translated in Middleton, op. cit., pp. 221–225.

4. The experiments on magnetism and electricity are recorded in *Saggi di Naturali Esperienze*, CCXVII–CCXXVII, translated in Middleton, op. cit., pp. 225–233.

5. This is in his later work, *De mundo nostro sublunari philosophia nova*, discussed in Bronfenbrenner, *The Role of Scientific Societies in the Seventeenth Century*, p. 23.

6. Aristotle, *Physics*, IV.8 (214b17–27). For Aristotle, nothing can be *in* a void 'in the proper local sense of "in"' (214b25). The same argument, that there is a belonging together of place and what is naturally 'in' a given place, is reiterated in *Asclepius*, one of the most influential Hermetic texts available during the Middle Ages. In *Asclepius* 34, in Brian P. Copenhaver, *Hermetica* (Cambridge: Cambridge University Press, 1992), p. 88, we read:

> If you assume a place apart from that of which it is the place, it would seem to be an empty place, which, I believe, cannot exist in the world.

Another common argument against the void was based on the so-called 'principle of plenitude', according to which the universe is *completely filled* with living beings, as *Asclepius* 33, in Copenhaver, op. cit., p. 87, explains:

> There is no such thing as a void, nor can there have been, nor will there ever be. For all the members of the world are completely full so that the world is complete and filled with bodies diverse in quality and form ...

This perspective is articulated at length in Plato, *Timaeus*, 29A–37D. The *Timaeus* exercised a powerful influence on medieval natural philosophers. For a discussion of the principle of plenitude, see Arthur O. Lovejoy, *The Great Chain of Being* (Cambridge MA: Harvard University Press, 1978), pp. 52–55.

7. Otto von Guericke, *Experimenta Nova (ut vocantur) Magdeburgica de Vacuo Spatio* (Amsterdam, 1672), Book III, Chapters 12–16, translated in Thomas E. Conlon,

Thinking About Nothing: Otto Von Guericke and the Magdeburg Experiments (London: St. Austin Press, 2011), pp. 237–245.

8. Ibid., p. 267f.

9. When the air is removed from inside the receptacle, the pressure of the surrounding atmosphere is 10 tonnes per square meter, with no compensating pressure on the inside to offset this. Were the receptacle made of a weaker substance than brass it would undoubtedly have collapsed. See Frank Close, *The Void* (New York: Sterling, 2009), pp. 15–17.

10. For the medieval religious sensibility, exemplified by Thomas Aquinas, *Summa Theologica*, 1a, 8, 2, 'God is everywhere and in every place, first because he is in all things giving them substance, power and operation. Since place is real he is present there.' A vacuum, however, is no longer strictly speaking a 'place', as Aristotle argued in *Physics*, IV, 8. See note 6 above. The implication is that were a void to be created, then it would be a non-place, from which God would be absent—an appalling notion, and for the medieval mind a theological nightmare. The idea of 'subnature' was reintroduced in modern times by Rudolf Steiner, who specifically linked it to the force of electricity. See, for example, Rudolf Steiner, *Anthroposophical Leading Thoughts* (Forest Row: Rudolf Steiner Press, 1973), pp. 216–219.

11. Martha Ornstein Bronfenbrenner, *The Role of Scientific Societies in the Seventeenth Century* (New York: Arno Press, 1975), p. 85f and p. 153.

12. The phrase 'nature abhors a vacuum' (*natura abhorret vacuum*) was common from the thirteenth century onwards, deriving from the arguments put forward by Aristotle in his *Physics*, IV, 6–9 and *On the Heavens*, IV, 5. See Edward Grant, 'Medieval Explanations and Interpretations of the Dictum that "Nature Abhors a Vacuum"', *Traditio*, 29 (1973), pp. 327f with n. 4. That nature should abhor a vacuum follows from the generally accepted 'principle of plenitude', or fullness of the universe, created through the divine outpouring into materiality, for which see note 6 above.

13. Otto von Guericke, op. cit., IV, Chapter 15.

14. Park Benjamin, *A History of Electricity*, p. 395.

15. *Philosophical Transactions* 25 (1706–7), pp. 2277–2282, discussed in J. L. Heilbron, *Electricity in the 17th and 18th Centuries: A Study of Early Modern Physics* (Berkeley: University of California Press, 1979), p. 230.

16. Lehrs, *Man or Matter*, pp. 64–70.

17. Book of Genesis, 3:8.

18. Lehrs, *Man or Matter*, p. 54.

19. Between 1667 and 1866 (the year the version with Gustave Doré's engravings was published) over 70 illustrated editions of the poem were issued, an average of roughly one every three years. See Robert Wolf, et al., *Paradise Lost: The Poem and its Illustrators* (Grasmere: The Wordsworth Trust, 2004), p. 16.

20. Milton records this meeting in his *Areopagitica*. For a discussion, see Allan H. Gilbert, 'Milton and Galileo', *Studies in Philology*, 19, 2 (1922), pp. 152–185. See also, George F. Butler, 'Milton's Meeting with Galileo: A Reconsideration', *Milton Quarterly*, 39, 3, (2005), pp. 131ff.

21. Milton, *Paradise Lost*, Book One, lines 287–291. For Galileo and Milton, see Allan H. Gilbert, op. cit., pp. 152–185.

22. Milton, *Paradise Lost*, Book One, lines 249–252.

23. Milton, *Paradise Lost*, Book One, lines 261–263.

24. Milton, *Paradise Lost*, Book Six, lines 490–491.

Chapter Thirteen

1. For a discussion of 'the quantitative way of seeing', see Henri Bortoft, *The Wholeness of Nature: Goethe's Way of Science* (Edinburgh: Floris Books, 1996), pp. 173–179.

2. Francis Bacon, *The Great Instauration*, Preface, in Rose-Mary Sargent, ed., *Francis Bacon: Selected Philosophical Works* (Indianapolis: Hackett Publishing Inc., 1999), p. 75.

3. For the *intellectus* and the *ratio*, see Chapter 7, pp. 61–63.

4. *Novum Organum* 1, §104, p. 132. Page numbers for this and subsequent citations of Bacon's writings are from Rose-Mary Sargent, ed., *Francis Bacon: Selected Philosophical Works*.

5. Francis Bacon, *Novum Organum* 1, §100, p. 131.

6. Ibid., Preface, p. 87 (italics added).

7. Francis Bacon, *Novum Organum* 1, §122, p. 142.

8. Ibid., §104, p. 132. The statement in full is:

   > The understanding must not therefore be supplied with wings but rather hung with weights to keep it from leaping and flying.

9. Ibid., §115, p. 137, §116, p. 138 and §129, p. 147.

10. Ibid., §102, p. 131.

11. Ibid., §103, p. 132:

    > For our road does not lie on a level, but ascends and descends—first ascending to axioms, then descending to works.

12. Ibid., §124, p. 143.

13. Ibid., 2, §52, p. 189. For Bacon's project of restoring Paradise on earth, see Peter Dawkins, *Building Paradise* (Warwick: The Francis Bacon Research Trust, 2001), pp. 62–130.

14. Francis Bacon, *Novum Organum* 1, §84, p. 119f.

15. I owe this insight to Paul Emberson, *From Gondhishapur to Silicon Valley* (Tobermory: Etheric Dimensions Press, 2009), p. 96f.

16. Francis Bacon, *New Atlantis*, in Rose-Mary Sargent, ed., op. cit., p. 267f.

17. Ibid., p. 261f.

18. Ibid., pp. 263–266.

19. Paul Emberson, op. cit., p. 99–101. See also Peter Dawkins, *Building Paradise* (Warwick: The Francis Bacon Research Trust, 2001), p. 182.

20. Francis Bacon, *De Augmentis Scientiarum* (1620), Book Six, in James Spedding, et al., eds., *The Works of Francis Bacon*, vol. 1 (London: Longman, 1858), pp. 660–661.

21. Ibid., p. 661.

22. Bacon's significance in this respect was recognized by the American cryptologist Colonel W.F. Friedman, whose long and distinguished career culminated in his serving as chief cryptologist of the U.S. Government's National Security Agency between 1952 and 1956. He wrote: 'Bacon was, in fact, the inventor of the binary code that forms the basis of modern ... computers.' Quoted in William H. Sherman, 'How to Make Anything Signify Anything' in *Cabinet Magazine*, 40 (2010/11).

Chapter Fourteen

1. See Chapter Three, p. 27, with note 10.

2. For example, Moderatus, Eudorus and Nicomachus. See Charles H. Kahn, *Pythagoras and the Pythagoreans* (Indianapolis: Hackett Publishing Company, 2001),

pp. 107–109 and 116f. The Pythagorean origins of the view of the Monad as first principle are attested in Aetius, 1.7.18, quoted in Guthrie, *History of Greek Philosophy*, vol. 1 (Cambridge: Cambridge University Press, 1962), p. 248, as follows:

> Of the principles, Pythagoras said that the Monad was God and the good, the true nature of the One, Mind itself...

For the One in Neoplatonism, see A. H. Armstrong, *An Introduction to Ancient Philosophy* (Boston: Beacon Press, 1963), pp. 180–188, and for the One as 'the light before all light' in Plotinus, see the discussion in Frederic M. Schroeder, *Form and Transformation: A Study in the Philosophy of Plotinus* (Montreal and Kingston: McGill-Queen's University Press, 1992), p. 45f. For the Monad in the Hermetic tradition, see *Corpus Hermeticum*, IV.10 in Brian P. Copenhaver, *Hermetica* (Cambridge: Cambridge University Press, 1992), p. 17.

3. So, for example, Aetius, 1.7.18 in Guthrie, op. cit., p. 248 says that Pythagoras said 'the indefinite Dyad is a daimon and evil, concerned with material plurality'. See also Porphyry, *Life of Pythagoras*, 50, in David Fideler, ed., *The Pythagorean Sourcebook* (Grand Rapids, Michigan: Phanes Press, 1988), p. 133. Philo, *Questions on Exodus*, discussed in John Dillon, *The Middle Platonists* (London: Duckworth, 1996), p. 173 refers obliquely to the Dyad as an 'evil principle, akin to the Persian Ahriman and to the Evil Soul of [Plato's] *Laws* X'. Plutarch's view of the Dyad as a principle of confusion, a 'maleficent world soul' and 'evil principle' is discussed in Dillon, op. cit., pp. 204–206. See also *Corpus Hermeticum*, IV.10 in Copenhaver, *Hermetica*, p. 17, for divisibility associated with imperfection.

4. According to Porphyry, *Life of Pythagoras*, 51, in Fideler, *The Pythagorean Sourcebook*, p. 133, the Pythagoreans said that 'if anything was perfect it would make use of this principle, and be adorned according to it'. The Platonically inclined Neopythagoreans saw the heavenly bodies as held together by the Triad, and removing the influence of the Dyad, for which see Dillon, *The Middle Platonists*, p. 356f. In the Hermetic tradition the Triad of God, the cosmos and the human being is frequently reiterated, for example in *Corpus Hermeticum*, X.14; in *Corpus Hermeticum*, XVI.4, Heaven, Earth and Sun are presented as a Triad, with the Sun as the third principle perfecting the whole.

5. The first printed version of Ficino's Latin translation of the *Corpus Hermeticum* appeared in 1471, the dialogues of Plato in 1484, the *Enneads* of Plotinus in 1492 and the writings of Iamblichus, Proclus, Porphyry and other Neo-Platonists in 1497. See Kocku von Stuckrad, *Western Esotericism: A Brief History of Secret Knowledge* (Durham: Acumen Publishing, 2013), pp. 53–56.

6. The resurgence of interest in the Kabbalah was greatly stimulated by the expulsion of the Jews from Spain in 1492, many of whom went to Italy where the ferment of interest in Neoplatonism, Hermeticism and Pythagoreanism was strongest. See Frances A. Yates, *The Occult Philosophy* (Routledge and Kegan Paul, 1983), p. 14. For the Pythagorean revival in the Renaissance, see Kahn, *Pythagoras and the Pythagoreans*, p. 167.

7. Dorn's refers frequently to the Binarius in his tracts. The spiritual biography of the Binarius that follows is based on *De Tenebris Contra Naturam* ('Of Darkness Against Nature'), *De Spagirico Artificio Johannes Trithemii Sententia* ('On the Propositions of Johannes Trithemius concerning the Spagyric Art') and *De Duello Animi cum Corpore* ('Of the Conflict of the Soul and the Body') reproduced in L. Zetzner (ed.), *Theatrum Chemicum* (Ursellis,1659), volume 1. All three treatises have been translated by Paul Ferguson in 2014, and are available online at

<http://www. independent.academia.edu/PaulFerguson2>. A further short but visually important treatise is Dorn's *The Monarchy of the Ternary in Union Versus the Monomachia of the Dyad in Confusion*, translated from the French by Daniel Willens in *Alexandria* 2 (1993), pp. 214–231. For a brief summary of Dorn's spiritual biography of the Binarius, see C. G. Jung, *Psychology and Religion: West and East*, C.W. 11, trans. R. F. C. Hull (London: Routledge and Kegan Paul, 1958), p. 60, n. 47.

8. Gerhard Dorn, *De Tenebris contra Naturam*, in Zetzner op. cit., p. 263.

9. Henry Cornelius Agrippa, *Three Books of Occult Philosophy*, vol. 2, chap. 5, ed. Donald Tyson (St. Paul, MN: Llwellyn Publications, 1993), p. 245f. Translation adapted.

10. See notes 1–4 above.

11. *De Duello Animi cum Corpore* in L. Zetzner, op. cit., p. 282. The Pythagorean number symbolism that Dorn adopts is explained in *De Spagirico Artificio Johannes Trithemii Sententia*, in Zetzner, op. cit., p. 388ff.

12. Dee (1527–1608), was an almost exact contemporary of Dorn (ca. 1530–1584), and both were a generation older than Francis Bacon (1561–1626), who it can safely be said must have loathed everything they stood for. As well as being a Pythagorean and Christian Hermeticist, Dee was an astrologer and magician, renowned in his time as a great magus, and as someone who had insight into the world of spirit. Despite moving in the same court circles, and also having a shared interest in cryptology, it seems that the contact between Dee and Bacon was minimal. For Dee's geometrical symbolism, see *Monas Hieroglyphica*, Theorem 2 and Theorem 20, in C. H. Josten, 'A Translation of John Dee's *Monas Hieroglyphica* (Antwerp, 1564), with an Introduction and Annotations' in *Ambix* 12 (1964), p. 154f and pp. 180–186. Dorn used Dee's 'monas' symbol on the title page of his book *Chymisticum Artificium Naturae* (1568), and was clearly indebted to him. See Peter J. Forshaw, 'The Early Alchemical Reception of John Dee's *Monas Hieroglyphica*' in *Ambix*, 52, 3 (2005), p. 256f.

13. Gerhard Dorn, *De Duello Animi cum Corpore*, in Zetzner, op. cit., pp. 482–483.

14. The geometrical representation of the twofold serpent (Fig. 13.3) can be found in Gerhard Dorn, *The Monarchy of the Ternary*, p. 222.

15. Gerhard Dorn, *The Monarchy of the Ternary*, p. 222.

16. Gerhard Dorn, *De Duello Animi cum Corpore*, in Zetzner, op. cit., p. 480.

17. Gerhard Dorn, *De Duello Animi cum Corpore*, in Zetzner, op. cit., p. 476. Translation by Paul Ferguson (see note 7 above).

18. Gerhard Dorn, *De Tenebris*, in Zetzner, op. cit., p. 465. This identification of the number two with the female goes back to the Pythagorean tradition. According to the fourth century BC Platonist, Xenocrates, the Monad is male and the Dyad is female—a doctrine in all likelihood emanating from Pythagoras. See Kahn, *Pythagoras and the Pythagoreans*, p. 115f. See also notes 2, 3 and 4 above.

19. *De Spagirico Artificio Jo. Trithemii Sententia*, in Zetzner, op. cit., p. 393.

20. Thus, according to Thomas Vaughan, 'Anima Magica Abscondita' in Arthur Edward Waite (ed.), *The Works of Thomas Vaughan* (London: Theosophical Publishing House, 1919), p. 92, 'the triad, joined to the duad, maketh all things one'. This is accomplished through the purification of the tetrad or Quaternarius (corresponding to the four elements). Since Dorn regarded the Quaternarius as the offspring of the Binarius, it was for him as irredeemable as the Binarius. Trithemius of Sponheim's more positive view of the Quaternarius is stated in his letter to Joachim Margrave of Brandenberg, 24 August, 1605, in which the progression is from the ternary and the quaternary to the monad, for which see Forshaw, 'The Early Alchemical Reception

of John Dee's *Monas Hieroglyphica*', p. 249. John Dee echoes the notion that the magus proceeds through the Binary and Ternary to the restoration of Unity in a purified Quaternity. In *Monas Hieroglyphica*, Theorem 20, translated in Josten, 'A Translation of John Dee's *Monas Hieroglyphica*', p. 185, Dee writes: 'This is the way proceeding along which, through binary and ternary, our monad is to be restored to its oneness in a purified quaternary . . .'. The long tradition of not rejecting but integrating the Binarius goes back at least as far as Thierry of Chartres, for which see Theodore Silverstein, 'The Fabulous Cosmology of Bernardus Silvestris', in *Modern Philology*, 46 (1948–9), p. 101f.

Chapter Fifteen

1. Adrien Baillet, *La Vie de M. Descartes* (1691), vol. 1, p. 39, quoted in Marie Louise von Franz, *Dreams* (Boston: Shambhala, 1998), p. 118.
2. Baillet's account is translated in Marie Louise von Franz, *Dreams* (Boston: Shambhala, 1998), pp. 118–121), According to von Franz, op. cit., p. 156, the unknown figure would best be interpreted as the trickster or the Spirit Mercurius. Jacques Maritain, *The Dream of Descartes* (London: Nicholson and Watson, 1946), p. 11, pondering on the identity of the 'spirit of truth', concluded that it was most likely the same spirit as appeared in the *Meditations*, 1, as the 'evil genius' or spirit of deceit.
3. Descartes, *Treatise on Man* (1632), in Ralph M. Eaton, ed., *Descartes: Selections* (London: Charles Scribner's Sons, 1927), p. 350 and p. 354.
4. Descartes, *Meditations*, 6, in Elizabeth Anscombe and Peter Thomas Geach *Descartes: Philosophical Writings* (Sunbury-on-Thames: Thomas Nelson and Sons, 1970), p. 120.
5. Descartes, *Principles of Philosophy* 4, 103 in Anscombe and Geach, *Descartes: Philosophical Writings*, p. 236.
6. See, for example, C. G. Jung, 'Spirit and Life' in *The Structure and Dynamics of the Psyche*, CW 8 (London: Routledge and Kegan Paul, 1970), pars. 618f.
7. *Rules for the Direction of the Mind* in Anscombe and Geach, *Descartes: Philosophical Writings*, pp. 177–180.
8. Thomas Hobbes, *Leviathan* (1651; London: J. M. Dent and Sons, 1973), Ch. 5, p. 18 (spelling updated).
9. Descartes, letter to Mersenne, 20 November, 1629, in *Descartes: Philosophical Letters*, translated and edited by Anthony Kenny (Oxford: Clarendon Press, 1970), p. 5.
10. Ibid.
11. Bacon, *Novum Organum*, 1.38ff, p. 95f.
12. Ibid., 1.43, p. 96.
13. Gerhard Dorn, *De Duello Animi cum Corpore*, in L. Zetner (ed.), *Theatrum Chemicum* (Ursellis, 1659), volume 1, p. 476. See also Chapter 14, p. 5 above. From the perspective of depth psychology, archetypes may rise up from the collective unconscious and manifest in the form of gods or other archetypal figures. As such they may legitimately be regarded as '*daimones* or personal agencies', for which see C. G. Jung, *Symbols of Transformation*, CW.5 (London: Routledge and Kegan Paul, 1956), para. 388. From the esoteric perspective, there are objectively existing spiritual beings that exert an influence upon the psyche from the spiritual world. See Rudolf Steiner, *The Incarnation of Ahriman* (Forest Row: Rudolf Steiner Press, 2006), pp. 19–23.
14. Pascal, *Pensées*, VI.340, in Blaise Pascal, *Thoughts, Letters and Minor Works*, selected by Charles W. Eliot (New York: P. F. Collier and Son, 1910).

15. Pascal, *Pensées* IV.§277, translated by A. J. Krailsheimer (Harmondsworth: Penguin, 1966), p. 154; in the following section, §278, Pascal writes: 'It is the heart which perceives God and not the reason.'
16. See Chapter 7, pp. 61–64.

Chapter Sixteen

1, For the universe as a giant clock, see Leibniz's second letter to Clarke in Leroy E. Loemker, ed., *Gottfried Wilhelm Leibniz: Philosophical Papers and Letters* (Dordrecht: D. Reidel, 1969), p. 679, where he states that the universe is a machine or watch that God has created but, unlike watches manufactured by human beings, it is 'a watch that goes without wanting to be mended'. For natural organisms as machines see, for example, *A New System of the Nature and the Communication of Substances* (1695) in Loemker, op. cit., p. 455, where animals are described as 'organic machines'; and *Monadology* (1714) §64, where 'the machines of nature, that is to say living bodies, are still machines in the least of their parts *ad infinitum*' in G.H.R. Parkinson, *Leibniz: Philosophical Writings* (London and Toronto: J. M. Dent and Sons, 1973), p. 189.
2. For example, *The Principles of Nature and Grace* (1714), in Loemker, op. cit., p. 637, where we read: 'Such a living being is called an *animal*, as its monad is called a *soul*. When this soul is raised to the level of *reason*, it is something more sublime and is counted among the spirits . . .'
3. Leibniz, *The Art of Combination* (1666) in G. H. R. Parkinson ed., *Leibniz: Logical Papers* (Oxford: Clarendon Press, 1966), pp. 1–11.
4. George Dalgarno, *Ars Signorum* (London: 1661); John Wilkins, *Essay Towards a Real Character and a Philosophical Language* (London: 1668). Leibniz's *The Art of Combination* was written in 1666.
5. Leibniz, *Preface to a Universal Characteristic* (1678–9) in R. Ariew and D. Garber, eds., *Leibniz: Philosophical Essays* (Indianapolis: Hacket Publishing Co., 1989), p. 6.
6. Leibniz, *Of the Art of Combination* (1666) in Parkinson, *Leibniz: Logical Papers*, p. 3.
7. To put it more precisely, in a true proposition, the meaning of the predicate term is contained in the meaning of the subject term. See William and Martha Kneale, *The Development of Logic* (Oxford: Oxford University Press, 1962), p. 338.
8. P. H. Nidditch, *The Development of Mathematical Logic* (London: Routledge and Kegan Paul, 1962), p. 19.
9. Leibniz, *Elements of a Calculus* (1679) in Parkinson, *Leibniz: Logical Papers*, p. 17.
10. Ibid., p. 22.
11. In Latin, *Regulae de bonitate consequentiarum*. The full title (in English) is *Rules from which a decision can be made, by means of numbers, about the validity of inferences and about the forms and moods of categorical syllogisms*, reproduced in Parkinson, *Leibniz: Logical Papers*, pp. 25–32.
12. Leibniz, *Elements of a Calculus* (1679) in Parkinson, *Leibniz: Logical Papers*, p. 17f. In his *The Art of Discovery* (1685), in Philip P. Weiner, ed., *Leibniz: Selections* (New York: C. Scribner's Sons, 1951), p. 51, Leibniz writes:

> The only way to rectify our reasonings is to make them as tangible as those of the Mathematicians, so that we can find our error at a glance, and when there are disputes among persons, we can simply say: Let us calculate (*calculemus*), without further ado, to see who is right.

13. Leibniz's letter to Nicolas Remond, January 10, 1714, in Loemker, *Gottfried Wilhelm Leibniz: Philosophical Papers and Letters*, p. 654.
14. Leibniz, *Preface to a Universal Characteristic* in R. Ariew and D. Garber, op. cit., p. 10.
15. See Leibniz's letter of 1 March, 1673 to the editor of the *Journal des Savants*, quoted in L. Couturat, *La Logique de Leibniz d'Après des Documents Inédits* (Paris: F. Alcan, 1901), p. 296, in which Leibniz explains that his calculating machine demonstrates 'a purely mechanical act of thought (*reflexion*)' in a way that 'is obvious to our senses'.
16. L. Couturat, op. cit, p. 115f.
17. Leibniz, *De Progressione Dyadica*, translated in F. L. Bauer, *Origins and Foundations of Computing* (Heidelberg: Springer-Verlag, 2010), p. 15.
18. In *De Progressione Dyadica*, Leibniz wrote the following description of his binary calculating machine, translated in F. L. Bauer, *Origins and Foundations of Computing* (Heidelberg: Springer-Verlag, 2010), p. 15:

> The following method would certainly be very easy and without effort: a container should be provided with holes in such a way that they can be opened and closed. They are to be open at those positions that correspond to a 0. The open gates permit small cubes or marbles to fall through into a channel; the closed gates permit nothing to fall through. They are moved and displaced from column to column as called for by the multiplication. The channels should represent the columns, and no ball should be able to get from one channel to another except when the machine is in motion. Then all the marbles run into the next channel, and whenever one falls into an open hole it is removed.

19. 'Of an *Organum* or *Ars Magna* of Thinking', reproduced in Parkinson, *Leibniz: Philosophical Writings*, p. 3.
20. Letter to Duke Rudolf August, January 2, 1697, quoted in Florian Cajori, 'Leibniz's "Image of Creation"', in *The Monist*, 26.4 (1916), p. 561.
21. This explanation is given in a letter of Leibniz to Bouvet in 1703, translated in Frank J. Swetz, 'Leibniz, the Yijing, and the Religious Conversion of the Chinese' in *Mathematics Magazine*, 76.4 (2003), p. 285f, and summarized in Yueng-Ting Lai, 'Leibniz and Chinese Thought' in A. P. Coudert, et al., eds., *Leibniz, Mysticism and Religion* (Dordrecht: Kluwer Academic Publishers, 1998), p. 151.
22. 'Explanation of Binary Arithmetic' ('*Explication de l'arithmétique binaire*', 1703) translated in Anton Glaser, *History of Binary and Other Nondecimal Numeration* (Los Angeles: Tomash Publishers, 1981), pp. 42–43.
23. Robert Batchelor, 'Binary as Transcultural Technology: Leibniz, *Methesis Universalis*, and the *Yijing*' in Michelle R. Warren and David Glimp, eds., *Arts of Calculation: Quantifying Thought in Early Modern Europe* (New York: Palgrave Macmillan, 2004), p. 247.
24. Rudolf Steiner discusses the entry into modern consciousness of the 'delusion of the duad' and the withholding of 'the truth of the triad' in *The Mission of the Archangel Michael* (New York: Steiner Books, 1961), Lecture 1. Like Steiner, Jung also well understood the saving power of threeness, which he linked to the imaginative consciousness of living symbols that transcend the binary oppositions of logical thought. See, for example, C. G. Jung, *Alchemical Studies*, CW13 (Princeton: Princeton University Press, 1967), §199, p. 162:

In accordance with its paradoxical nature, [the symbol] represents the 'tertium' that in logic does not exist, but which in reality is the living truth.

Chapter Seventeen

1. Park Benjamin, *A History of Electricity*, p. 467f. Hauksbee explained this in terms of invisible (but nevertheless material) 'effluvia' emanating from the electrified body:

   > That the Effluvia pass along, as it were, in so many *Physical Lines or Rays*; and all the Parts that compose them, adhere and joyn to one another, in such manner, that when any of 'em are push'd, all in the same Line are affected by the Impulse given to others. (Quoted in Park Benjamin, op. cit., p. 468.)

2. Brian Baigrie, *Electricity and Magnetism: A Historical Perspective* (Westport, Connecticut: Greenwood Press, 2007), p. 27. See also Herbert W. Meyer, *A History of Electricity and Magnetism* (Norwalk, Connecticut: Burndy Library, Massachusetts Institute of Technology, 1971), p. 17.
3. Benjamin, *A History of Electricity*, p. 479. Note that William Gilbert had extended the range of substances that could be classified as 'electrics', but had not reached the idea that electricity was a force in its own right. For Robert Boyle, see Park Benjamin, op. cit., p. 419.
4. Benjamin, *A History of Electricity*, p. 482.
5. For example C. A. Hausen and G. M. Bose in Germany, and J. A. Nollet in France.
6. Patricia Fara, *An Entertainment for Angels: Electricity in the Enlightenment* (Duxford: Icon Books, 2002), Chapter 3.
7. Heilbron, *Electricity in the 17th and 18th Centuries*, p. 233, who notes that it is the pull on the experimenter's own hairs that causes this sensation. For Von Guericke, see Meyer, *A History of Electricity and Magnetism*, p. 13.
8. Benjamin, *A History of Electricity*, p. 483.
9. Ibid., pp. 494–496.
10. Van Musschenbroek, *Mémoire de l'Académie Royale des Sciences*, 1746, quoted in Benjamin, *A History of Electricity*, p. 519. For a full description of the experiment, see Heilbron, *Electricity in the 17th and 18th Centuries*, p. 313f.
11. Benjamin, op. cit., p. 519.
12. *Philosophical Transactions of the Royal Society*, 44 (1746), pp. 211–212, quoted in Fara, *An Entertainment for Angels*, p. 55.
13. Baigrie, *Electricity and Magnetism*, p. 32.
14. Benjamin, *A History of Electricity*, p. 524. See also Fara, *An Entertainment for Angels*, p. 56f.
15. Heilbron, *Electricity in the 17th and 18th Centuries*, p. 334.
16. Meyer, *A History of Electricity and Magnetism*, p. 13.
17. Heilbron, *Electricity in the 17th and 18th Centuries*, p. 330.
18. Heilbron, *Electricity in the 17th and 18th Centuries*, pp. 330–331, summarizes Franklin's explanation as follows:

   > Give an uncharged jar to a man standing on the floor. Rub the tube and touch it to the bottle's internal wire or hook, thereby throwing, say, one unit of electrical fire into the jar. None crosses the bottle's impermeable bottom: all accumulates within. Where, then, is the minus electrification necessarily associated with this unit of positive electricity? It resides, Franklin says, in the outer coating of the jar: 'At the same time that the wire at the top of the bottle, etc., is electrized *positively* or *plus*,

the bottom of the bottle is electrized *negatively* or *minus*, in exact proportion.' The unit of electrical fire within the bottle drives an equal unit from the other side, through the man holding the coating and into the ground; the ejected unit compensates for the fire pumped *from* the ground, via the tube, into the bottle. Rub the tube again and present it to the hook: a second unit enters the bottle, and another exits from it. The process can be repeated until the external surface of the jar has surrendered all its fire. Then the operation ceases.

Franklin understood that the accumulation of positive charge inside the jar could only occur when the outer surface of the jar was in contact with the earth (through the person holding it), for only then could it surrender all its 'fire'.

19. Benjamin, *A History of Electricity*, p. 554.
20. Fara, *An Entertainment for Angels*, p. 114; see also Michael Brian Schiffer, *Draw the Lightning Down: Benjamin Franklin and Electrical Technology in the Age of Enlightenment* (Berkeley: University of California Press, 2003), p. 283, n. 62.
21. Schiffer, *Draw the Lightning Down*, p. 163f.
22. Ibid., p. 167f.
23. Ibid., p. 162. See also Heilbron, *Electricity in the 17th and 18th Centuries*, p. 342.
24. Fara, *An Entertainment for Angels*, p. 75.
25. William Shakespeare, *All's Well That Ends Well*, II.3.1.
26. Franz Aepinus, *Tentamen Theoriae Electricitatis et Magnetismi* or 'An Attempt at a Theory of Electricity and Magnetism' (St. Petersburg, 1759).
27. The torsion balance enabled Coulomb to demonstrate that the force between two electrical charges is proportional to their product and inversely proportional to the square of the distance between them.
28. Baigrie, *Electricity and Magnetism*, pp. 43–46.
29. Newton, *Principia*, B.iii, quoted in Benjamin, *A History of Electricity*, p. 511:

> ... by the force and action of which spirit [i.e. ether], the particles of bodies mutually attract one another at small distances and cohere when in contact, and electric bodies operate at greater distances as well as by repelling and attracting the neighboring corpuscle, and by which light is emitted.

30. Heilbron, *Electricity in the 17th and 18th Centuries*, p. 300. See also Benjamin, *A History of Electricity*, pp. 53–535.
31. Such views can be found in the work of the eighteenth century German Pietist thinkers, Friedrich Christoph Oetinger, Prokop Divisch and Johann Ludwig Fricker, for which see Ernst Benz, *The Theology of Electricity* (Eugene, Or: Pickwick Publications, 1989), Chapters 3 and 4. See also Nicholas Goodrick-Clarke, 'The Esoteric Uses of Electricity: Theologies of Electricity from Swabian Pietism to Ariosophy', pp. 70–73.
32. Joseph Priestley, *The History and Present State of Electricity* (1775), II.3, quoted in A. Wolf, *A History of Science, Technology and Philosophy in the 18th Century* vol. 1 (London: George Allen and Unwin, 1962), p. 240.

Chapter Eighteen

1. The quotation is from Benjamin Franklin, 'Information to Those who would Remove to America' (1782), quoted in Liah Greenfield, *Nationalism: Five Roads to Modernity* (Cambridge, MA: Harvard University Press, 1992), p. 408. For Franklin's

mechanistic Deism, see Kerry S. Walters, *Benjamin Franklin and His Gods* (Urbana: University of Illinois Press), 1999), p. 52, who observes that Franklin was one of 'those deists who accepted the mechanistic teachings of modern science but took pride in human freedom'.

2. Julien Offray de la Mettrie, *Man a Machine* trans. Richard A. Watson and Maya Rybalka (Cambridge, Indianapolis: Hackett, 1994), p. 72:

> Thought is so far from being incompatible with organized matter that it seems to me to be just another of its properties, such as electricity...

3. La Mettrie, *Man a Machine*, p. 43f.
4. Ibid, p. 65.
5. The last book he wrote, published in the year of his death, 1751, was called *L'art de juir ou l'école de la volupté* ('The Art of Enjoyment or the School of Pleasure').
6. See, for example, Turing's essay 'Can Digital Computers Think?' (1951), in B. Jack Copeland, ed., *The Essential Turing* (Oxford: Clarendon Press, 2004), where he asks the question: 'Is there some particularly human characteristic that could never be imitated by a machine?' and answers: 'I believe no such bounds can be set.' See also his essay 'Intelligent Machinery' (1948) in the same volume. Von Neumann's views on the subject were put forward in his 1953 lectures at Princeton University entitled 'Machines and Organisms', which were summarized by John Kemeny in 'Man Viewed as a Machine' in *Scientific American*, 192 (April, 1955), pp. 58–67. According to von Neumann, the difference between the computer and the human brain is that although computers can calculate faster they lack the compactness of the brain and the ability to deal with complexity. This is a technical issue that will be resolved in time.
7. For the archetype as 'an autonomous factor' see Jung, *Psychology and Religion*, p. 469, para.758:

> The archetypes in question are not mere objects of the mind, but are also autonomous factors, i.e. living subjects...

This characterization of the archetypes enables us to link the thought of Jung to that of Rudolf Steiner, who gave to the 'living subject' in question the name of the ancient Persian adversary power, Ahriman. See note 19 below.

8. A. Wolf, *A History of Science, Technology and Philosophy in the 18th Century*, vol. 2, p. 654 observes:

> The calculating machines of the eighteenth century ... were all in the nature of experiments; most of them were intolerably complicated in their modes of operation, and, despite the claims of their respective inventors, it is doubtful whether any of them ever consistently maintained a satisfactory standard of performance.

The main inventors were Lépine, Hillerin de Boistissandeau and Leupold in the earlier part of the eighteenth century and Stanhope, Hahn and Müller in the later part, for which see Wolf, op. cit., pp. 654–660.

9. Sigvard Strandh, *A History of the Machine*, p. 211.
10. Justin Leiber, in La Mettrie, *Man a Machine*, p. 4.
11. J. D. Bernal, *Science in History*, vol. 2 (Harmondsworth: Penguin, 1969), p. 520.
12. Peter Mathias, *The First Industrial Nation* (London: Routledge, 1983), p. 121.
13. James Phillips Kay, *The Moral and Physical Condition of the Working Classes Employed in the Cotton Manufacture in Manchester* (London: James Ridgway, 1832), p. 10.

14. As René Guénon, *The Reign of Quantity and the Signs of the Times* (London: Luzac, 1953), p. 76f observed:

> The workman in industry cannot put into his work anything of himself, and a lot of trouble would even be taken to prevent him if he had the least inclination to try to do so . . . Servant of the machine, the man must become a machine himself, and thenceforth his work has nothing really human in it, for it no longer implies the putting to work of any of the qualities which really constitute human nature.

15. Peter Ackroyd, *Blake* (London: Sinclair-Stevenson, 1995), p. 130.
16. William Blake, *There is No Natural Religion* (1788) in Geoffrey Keynes, ed., *Poetry and Prose of William Blake* (London: Nonesuch Press, 1967), p. 148. The phrase 'dark, satanic mills' comes from Blake's poem *Milton* in Keynes, op. cit. p. 375.
17. It was not until the second half of the century, with the publication of Darwin's *On the Origin of Species* (1859), that the Church was finally forced onto the defensive.
18. See Friedrich Engels, *The Condition of the Working Class in England* (New York: Cosimo, 2008), pp. 134–187. First published in Leipzig in 1844, Engels' study gives a detailed account of the horrors of life in the mills.
19. See Rudolf Steiner, *The Incarnation of Ahriman*, pp. 17–18, where he characterizes this mechanical intelligence, to which he assigns the name Ahriman, as one of the greatest, most comprehensive and penetrating intelligences in the cosmos, but it is at the same time calculating and utterly cold. It reduces everything with which it works to measure, weight and number, and it operates in a thoroughly mechanistic and deterministic way, entrapping humanity in a narrowly materialistic view of the world. See also the seminal essay by David B. Black, *The Computer and the Incarnation of Ahriman* (New York: St. George Publications, 1981), p. 19.
20. See Chapter 10, pp. 102–103. The phrase, 'the spiritual catastrophe of the Reformation' is from C. G. Jung, *The Structure and Dynamics of the Psyche*, CW 8, p. 338, para. 649.
21. Thus the basis of religious belief for the educated eighteenth century Deist was reason and evidence, rather than revelation and actual spiritual experience. The God of Deism thereby became little more than a philosophical convenience, and the Deists became allies by default of the atheists.
22. Quoted in King, *Western Monasticism*, p. 319.
23. King, ibid., p. 322. For the emperor Joseph II, see King, op. cit., p. 321.
24. King, ibid., pp. 336–337. For the confiscation of church and monastic property in Germany during this period, see Kristina Krüger, *Monasteries and Monastic Orders* (Potsdam: Ullmann, 2008), p. 400; see also Peter King, op. cit. p. 337.
25. See William Durant, *The Age of Napoleon* (New York: Simon and Schuster, 1953), p. 73. Jung on more than one occasion referred to the profound symbolic importance of this event. See C. G. Jung, *Aion*, CW 9.ii, (Princeton: Princeton University Press, 1959), par. 155, and *Civilization in Transition*, CW 10 (London: Routledge, 1964), pars. 174f.
26. Jung, *Civilization in Transition*, par. 174.
27. In France, the brilliant invention by Dr Guillotine of his death-dealing machine, which had the capacity to execute with such chilling efficiency hundreds in a single day, well exemplifies this kind of applied reasoning. The daily stream of executions could even be thought of as a form of grim public worship of the Goddess Reason.

28. Roy Porter, in John Haslam, *Illustrations of Madness* (1810) edited with an introduction by Roy Porter (London: Routledge, 1988), p. xii.
29. It was first properly studied in Viktor Tausk's seminal paper, 'On the Origin of the "Influencing Machine" in Schizophrenia' (1919), translated in *Psychoanalytic Quarterly* (1933), 2: pp. 519–556.
30. Haslam, *Illustrations of Madness*, pp. 31–38.
31. Ibid., p. 39ff.
32. Ibid., p. 35.
33. Babbage's machine did this by translating them into algebraic form. Then it would, as Ada Lovelace explained, 'weave algebraical patterns just as the Jacquard loom weaves flowers and leaves.' See Ada Lovelace, 'Notes Upon the Memoir' appended to L. F. Menabrea, *Sketch of the Analytical Engine Invented by Charles Babbage*, in Philip Morrison and Emily Morrison, eds., *Charles Babbage and His Calculating Engines: Selected Writings by Charles Babbage and Others* (New York: Dover, 1961), p. 273.

Chapter Nineteen
1. Haslam, *Illustrations of Madness*, p. 51.
2. Alfred Barlow, *The History and Principles of Weaving by Hand and Power* (London: Sampson Low, Marston, Searle and Rivington, 1879), p. 130, gives the date as 1567.
3. James Essinger, *Jacquard's Web* (Oxford: Oxford University Press, 2004), p. 11.
4. Essinger, *Jacquard's Web*, p. 30.
5. This description is based on Eric Broudy, *The Book of Looms* (London: Cassell, 1979), p. 134, and the fuller, admirably clear description of Emberson, *From Gondishapur to Silicon Valley*, pp. 159–162.
6. Essinger, *Jacquard's Web*, p. 30.
7. For the influence of Vaucanson on Jacquard, see Anthony Hyman, *Charles Babbage: Pioneer of the Computer* (Oxford: Oxford University Press, 1982), p. 166.
8. Strandh, *A History of the Machine*, p. 226.
9. Essinger, *Jacquard's Web*, p. 42.
10. In England the first power-loom patents were taken out by Edmund Cartwright between 1786 and 1788. See Mathias, *The First Industrial Nation*, p. 117.
11. John Ruskin, *The Stones of Venice*, vol. 3 (1853), XI.64, in Robert L. Herbert, ed., *The Art Criticism of John Ruskin* (New York: Plenum, 1964), p. 20.
12. Doron Swade, *The Cogwheel Brain* (London: Little, Brown and Company, 2000), p. 28f. The method of differences can be explained very simply as follows. Suppose we wish to record the values of the expression $N^2 + N + 41$, when N is a given number from 0 to 5. We can set this out as a table with four columns:

N	N^2+N+41	D_1	D_2
0	41		
1	43	2	
2	47	4	2
3	53	6	2
4	61	8	2
5	71	10	2

In this table column D_1 consists in the differences between the successive entries in the column headed N^2+N+41, while the column headed D_2 shows the differences between successive entries in the column headed D_1. Since all the differences in D_2 are the same, the table can be extended *ad infinitum* without having to do any multiplication or division, for each successive result in the column headed N^2+N+41 can be arrived at by adding the entries in columns D_1 and D_2. For a fuller version of this table, and a fuller explanation, see Hyman, *Charles Babbage: Pioneer of the Computer*, pp. 17–19.

13. Swade, *The Cogwheel Brain*, p. 30.
14. Ibid., p. 48.
15. A smaller and much cruder Difference Engine *was* successfully constructed quite independently by Georg and Edvard Scheutz in Sweden in 1843, followed by a much improved version in 1853. See Swade, *The Cogwheel Brain*, p. 197ff.
16. This, too, can be viewed in the Science Museum in London.
17. When he finally did obtain one (in 1840) he hung it in his drawing room, and proudly showed it to his guests, delighting in the fact that they invariably mistook it for an engraving.

Chapter Twenty
1. Charles Babbage, *Passages from the Life of a Philosopher* (1864) reproduced in Philip Morrison and Emily Morrison, eds., *Charles Babbage and His Calculating Engines: Selected Writings by Charles Babbage and Others* (New York: Dover, 1961), p. 16.
2. Babbage, *Passages from the Life of a Philosopher*, p. 22f.
3. Ibid., p. 33.
4. Hyman, *Charles Babbage*, p. 109, p. 156 and p. 219.
5. Charles Babbage, 'Calculating Engines' (1851) in Morrison and Morrison, *Charles Babbage and His Calculating Engines*, p. 322:

   It is not a bad definition of *man* to describe him as a *tool-making animal*. His earliest contrivances to support uncivilized life were tools of the simplest and rudest construction. His latest achievements in the substitution of machinery, not merely for the skill of the human hand, but for the relief of the human intellect, are founded on the use of tools of a still higher order.

6. According to Hyman, *Charles Babbage*, p. 227, 'Babbage was the outstanding cryptologist of his age, wholly without rival.'
7. *The Ninth Bridgewater Treatise* (London: John Murray,1838), p. 30f, discussed in Hyman, *Charles Babbage*, p. 138f.
8. Nidditch, *The Development of Mathematical Logic*, p. 23.
9. John Passmore, *A Hundred Years of Philosophy* (Harmondsworth: Penguin Books, 1968), p. 122. The example cited is known as the rule of commutation. The same principle applies to the rule of association, that $(a + b) + c = a + (b + c)$ and the rule of distribution, that $a (b + c) = ab + ac$.
10. Hyman, *Charles Babbage*, p. 244.
11. George Boole, *The Mathematical Analysis of Logic* (1847), quoted in William Kneale and Martha Kneale, *The Development of Logic* (Oxford: Oxford University Press, 1962), p. 405:

    That to the existing forms of Analysis a quantitative interpretation is assigned, is the result of circumstances by which those forms were determined, and is not to be construed into a universal condition of Analysis. It is upon the foundation of this general principle, that I purpose to establish the Calculus of Logic . . .

    See also George Boole 'The Calculus of Logic' in *Cambridge and Dublin Mathematical Journal*, 3 (1848).

12. H. W. Buxton, *Memoir of the Life and Labours of the Late Charles Babbage Esq., F.R.S.* (Cambridge, MA: MIT Press, 1988), p. 155:

> It is manifest that the language of algebra is more simple and precise than the symbols of language expressed by sound, and it would seem therefore within the range of our intelligence to be able to reduce our thoughts in most cases into the form of mathematical language, and thus adapt the subject of our enquiry to the operations of the Analytical Engine.

Once reduced to algebra, the 'calculus of logic' (as Boole would have called it) can be applied to human thought by a machine. See Hyman, *Charles Babbage*, p. 244.

13. *Passages from the Life of a Philosopher* (1864) in Morrison and Morrison, *Charles Babbage and His Calculating Engines*, p. 55.
14. Ibid., p. 59.
15. Ada Lovelace, 'Notes Upon the Memoir' appended to L. F. Menabrea, *Sketch of the Analytical Engine Invented by Charles Babbage*, in Morrison and Morrison, *Charles Babbage and His Calculating Engines*, p. 273.
16. Ibid., p. 251.
17. Ibid., p. 249.
18. Ibid., p. 252.
19. The following description is extremely simplified. For the reader interested in a more detailed account, see Allan G. Bromley, 'Charles Babbage's Analytical Engine, 1838' in *Annals of the History of Computing* (July 1982), 4.3, pp. 196–217.
20. The memory capacity was to be vast—at one point Babbage thought it should be 1,000 numbers of 50 digits each. This would have occupied a lot of space! Swade, *The Cogwheel Brain*, p. 115.
21. H. P. Babbage, 'The Analytical Engine' in Morrison and Morrison, *Charles Babbage and His Calculating Engines*, p. 334.
22. Ibid. H. P. Babbage listed the series of operations in the table below. Note that the Number Cards and the Variable Cards are together referred to by him as 'Directive Cards'.

Directive Card	Operation Card	
1st	...	Places a on column 1 of Store
2nd	...	,, b ,, 2 ,,
3rd	...	,, c ,, 3 ,,
4th	...	,, d ,, 4 ,,
5th	...	Brings a from Store to Mill
6th	...	,, b ,, ,,
...	1	Multiplies a and $b = p$
7th	...	Takes p to column 5 of Store where it is kept for use and record
8th	...	Brings p into Mill
9th	...	Brings c into Mill
...	2	Adds p and $c = q$
10th	...	Takes q to column 6 of Store
11th	...	Brings d into Mill
12th	...	,, q ,,
...	3	Multiplies $d \times q = p_2$
13th	...	Takes p_2 to column 7 of Store
14th	...	Takes p_2 to printing or stereo-moulding apparatus

23. See Christopher D. Green, 'Was Babbage's Analytical Engine Intended to be a Mechanical Model of the Mind?' in *History of Psychology*, 8, 1 (2005), pp. 35–45. See also Swade, *The Cogwheel Brain*, p.-114.

24. Hyman, *Charles Babbage*, p. 172.

25. In a letter of 1843, quoted in Swade, *The Cogwheel Brain*, p. 165. The statements of Lovelace scrutinized and corrected by Babbage were appended in the form of 'Notes' to Menabrea's *Sketch of the Analytical Engine*. See Swade, op. cit., pp. 160–161.

26. Swade, *The Cogwheel Brain*, p. 252.

27. In awe of its capabilities, but taking care not to exaggerate them, the Italian mathematician and friend of Babbage, L. F. Menabrea, *Sketch of the Analytical Engine Invented by Charles Babbage* (1842) in Morrison and Morrison, *Charles Babbage and His Calculating Engines*, p. 243, remarked:

> Thus, although it is not itself the being that reflects, it may yet be considered as the being which executes the conceptions of intelligence.

Menabrea's reference to the Engine as a 'being' is particularly striking.

28. Charles Babbage, *Passages from the Life of a Philosopher*, in Morrison and Morrison, *Charles Babbage and His Calculating Engines*, p. 68. See also Hyman, *Charles Babbage*, p. 244.

29. Lovelace, 'Notes Upon the Memoir' in Morrison and Morrison, op. cit., p. 252.

30. Hyman, *Charles Babbage*, p. 172, note 18; and p. 227, note 3. Babbage was a friend of the telegraphic pioneer Charles Wheatstone, and it was after seeing models and drawings of some of Wheatstone's telegraphic equipment that he considered using an electromechanical switching system in his machine.

Chapter Twenty-One

1. J. L. Fricker, 'Kurzen Auszug' ('Brief Excerpt'), p. 71, published as an Appendix to Prokop Divis, *Theorie von der meteorologischen Electricité* (1765), quoted in Benz, *The Theology of Electricity*, p. 46, to which Benz adds the comment:

> This electrical fire of nature, added to matter itself, is the life principle that again and again rushes into new forms, that wants to manifest itself again and again in new living shapes—it is, strictly speaking, the principle of evolution itself that was part of Creation from the beginning . . .

2. Quoted in Alison Muri, *The Enlightenment Cyborg: A History of Communications and Control in the Human Machine, 1660–1830* (Toronto: University of Toronto Press, 2007), p. 71.

3. Adam Walker, *A System of Familiar Philosophy* (London, 1802), vol. 2, p. 7. Walker gives no sources for this claim but is presumably drawing on the earlier experiments of Nollet. Nollet had observed that seeds germinated more swiftly in an electrified environment, and that such an environment also had the effect of accelerating growth. However, he also observed that the plants thus forced were less hardy than those grown under natural conditions. Animals kept in electrified environments were observed to lose weight. For these and other experiments of Nollet, see Benjamin, *A History of Electricity*, pp. 527–530.

4. Walker, *A System of Familiar Philosophy*, pp. 42–44.

5. Ibid., p. 74. See also the discussion in Muir, *The Enlightenment Cyborg*, pp. 71–79.

6. Luigi Galvani, *Commentary on the Effect of Electricity on Muscular Motion (De Viribus*

Electricitatis in Motu Musculari Commentarius) (1791), translated by Robert Montraville Green (Cambridge MA: Elizabeth Licht, 1953).

7. Galvani, op. cit., p. 36.
8. See Lehrs, *Man or Matter*, p. 52.
9. According to Galvani, op. cit., p. 68, the inside of the muscles have a positive charge and their exterior a negative charge. The Leyden jar comparison is given in Galvani, op. cit., p. 61:

> It would perhaps not be an inept hypothesis and conjecture, nor altogether deviating from the truth, which should compare a muscle fibre to a small Leyden jar, or other similar electric body, charged with two opposite kinds of electricity; but should liken the nerve to the conductor, and therefore compare the whole muscle with an assemblage of Leyden jars.

10. Galvani, op. cit., p. 70f.
11. Fara, *An Entertainment for Angels*, p. 165f.
12. The quantitative approach had already been begun by Franklin in the 1740s and was then taken forward by Aepinus in the 1750s and Coulomb in the 1770s and following decades.
13. Coleridge drew on the medieval distinction between *Natura naturans and Natura naturata*, the former referring to the 'pre-phenomenal' creative powers of nature and the latter to the observable phenomena produced by them. For Coleridge's use of this distinction, see Owen Barfield, *What Coleridge Thought* (San Rafael, CA: Barfield Press, 1971), Chapter 2.
14. As Coleridge, quoted in Barfield, op. cit., p. 204, put it:

> The term, matter, therefore, taken separately, should be confined to the Phaenomena—i.e. to Length and Breadth without Depth—now as in bodies the only universal Evidence of Depth is Weight, therefore matter but not body should be attributed to imponderable Phaenomena—Light, Heat, Magnetism, Electricity are material but not corporeal.

For Coleridge's reflections on electricity, see Barfield, op. cit., pp. 37–40.
15. While this is left open in the novel, in Mary Shelley's Introduction to *Frankenstein* she makes clear that at the back of her mind the life-injecting machine is an electric battery. See Mary Shelley, *Frankenstein* (Harmondsworth: Penguin Books, 1985), p. 54f, and n. 5 on p. 263f.
16. Shelley, Frankenstein, p. 101.
17. Humphry Davy, *A Discourse Introductory to a Course of Lectures on Chemistry* (1802) in *The Collected Works of Sir Humphry Davy*, ed. John Davy (London: Smith, Elder and Co., 1839) Volume 2, p. 319 [emphasis added]. Davy ominously continues:

> Science has done much for man, but is capable of doing still more; its sources of improvement are not yet exhausted; the benefits that it has conferred ought to excite our hopes of its capability of conferring new benefits...

Mary Shelley is known to have attended Humphry Davy's lectures at the Royal Institution in 1812, and was reading his *Discourse* (or possibly a later publication by Davy) about six months before she began work on *Frankenstein*. It is likely that Professor Waldman in the novel, who inspired and encouraged Frankenstein, was modelled on Davy. The words of Waldman's lecture, which had such a profound impact on the young Frankenstein (*Frankenstein*, pp. 91–92) are similar in tone to Davy's *Discourse*. See Richard Holmes, *The Age of Wonder: How the Romantic Generation Discovered the Beauty and Terror of Science* (London: HarperCollins, 2008), pp. 325–330.

18. Wolf, *A History of Science, Technology and Philosophy in the 18th Century*, vol. 1, p. 260.
19. Baigrie, *Electricity and Magnetism*, p. 52.
20. Wolf, op. cit., p. 262.
21. In the case of Galvani's frogs, the convulsions of the frogs' legs were caused by the chemical action between the two metals and the juices of the frog tissues.
22. William Nicholson, 'Account of the new Electrical Apparatus of Sig. Alex. Volta, and Experiments performed with the same' in *A Journal of Natural Philosophy, Chemistry and the Arts*, vol. 4 (July, 1800), p. 181.
23. Carlisle and Nicholson conducted several experiments. In the first, the hydrogen revealed itself in bubbles at the silver end (negative charge), while the oxygen revealed itself through the oxidization of the brass wire at the zinc end (positive charge), which became tarnished and from which 'whitish filmy clouds' were emitted. Nicholson, op. cit., p. 182f. In the next experiment (illustrated in Fig. 21.4), two much finer wires of platina were inserted into a narrower tube containing water, and when the circuit was closed 'the silver side gave a plentiful stream of fine bubbles, and the zinc side also a stream less plentiful. No turbidness nor oxidation, nor tarnish appeared . . .' Nicholson, op. cit., p. 185.
24. His researches were preceded by those of Joseph Black, who discovered carbon dioxide in 1753, Daniel Rutherford who discovered nitrogen in 1772, and Carl Scheele who discovered oxygen—also in 1772.
25. William of Ockham, *Summulae*, 1, 14, quoted in Carré, *Realists and Nominalists*, p. 119. The full quotation is:

> It is impossible to have first matter (*prima materia*) without extension, for matter cannot exist without having parts distant from part. But this amounts to asserting that matter is extended, quantitative and has dimensions.

See also Chapter 10, pp. 64–67.

26. Thus the element Earth is characterized by the qualities of coldness and dryness, Water by coldness and moistness, Air by moistness and warmth, and Fire by warmth and dryness.
27. As Lehrs, *Man or Matter*, p. 278 pointed out:

> To think in this way [i.e. atomistically] is inherent in the spectator-mind, compelled as it is to proceed from the parts to the whole.

See also Bortoft, *The Wholeness of Nature*, p. 174f., quoted in Chapter 10 above, n. 36.

28. For the concept of the 'lower border' see Chapter 11, pp. 108–114. In this ambitious endeavour, Davy worked closely with John Dalton, who laid the foundations of atomism in chemistry, setting out the first list of 'atomic weights' of the chemical elements. John Dalton, *A New System of Chemical Philosophy* (1808), set out the Galilean intent of his work in the following words, quoted in F. Greenaway, *John Dalton and the Atom* (London: Heinemann, 1966), p. 132:

> . . . it is one great object of this work, to show the importance and advantage of ascertaining the relative weights of the ultimate particles both of simple and

compound bodies, the number of simple elementary particles which constitute one compound particle, and the number of less compound particles which enter into the formation of one more compound particle.

Some years before the publication of this book, Davy helped Dalton to hone the presentation of his atomistic model in his lectures at the Royal Institute in December 1803 and January 1804. See John Gribbin, *Science: A History, 1543–2001* (London: Penguin, 2003), p. 367.

Chapter Twenty-Two

1. For example, in a famous experiment conducted by William Watson, Henry Cavendish, a certain Dr Bevis and others in August, 1748, a circuit of 12, 276 feet was set up at Shooter's Hill, London, using wires supported by dry sticks, through which a Leyden jar was discharged. It was concluded that the transmission was instantaneous. See Meyer, *A History of Electricity and Magnetism*, p. 19.

2. The first detailed proposal for an alphabetical telegraph was put forward by an anonymous contributor to the *Scots' Magazine* in February 1753. It is reproduced in full in J. J. Fahie, *A History of Electric Telegraphy to the Year 1837* (London: E. and F. N. Spon, 1884), pp. 68–71. Subsequent to this, in 1767 Joseph Bozolus proposed an electric telegraph in which the spark was the active principle; in 1782 an anonymous letter to the *Journal de Paris* put forward another detailed proposal for a telegraph based on static electricity via the Leyden jar, which was taken up by George Louis Le Sage in the same year.

3. See Fahie, *A History of Electric Telegraphy*, pp. 91–93. Many more proposals followed, however, which continued to seek ways of using static electricity to send alphabetic messages, notably Jean Alexandre's so-called 'secret telegraph' of 1802, and the later proposals of Wedgwood, Ronald, Dyar and others. Despite their ingenuity, none of them received the necessary government backing, and so were never adopted. See Fahie, op. cit., Chapters 2–5.

4. For Francisco Salvá Campillo, see Fahie, *A History of Electric Telegraphy*, pp. 220–227; for Sömmerring, see Fahie, op. cit., pp. 227–243.

5. Ampère's paper 'Sur l'action mutuelle entre deux courans électrique' in *Annales de Chimie et de Physique* (1820), vol. 15, showed amongst other things, that two parallel wires carrying an electric current acted like magnets on each other: they were attracted to each other if the currents flowed in the same direction but were repelled from each other if the currents flowed in opposite directions. Since, in the case of static electricity, unlike charges attract and like charged repel, a quite different law was thereby shown to be operating with electric currents. See Baigrie, *Electricity and Magnetism*, pp. 68–70.

6. This was the great achievement of his masterwork, *Mémoire sur la théorie mathématique des phénomènes électrodynamique, uniquement déduite de l'expérience* (1827) ('Memoir on the Mathematical Theory of Electrodynamic Phenomena Deduced Solely from Experiment').

7. See Chapter 13, pp. 139–141 above.

8. Fahie, *A History of Electric Telegraphy*, pp. 309–311.

9. Ibid., pp. 312–319.

10. Ibid., p. 342. He also thought that two bells, each sounding a different tone, would work just as well. See Fahie, op. cit., pp. 336–342.

11. In England, at roughly the same time as the Morse/Vail system was introduced, the new telegraph system of Wheatstone and Cooke relied on the old-style technique of reversing the current in order to flick magnetized needles at the receiving end one way or the other.

12. Benjamin, *The Age of Electricity*, p. 231.

13. The dot was vocalized as 'di' except at the end of a letter when it became a 'dit'. For example, the letter 'B' would be vocalized as 'Dah-di-di-dit'.

14. The idea of nature embodying the language of God was a commonplace in the Middle Ages. For the medieval view of language, see Chapter 7, pp. 64–67.

15. Kathleen Ruppert, 'Gilded Age Art and Literature' in Kevin Hillstrom and Laurie Collier Hillstrom, eds., *The Industrial Revolution in America: Communications* (Santa Barbara: ABC-CLIO, 2007), pp. 205–207, records the praises lavished on the new electric telegraph by contemporaneous American poets who should have known better.

Chapter Twenty-Three

1. Boole was not a friend of Babbage's, although they did meet towards the end of Babbage's life, in 1862. See Hyman, *Charles Babbage*, p. 249.

2. George Boole, *An Investigation of the Laws of Thought* (1854; New York: Dover, 1958), pp. 29–34.

3. Ibid., p. 27.

4. For example, in his essay, 'The Calculus of Logic' in *Cambridge and Dublin Mathematical Journal*, vol. 3 (1848), pp. 183–185, Boole argued that well known algebraic laws such as those of association $[x + (y + w) = (x + y) + w]$, distribution $[x (y + w) = xy + xw]$, and commutation $[x + y = y + x]$ are as much logical as mathematical.

5. Boole, *An Investigation of the Laws of Thought*, p. 11.

6. Ibid., pp. 422–3.

7. Boole, 'The Calculus of Logic', p. 98.

8. Boole, *An Investigation of the Laws of Thought*, p. 37f.

9. Boole refers to the 'Universe of the Proposition' in *The Mathematical Analysis of Logic* (1847; Oxford; Oxford University Press, 1949), p. 49f. See also Kneale and Kneale, *The Development of Logic*, pp. 413–14. Boole refers to the 'law of duality' in *An Investigation of the Laws of Thought*, p. 51, where he claims that it is impossible for the human mind to go beyond thinking in dualities, although he does qualify this later in the book, on p. 423.

10. Strictly speaking, 1 and 0 were for Boole classes, the 1 denoting the universe or 'all', and the 0 denoting nothing (i.e. an empty set). However, applied to the universe of the proposition, if we say 'x is true', this can be translated into the equation '$x = 1$', just as 'x is false' can be translated into the equation '$x = 0$'.

11. See Chapter 7, pp. 66-67.

12. Boole, *An Investigation into the Laws of Thought*, p. 423.

13. Notably Peirce in the U.S.A. and Shröder in Germany, who both worked in the Boolean tradition of algebraic logic, and Frege, Peano and Russell who deviated significantly from it.

14. W. S. Jevons, *Principles of Science* (1874) quoted in Gardner, *Logic Machines and Diagrams*, p. 100. Gardner gives a detailed description of Jevons' logic machine in Chapter 5.

15. Although Boole himself did not explicitly formulate the idea of tabulating the truth

or falsity of logical propositions in Truth Tables, it is implicit in his thought, and Gottlob Frege developed the idea directly from Boole. See Kneale and Kneale, *The Development of Logic*, p. 420. It was subsequently taken up by Shröder, Peirce, Post and Wittgenstein.

16. Bernal, *Science in History*, vol. 1, pp. 612–616, and vol. 2, p. 729.

17. By Claude E. Shannon, in his paper 'A Symbolic Analysis of Relay and Switching Circuits' in *Transactions of the American Institute of Electrical Engineers*, 57 (1938).

18. Translated in Jean van Heijenoort *From Frege to Gödel: A Source Book in Mathematical Logic*, 1879–1931 (Cambridge, MA: Harvard University Press, 1990). For the esoteric significance of the year 1879, see Rudolf Steiner, *The Fall of the Spirits of Darkness* (Forest Row: Rudolf Steiner Press, 1993), Lectures 9 and 12. See also Black, *The Computer and the Incarnation of Ahriman*, p. 37f.

19. Gottlob Frege, *Begriffsschrift*, Preface, in van Heijenoort *From Frege to Gödel*, p. 7:

> If it is one of the tasks of philosophy to break the domination of the word over the human spirit by laying bare the misconceptions that through the use of language often almost unavoidably arise concerning the relations between concepts and by freeing thought from that with which only the means of expression of ordinary language, constituted as they are, saddle it, then my ideography, further developed for these purposes, can become a useful tool for the philosopher.

20. Gottlob Frege, 'What is a Function?' (1904), reproduced in Geach and Black, *Translations from the Philosophical Writings of Gottlob Frege* (Oxford: Basil Blackwell, 1966), p. 116. Frege believed that 'logic has hitherto always followed ordinary language and grammar too closely' (*Begriffsschrift*, Preface, in van Heijenoort *From Frege to Gödel*, p. 7).

21. Whereas signs are man-made and have but one unambivalent meaning, symbols are multivalent, reflecting and 'making present' transcendent realities. For a discussion of the traditional understanding of symbols and the symbolic consciousness, see S. H. Nasr, 'Scientia Sacra', in Martin Lings and Clinton Minnaar, eds., *The Underlying Religion: An Introduction to the Perennial Philosophy* (Bloomington, Indiana: World Wisdom, 2007), pp. 137–139.

22. John of Salisbury, *Metalogicon*, 1.22. See Chapter 7 above, p. 65.

23. For John of Salisbury, *Metalogicon*, I.22 'Poetry is the cradle of philosophy'. Heidegger held the same view. See, for example, his essay 'Poetically Man Dwells' (1951) in Martin Heidegger, *Poetry, Language and Thought* (New York: Harper-Collins, 1971).

24. Frege, 'On the Scientific Justification of a Conceptual Notation', quoted in Arthur Sullivan, *Logicism and the Philosophy of Language: Selections from Frege and Russell* (Toronto: Broadview Press, 2003), p. 120f. Frege's horror at the possibility of multiple levels of meaning is what drives his assault on language. He once wrote:

> ... when properly expressed, a thought leaves no room for different interpretations. We have seen that ambiguity [*Vieldeutigkeit*] simply has to be rejected.

Quoted in Warren Goldfarb, 'Frege's Conception of Logic' in Michael Beaney and Erich H. Reck, eds., *Frege's Philosophy of Logic* (London: Routledge, 2005), p. 53.

25. The subtitle of Frege's *Begriffsschrift* is 'a formula language, modelled upon that of arithmetic, for pure thought'.

26. The variable element is referred to as the 'argument'. One of Frege's most important

innovations was to substitute the mathematical distinction between argument and function for the traditional distinction between subject and predicate. If the argument is the variable or replaceable element in a sentence, then the 'function' is the fixed or constant element. As this distinction would take too long to explain further, the interested reader is referred to a very clear and succinct account given by Anthony Kenny, *A New History of Western Philosophy* (Oxford: Oxford University Press, 2010), pp. 833–836. For a fuller discussion, see Anthony Kenny, *Frege: An Introduction to the Founder of Modern Analytic Philosophy* (Oxford: Blackwell, 2000), Chapters 1 and 6.

27. The traditional understanding is well expressed in the pithy statement of Thomas Aquinas, *Commentary on Aristotle's 'De Interpretatione'*, 1.2, n. 3: 'Logic is ordered towards obtaining knowledge of things.' See Chapter 7, p. 67 above, with notes 20 and 21. For a modern critique of symbolic logic from this traditional standpoint, see G. R. G. Mure, *Retreat from Truth* (Oxford: Blackwell, 1958), pp. 123–142.

28. Martin Davis, *Engines of Logic* (New York: Norton, 2000), p. 53, according to whom Frege's Concept Script 'made it possible to exhibit logical inferences as purely mechanical operations, so-called rules of inference, having reference only to the patterns in which symbols are arranged'. Davis points out that in order to carry out the rule of inference, 'no understanding of what $\supset$ means is required'.

29. Martin Heidegger, 'Recollection in Metaphysics' in *The End of Philosophy* (New York: Harper and Row, 1973), p. 80. For Heidegger's critique of symbolic logic, see Albert Borgmann, 'Heidegger and Symbolic Logic' in M. Murray, ed., *Heidegger and Modern Philosophy* (New Haven and London: Yale University Press, 1978).

30. See, for example, Heidegger's 'Letter on Humanism' (1947) in Martin Heidegger, *Basic Writings* edited by D. F. Krill (New York: HarperCollins, 1993), where he writes (p. 246):

> Everything depends on this alone, that the truth of Being come to language and that thinking attain to this language.

This is exactly what the formula language of symbolic logic prevents.

31. See, for example, Kenny, *Frege: An Introduction to the Founder of Modern Analytic Philosophy*.

32. For example, A. J. Ayer, *Language, Truth and Logic* (Harmondsworth: Penguin, 1972), p. 46, writes:

> For we shall maintain that no statement which refers to a 'reality' transcending the limits of all possible sense-experience can possibly have any literal significance; from which it must follow that the labours of those who have striven to describe such a reality have all been devoted to the production of nonsense.

33. See Philip Wadler, 'Proofs are Programs: 19th Century Logic and 21st Century Computing' (November 2000), <http://www. home pages.inf.ed.ac.uk/wadler/papers/frege/frege.pdf> accessed 31 January, 2017.

34. Both Martin Heidegger and Alan Turing understood this intimate relationship between symbolic logic and computer technology. In *What is Called Thinking?* (New York: Harper and Row, 1968), p. 238, Heidegger described symbolic logic as 'an irrepressible development' that was 'meanwhile bringing about the electronic computer'. Heidegger pointed out that one reason for the widespread adoption of symbolic logic in Anglo-Saxon countries is that 'its results and procedures imme-

diately yield something definite towards the construction of the technological world. Hence, in America and elsewhere, symbolic logic, as the proper philosophy of the future, begins to assume its reign over the human spirit.' (op. cit., p. 21f). Turing said in a lecture given in 1947 to the London Mathematical Society:

> I expect digital computing machines will eventually stimulate a considerable interest in symbolic logic . . . The language in which one communicates with these machines . . . forms a sort of symbolic logic.

Quoted in Davis, *Engines of Logic*, p. 199.

Chapter Twenty-Four

1. Faraday's dates are September 22, 1791–August 25 1867; Babbage's dates are December 26, 1791–October 18, 1871. The year of their birth was, significantly, the year of the publication of Galvani's, *De Viribus Electricitatis*, which led to Volta's creation of the first 'pile' (or battery) capable of producing a continuous current of electricity nine years later.

2. Joseph Henry in the United States was working in a similar direction. His short paper 'On a Reciprocating Motion Produced by Magnetic Attraction and Repulsion' was published in *American Journal of Science and Arts*, 20 (July, 1831), pp. 340–343. In Russia, the physicist Heinrich Lenz formulated the law of electromagnetic induction in 1834. Many other scientists in Europe were working in the same field, but Faraday's researches led the way.

3. This had been done by the French scientist Dominique François Jean Arago. See Baigrie, *Electricity and Magnetism*, p. 68.

4. Michael Faraday, 'Experimental Researches in Electricity' in *Philosophical Transactions of the Royal Society*, 122 (1832), p. 131, pars. 27–28.

5. Ibid., p. 132, par. 30.

6. Ibid., p. 144, pars. 71 and 76.

7. Ibid., p. 134, par. 39.

8. Faraday built the first practical (although very simple) dynamo in 1834. See Peter Day, *The Philosopher's Tree: A Selection of Michael Faraday's Writings* (Bristol: Institute of Physics, 1999), p. 118f.

9. Although Faraday was aware of the changing magnetic flux in relation to the induced electric current, it fell to Heinrich Lenz to focus more closely on its significance, and in 1834 to formulate a law of electromagnetic induction now known as 'Lenz's Law': namely, that the electric current induced due to motion in a magnetic field will oppose the change in magnetic flux and exert a mechanical force opposing the motion.

10. Michael Faraday, *Experimental Researches in Electricity*, Vol. 1 (London: Taylor and Francis, 1839), Series 5, §517.

11. See the interesting discussion in Geoffrey Cantor, *Michael Faraday: Sandemanian and Scientist* (London: MacMillan, 1991), pp. 175–181, who argues that 'Faraday's universe was a universe of power, and that power was of divine origin' (p. 177). It is perhaps worth noting that to an ancient Egyptian priestly observer, the presence of the god Seth would undoubtedly have been recognized in these processes. Seth was the god of turmoil and strife, and his name means 'The Overpowering One'. He was also the instigator of division, and known as 'The Separator'. See H. Te Velde, *Seth, God of Confusion*, pp. 3–6. See also Chapter 2, p. 12 above, with n. 5.

12. Faraday, 'Experimental Researches in Electricity' in *Philosophical Transactions of the Royal Society* (1832), 122, p. 152, pars. 101 (metal strips) and 106 (copper wire).

13. Ibid,, p. 193, par. 256.

14. Ibid., p. 148, pars. 84–86.

15. Ibid., p. 154, par. 114. For the copper disk as a succession of radii, see ibid., p. 157, par. 119.

16. Martin Heidegger, 'Traditional Language and Technological Language' (1962) translated by Wanda Torres Gregory in *Journal of Philosophical Research*, 23 (1998), p. 137:

> It is not that natural science is the foundation of technology, but rather that modern technology is the main characteristic of modern natural science.

17. Letter to Ada Lovelace, 24 October, 1844, in Frank A. L. James ed., *The Correspondence of Michael Faraday, Volume 3: 1841–1848* (London: The Institution of Engineering and Technology, 1996), Letter 1631, p. 266. Faraday wrote this in reply to a letter of Ada Lovelace (16 October, 1844) declaring her belief that only 'through a high spiritual and moral development' could 'the more subtle and occult agents of nature' be attained, while at the same time declaring herself to be a Unitarian Christian with Swedenborgian, Roman Catholic and Rosicrucian sympathies! See James, op. cit., Letter 1620, p. 254. The correspondence between the two makes fascinating reading.

18. Cantor, *Michael Faraday: Sandemanian and Scientist*, pp. 177f:

> By contrast with the study of light, electricity was not only central to his scientific research, being the subject of numerous papers and lectures, but it was also for Faraday the most magnificent and potent of God's powers. In his view electricity was the highest power known to man.

19. Discussed in Cantor, op. cit., p. 177. Faraday's early biographer, J. H. Gladstone, *Michael Faraday* (New York: Harper and Brothers, 1872), p. 115, comments:

> No doubt his electrical knowledge added much to his interest in the grand discharges from the thunder-clouds, but it will hardly account for his standing long at a window watching the vivid flashes, a stranger to fear, with his mind full of lofty thoughts, or perhaps of high communings. Sometimes, too, if the storm was at a little distance, he would summon a cab, and, in spite of the pelting rain, drive to the scene of awful beauty.

Rudolf Otto, *The Idea of the Holy* (Harmondsworth: Penguin, 1959), p. 32ff, aptly describes thunder and lightning as manifestations of the awful majesty of the *mysterium tremendum*, often associated with the 'wrath of Yahweh'. Thus we gain the impression of Faraday as like an initiate of the ancient mysteries of the thunder gods, meeting the death-dealing force of electricity with an unshakeable commitment to living a life of utmost moral purity, this latter characteristic being noted by all his biographers.

20. Michael Faraday, *Experimental Researches in Electricity*, Vol. 3 (London: Taylor and Francis, 1855), Nineteenth Series, pp. 1–26. See also Day, *The Philosopher's Tree*, pp. 122–125, in which the sequence of experiments is briefly described, culminating in Faraday's declaration on 30 September, 1845:

I have at last succeeded in *illuminating a magnetic curve* or *line of force* and in magnetizing a ray of light.

On Faraday's relative lack of interest in light as such, see Cantor, *Michael Faraday: Sandemanian and Scientist*, pp. 177f.

21. Faraday first set out this view in his essay 'A Speculation touching Electric Conduction and the Nature of Matter' in *Philosophical Magazine*, Series 3, Vol. 24, Issue 157 (1844), pp. 136–144, republished in *Experimental Researches in Electricity*, Volume 2 (London: Richard and John Edward Taylor, 1844), pp. 284–293. He returned to the subject again in his lecture published as 'Thoughts on Ray Vibrations' in *Philosophical Magazine*, Series 3, Vol. 28, Issue 188 (1846), pp. 345–50. In the earlier essay ('A Speculation . . .') he writes:

> Matter is not merely mutually penetrable but each atom extends, so to say, throughout the whole of the solar system, yet always retaining its own centre of force. (Reprinted in *Experimental Researches in Electricity*, Volume 2, p. 293.)

22. He used the term 'magnetic field' for the first time in his laboratory notebook on November 7, 1845. See Frank A. J. L. James, *Michael Faraday: A Very Short Introduction* (Oxford: Oxford University Press, 2010), p. 81.

23. See Maxwell's Introduction to *A Treatise on Electricity and Magnetism*, Volume 1 (Oxford: Oxford University Press, 1873), where he also explains (p. x) what it was that so appealed to him in Faraday's thought:

> Faraday, in his mind's eye, saw lines of force traversing space where the mathematicians saw centres of force attracting at a distance: Faraday saw a medium where they saw nothing but distance: Faraday sought the seat of the phenomena in real actions going on in the medium . . .

24. Maxwell, 'A Dynamical Theory of the Electromagnetic Field' (1864) in W. D. Niven, ed., *The Scientific Papers of James Clerk Maxwell*, Volume 1 (Cambridge: Cambridge University Press, 1890), p. 556. Unlike Maxwell, Faraday did not embrace the concept of the aether, thinking it was dispensable.

25. Maxwell, 'A Dynamical Theory' in Niven, op. cit., p. 564.

26. James Clerk Maxwell, 'On Faraday's Lines of Force' in Niven, *The Scientific Papers of James Clerk Maxwell*, Volume 1, pp. 155–157.

27. Letter to André-Marie Ampère, 3 September, 1822, quoted in Day, *The Philosopher's Tree*, p. 96f, where Faraday compares himself to 'a timid navigator' who is 'afraid to leave sight of the shore because he understands not the power of the instrument that is to guide to him'. Faraday also famously wrote to Maxwell more than 30 years later in 1857, giving voice to the wish that mathematicians might express themselves in ordinary language and translate their formulae 'out of their hieroglyphics', for which see Lewis Campbell, *The Life of James Clerk Maxwell* (London: MacMillan, 1882), p. 290.

28. Quoted in Campbell, op. cit., p. 355f.

29. Ibid., p. 356.

30. As Baigrie, *Electricity and Magnetism*, p. 94, has well observed:

> By definition, a potential does not actually exist until it is manifested in experiment. The field around a magnet, for example, has a potential that is not realized until an electric charge is placed in that field. Only by interacting with something

can the potential of the field be manifested; until this time, it has only the potential to do something.

For Maxwell, the concept of potential was the basis of field theory. See, for example, James Clerk Maxwell, 'A Dynamical Theory' in Niven, op. cit., §15, p. 533. See also Maxwell, *A Treatise on Electricity and Magnetism*, Volume 1, p. xi.

31. In a letter to R. B. Litchfield, of 5 March, 1858, reproduced in Campbell, *The Life of James Clerk Maxwell*, pp. 304–306, Maxwell writes (p. 305):

> With respect to the 'material sciences', they appear to me to be the appointed road to all *scientific* truth, whether metaphysical, mental, or social. The knowledge which exists on these subjects derives a great part of its value from ideas suggested by analogies from the material sciences, and the remaining part, though valuable and important to mankind, is not *scientific* but aphoristic.

32. Maxwell, 'On Faraday's Lines of Force' in Niven, op. cit., p. 156.

33. The starkness of Maxwell's view of the scope of scientific enquiry is revealed in his paper 'On Physical Lines of Force' in Niven, op. cit., p. 451, where he states:

> In electric and magnetic phenomena, the magnitude and direction of the resultant force at any point is the main subject of investigation.

34. Acording to his biographer Campbell, *The Life of James Clerk Maxwell*, p. 170, n. 164, Maxwell was most strongly influenced by Scottish Calvinism, but he did not identify with any 'particular school of religious opinion', and scrupulously kept all 'religious opinion' out of his science. Thus in correspondence with C. J. Ellicott, Bishop of Gloucester and Bristol on the subject of light (in November 1876), published in Campbell, op. cit., p. 190f, the question of the spiritual nature of light is entirely subsumed by Maxwell's scientific conjectures on the aether.

35. Maxwell, 'On Physical Lines of Force' in Niven, op. cit., p. 500:

> Light consists in the transverse undulations of the same medium which is the cause of electric and magnetic phenomena.

> In 'A Dynamical Theory of the Electromagnetic Field', in Niven, op. cit., §97, p. 580, Maxwell made the more definitive statement:

> Light and magnetism are affections of the same substance, and ... light is an electromagnetic disturbance propagated through the field according to electromagnetic laws.

36. Hertz's first experimental work showed that electricity had no inertia. See Heinrich Hertz, *Miscellaneous Papers*, translated by D. E. Jones and G. A. Schott (London: MacMillan, 1896), pp. xi–xvi, and Chapter One.

37. Hertz used a Ruhmkorff coil, which actually consists of two coils, one inside the other. The primary coil has relatively few turns of insulated copper wire, but the secondary has many thousands of turns, so it acts as a transformer, vastly amplifying the voltage coming from the battery. It is connected to a mechanism called an 'interrupter' that repeatedly interrupts the Direct Current from the batteries (he used six Bunsen cells), thereby collapsing the magnetic field and at once re-establishing it, thus enabling induction to take place in the coil. He describes the apparatus in his paper, 'On Very Rapid Electric Oscillations' (1887) in Heinrich Hertz, *Electric*

Waves: Researches on the Propagation of Electric Action with Finite Velocity Through Space, translated by D. E. Jones (London: MacMillan, 1893), p. 31f.

38. Baigrie, *Electricity and Magnetism*, pp. 106–107.
39. It was also possible to make adjustments to the oscillator, which terminated at either end in a zinc plate, by adding further plates that made it act as a capacitor, the distance between the original and the additional plate affecting the intensity of the sparks and frequency of oscillations. See 'On the Finite Velocity of Electromagnetic Actions' (1888) in Hertz, *Electric Waves*, p. 111.
40. 'The Forces of Electric Oscillations, Treated According to Maxwell's Theory' in Hertz, *Electric Waves*, pp. 144–149.
41. 'The Forces of Electric Oscillations' in Hertz, *Electric Waves*, p. 146. See also op. cit., p. 15: 'The seat and field of action of the waves is not in the interior of the conductor, but rather in the surrounding space.' And again (op. cit., p. 19): 'The electric forces can disentangle themselves from material bodies, and can continue to subsist as conditions or changes in the state of space.'
42. In his lecture 'On the Relations Between Light and Electricity' (1889) in Hertz, *Miscellaneous Papers*, p. 313, Hertz states:

> Light of every kind is an electrical phenomenon—the light of the sun, the light of a candle, the light of a glow-worm. Take away from the world electricity, and light disappears . . .

43. 'On the Relations Between Light and Electricity' (1889) in Hertz, *Miscellaneous Papers*, p. 326. Italics added.
44. In the Introduction to *Electric Waves*, p. 20, Hertz wrote:

> The object of these experiments was to test the fundamental hypotheses of the Faraday-Maxwell theory, and the result of the experiments is to confirm the fundamental hypotheses of the theory.

45. 'On the Relations Between Light and Electricity' (1889) in Hertz, *Miscellaneous Papers*, p. 318.
46. Ibid. p. 21.

Chapter Twenty-Five

1. Gribbin, *Science: A History*, p. 505. This means the diameter of an atom is 0.1 nanometres. The diameters of atoms do vary, ranging between 0.1 and 0.5 nanometres.
2. Neils Bohr, *Atomic Physics and the Description of Nature* (London: Cambridge University Press, 1934), p. 57.
3. J. R. Oppenheimer, *Science and the Common Understanding* (Oxford: Oxford University Press, 1954), p. 42f. F. Sherwood Taylor, *The World of Science* (London: Heinemann, 1950), p. 276, in attempting to summarize the new science, hazarded the following view of the electron:

> The likeliest view is that an electron is a portion of energy: we cannot really picture this, but anything we *can* picture about the electron seems to be untrue.

4. Thomas Aquinas, *The Principles of Nature (De Principiis Naturae)*, I.2, translated by Robert P. Goodwin (Indianapolis: Bobbs-Merrill Company, Inc, 1965), p. 8:

> Likewise, properly speaking, what is in potency to exist substantially is called *prime matter*...

5. Faraday, 'A Speculation Touching Electric Conduction' in *Experimental Researches in Electricity*, Vol. 2 contains some of his most interesting statements, for example the following on p. 289f:

> The safest course appears to be to assume as little as possible, and in that respect the atoms of Boscovich appear to me to have a great advantage over the more usual notion. His atoms, if I understand alright, are mere centres of forces or powers, not particles of matter, in which the powers themselves reside.

6. For a comprehensive review of research on the health effects of electromagnetic radiation, see *The BioInitiative Report* (2012), available online at http:/www.bioinitiative.org. See also Katie Singer, *An Electronic Silent Spring* (Great Barrington, MA: Portal Books, 2014) and Arthur Firstenberg, *The Invisible Rainbow: A History of Electricity and Life* (Santa Fe: AGB Press, 2017).

7. Hertz, 'On the Relations Between Light and Electricity' (1889) in *Miscellaneous Papers*, p. 313. Italics added.

8. Rudolf Steiner (1861–1925), philosopher and esotericist, who had a very thorough knowledge of the physical sciences of his day, while at the same time developing faculties of spiritual perception that enabled him to gain profound insights into the spiritual forces at work within nature, characterized electricity in the following way at the end of his lecture, *The Etherisation of the Blood* (London: Rudolf Steiner Press, 1971) given in 1911:

> When light is thrust down into the sub-material state—that is to say, a stage deeper than the material world—electricity arises.

9. Steiner, *Anthroposophical Leading Thoughts*, p. 218. Translation adapted.

10. A good example of the 'biological computer' view of the human being is to be found in Richard Dawkins, *The God Delusion* (London: Transworld Publishers, 2007), pp. 411–412, where Dawkins refers to human beings as 'chunks of complex matter' and the brain as an 'on-board computer'. For the outlook of 'dataism', raised to the status of a new religion, see Yuval Noah Harari, *Homo Deus* (London: Penguin Random House, 2016), Chapter 12.

11. See Stephen Talbott's illuminating essay, 'Awakening from the Primordial Dream' in Stephen Talbott, *The Future Does Not Compute* (Sebastopol, CA: O'Reilly and Associates, 1995).

12. Ray Kurzweil, *The Singularity is Near* (London: Duckworth, 2005), p. 313f. Kurzweil is one of the foremost advocates for a merger of human beings and machines, for which see also Ray Kurzweil, *The Age of Spiritual Machines* (London: Penguin, 2000), Chapter 12.

13. *The Gospel of Luke*, 2:19 and 2:51.

BIBLIOGRAPHY

Abelson, Paul. *The Seven Liberal Arts: A Study in Medieval Culture*. New York: Columbia University Press, 1906.

Ackroyd, Peter. *Blake*. London: Sinclair-Stevenson, 1995.

Aetius, *Placita Philosophorum*. In Pseudo-Plutarch, *Sentiments Concerning Nature With Which Philosophers Were Delighted* in *The Complete Works of Plutarch: Essays and Miscellanies*. New York: Crowell, 1909.

Agricola, Georg. *De Re Metallica* (1556). New York: Dover, 1950.

Agrippa, Henry Cornelius. *Three Books of Occult Philosophy*. Edited by Donald Tyson. St Paul, MN: Llwellyn Publications, 1993.

Albertus Magnus. *The Book of Minerals (De Mineralibus)*. Translated by Dorothy Wyckoff. Oxford: Oxford University Press, 1967.

Allen, T. G. *Occurrences of Pyramid Texts*. Chicago: University of Chicago Press, 1951.

Andeli, Henri d'. *The Battle of the Seven Arts*. Translated by Louis John Paetow. Berkeley: University of California Press, 1914.

Anderson, William. *Dante the Maker*. London: Hutchinson, 1983.

Appleyard, Rollo. 'Pioneers of Electrical Communication—Heinrich Rudolf Hertz' in *Electrical Communication*, 6, 2 (1927).

Archimedes. *On the Equilibrium of Planes*. Translated in Morris R. Cohen and I. E. Drabkin, *A Sourcebook in Greek Science*. Cambridge, Massachusetts, 1948.

Aquinas, Thomas. *Commentary on the Gospel of John: Chapters 6–12*. Translated by Fabian R. Larcher and James A. Weisheipl. Washington, DC: Catholic University of America Press, 2010.

—— *Commentary on Aristotle's Nicomachean Ethics*. Translated by C. I. Litzinger. Chicago: Henry Regnery Company, 1964.

—— *Questiones Disputatae de Potentia Dei, (On the Power of God)*. Translated by the English Dominican Fathers. Westminster, Maryland: The Newman Press, 1952.

—— *Selected Philosophical Writings,* translated by Timothy McDermott. Oxford: Oxford University Press, 1993.

—— *Summa Theologiae: A Concise Translation*. Edited by Timothy McDermott. London: Eyre and Spottiswoode, 1989.

—— *The Principles of Nature (De Principiis Naturae)*. Translated by Robert P. Goodwin. Indianapolis: Bobbs-Merrill Company, Inc, 1965.

Aristotle, *De Anima (On the Soul)*. Translated by W. S. Hett. London: William Heinemann, 1975.

—— *Metaphysics*. Translated by Hugh Tredennick. London: William Heinemann, 1975.

—— *Meteorologica*. Translated by H. D. P. Lee. London: William Heinemann, 1952.

—— *Nicomachean Ethics*. Translated by H. Rackham. London: William Heinemann, 1975.

—— *Physics*. Translated by P. H. Wickstead and F. M. Cornford. London: William Heinemann, 1934.

—— *Prior Analytics*. Translated by Hugh Tredennick. London: William Heinemann, 1973.

Armstrong, A. H. *An Introduction to Ancient Philosophy*. Boston: Beacon Press, 1963.

Augustine. *The Confessions of St. Augustine*. Translated by F. J. Sheed. London: Sheed and Ward, 1948.

—— *The Trinity (De Trinitate)*. Translated by Edmund Hill. Brooklyn, New York: New City Press, 1991.

Ayer, A. J. *Language, Truth and Logic*. Harmondsworth: Penguin, 1972.

Babbage, Charles. *The Ninth Bridgewater Treatise*. London: John Murray, 1838.

—— *Passages from the Life of a Philosopher* (1864). In Philip Morrison and Emily Morrison (eds.), *Charles Babbage and His Calculating Engines: Selected Writings by Charles Babbage and Others*. New York: Dover, 1961.

Babbage, H. P. 'The Analytical Engine'. In Philip Morrison and Emily Morrison (eds.), *Charles Babbage and His Calculating Engines: Selected Writings by Charles Babbage and Others*. New York: Dover, 1961.

Bacon, Francis. *The Works of Francis Bacon*, vol.1. Edited by James Spedding, et al. London: Longman, 1858.

—— *Selected Philosophical Works*. Edited by Rose-Mary Sargent. Indianapolis: Hackett Publishing Inc., 1999.

Baigrie, Brian. *Electricity and Magnetism: A Historical Perspective*. Westport, Connecticut: Greenwood Press, 2007.

Barfield, Owen. *Saving the Appearances: A Study in Idolatry*. London: Faber and Faber, 1957.

—— *History in English Words*. Edinburgh: Floris, 1985.

—— *What Coleridge Thought*. San Rafael, CA: Barfield Press, 1971.

Barlow, Alfred. *The History and Principles of Weaving by Hand and Power*. London: Sampson Low, Marston, Searle and Rivington, 1879.

Barnes, Jonathan. *Early Greek Philosophy*. London: Penguin, 1987.

Batchelor, Robert. 'Binary as Transcultural Technology: Leibniz, *Mathesis Universalis*, and the *Yijing*' in Michelle R. Warren and David Glimp (eds.), *Arts of Calculation: Quantifying Thought in Early Modern Europe*. New York: Palgrave Macmillan, 2004.

Bauer, F. L. *Origins and Foundations of Computing*. Heidelberg: Springer-Verlag, 2010.

Beazley, J. D. *Attic Red-Figure Vase-Painters*. Oxford: Oxford University Press, 1942.

Bedini, Silvio A. and Maddison, Francis R. *Mechanical Universe: The Astrarium of Giovanni de' Dondi*. Philadelphia: American Philosophical Society, 1966.

Benjamin, Park. *A History of Electricity (The Intellectual Rise in Electricity): From Antiquity to the Days of Benjamin Franklin*. New York: John Wiley, 1898.

—— *The Age of Electricity*. New York: Charles Scribner's, 1892.

Benz, Ernst. *The Theology of Electricity*. Eugene, Or: Pickwick Publications, 1989.

Bernal, J. D. *Science in History*. Four volumes. Harmondsworth: Penguin, 1969.

BioInitiative Report (2012). <http://www. bioinitiative.org>

Black, David B. *The Computer and the Incarnation of Ahriman*. New York: St George Publications, 1981.

Black, Jeremy and Green, Anthony. *Gods, Demons and Symbols of Ancient Mesopotamia*. London: British Museum Press, 1992.

Blake, William. *Poetry and Prose*. Edited by Geoffrey Keynes. London: Nonesuch Press, 1967.

Boethius. *The Consolation of Philosophy*. Translated by S. J. Tester. London: Harvard University Press, 1973.

Bohr, Neils. *Atomic Physics and the Description of Nature*. London: Cambridge University Press, 1934.

Boole , George. *The Mathematical Analysis of Logic* (1847). Oxford; Oxford University Press, 1949.

—— 'The Calculus of Logic' in *Cambridge and Dublin Mathematical Journal,* 3 (1848).

—— *An Investigation of the Laws of Thought* (1854). New York: Dover, 1958.

Borgmann, Albert. 'Heidegger and Symbolic Logic'. In M. Murray (ed.), *Heidegger and Modern Philosophy.* New Haven and London: Yale University Press, 1978.

Bortoft, Henri. *The Wholeness of Nature: Goethe's Way of Science.* Edinburgh: Floris Books, 1996.

Brodrick, M. and Morton, A. A. *A Concise Dictionary of Egyptian Archaeology.* Chicago: Ares Publishers, 1924.

Bromley, Allan G. 'Charles Babbage's Analytical Engine, 1838.' *Annals of the History of Computing* (July 1982), 4.3.

Bronfenbrenner, Martha Ornstein. *The Role of Scientific Societies in the Seventeenth Century.* New York: Arno Press, 1975.

Brooke, Christopher. *The Monastic World: 1000–1300.* London: Paul Elek, 1974.

Broudy, Eric. *The Book of Looms.* London: Cassell, 1979.

Bruton, Eric. *The History of Clocks and Watches.* London: Orbis Publishing, 1979.

Budge, E. A. Wallis. *The Book of the Dead* (1923). London: Arkana 1985.

Burkert, Walter. *Greek Religion.* Oxford: Basil Blackwell, 1987.

Burnham, Dorothy. *Warp and Weft: A Textile Terminology.* Toronto: Royal Ontario Museum, 1980.

Burtt, E. A. *The Metaphysical Foundations of Modern Physical Science.* London: Routledge and Kegan Paul, 1932.

Butler, George F. 'Milton's Meeting with Galileo: A Reconsideration'. *Milton Quarterly,* 39, 3, (2005).

Buxton, H. W. *Memoir of the Life and Labours of the Late Charles Babbage Esq., F.R.S.* Cambridge, MA: MIT Press, 1988.

Cajori, Florian. 'Leibniz's "Image of Creation" ', in *The Monist,* 26.4 (1916).

Campbell, Lewis. *The Life of James Clerk Maxwell.* London: MacMillan, 1882.

Cantor, Geoffrey. *Michael Faraday: Sandemanian and Scientist.* London: MacMillan, 1991.

Capella, Martianus. *De Nuptiis Philologiae et Mercurii.* Translated in W. H. Stahl and R. Johnson with E. L. Burge in *Martianus Capella and the Seven Liberal Arts,* vol. 2. New York: Columbia University Press, 1977.

Carré, Meyrick H. *Realists and Nominalists.* Oxford: Oxford University Press, 1946.

Cicero, Marcus Tullius. *De Republica.* Translated and edited by Niall Rudd and Jonathan G. F. Powell in Cicero, *The Republic and the Laws.* Oxford: Oxford University Press, 2016.

—— *De Divinatione.* Translated by W. A. Falconer in Cicero, *On Divination.* Cambridge, Mass: Harvard University Press, 1923.

—— *On the Nature of the Gods.* Translated by Horace C. P. McGregor. Harmondsworth: Penguin, 1972.

Clagett, Marshall, et al, (eds). *Twelfth Century Europe and the Foundations of Modern Society.* Madison: University of Wisconsin Press, 1961.

Clark, Mary T. *Augustine.* London: Continuuum, 1994.

Clark, R. T. Rundle. *Myth and Symbol in Ancient Egypt.* London: Thames and Hudson, 1978.

Close, Frank. *The Void.* New York: Sterling, 2009.

Cobbett, William. *A History of the Protestant Reformation*. Dublin: James Duffy, 1826.

Cohen, Morris R. and Drabkin, I. E. *A Sourcebook in Greek Science*. Cambridge, Massachusetts, 1948.

Conlon, Thomas E. *Thinking About Nothing: Otto Von Guericke and the Magdeburg Experiments*. London: St. Austin Press, 2011.

Copenhaver, Brian P. *Hermetica*. Cambridge: Cambridge University Press, 1992.

Copeland, B. Jack (ed.). *The Essential Turing*. Oxford: Clarendon Press, 2004.

Copleston, Frederick. *A History of Medieval Philosophy*. London: Methuen, 1972.

Cornford, F. M. *Principium Sapientiae: A Study of the Origins of Greek Philosophical Thought*. Gloucester Mass: Peter Smith, 1971.

Cotterell, Brian and Kamminga, Johan. *Mechanics of Pre-industrial Technology*. Cambridge: Cambridge University Press, 1990.

Coudert, A. P. et al. (eds.). *Leibniz, Mysticism and Religion*. Dordrecht: Kluwer Academic Publishers, 1998.

Courtenay, William J. '*Antiqui* and *Moderni* in Late Medieval Thought'. *Journal of the History of Ideas*, 48, 1 (Jan–March,1987).

Couturat, L. *La Logique de Leibniz d'Après des Documents Inédits*. Paris: F. Alcan, 1901.

Crombie, A. C. *From Augustine to Galileo*. London: Heinemann, 1957.

Crosby, Alfred W. *The Measure of Reality: Quantification and Western Society*. Cambridge: Cambridge University Press, 1997.

Curtius, Ernst Robert. *European Literature and the Latin Middle Ages*. London: Routledge and Kegan Paul, 1979.

Cushman, Robert Earl. *Therapeia: Plato's Conception of Philosophy*. New Brunswick, NJ: Transaction Publishers, 2002.

Daiber, Hans. 'The Meteorology of Theophrastus in Syriac and Arabic Translation'. In William Wall Fortenbaugh and Dimitri Gutas, *Theophrastus: His Psychological, Doxographical and Scientific Writings*. New Brunswick, NJ: Translation Publishers, 1992.

Dalley, Stephanie. *Myths from Mesopotamia*. Oxford: Oxford University Press, 1991.

Davis, Martin. *Engines of Logic*. New York: Norton, 2000.

Davies, Norman de G. *Two Ramesside Tombs at Thebes*. NY: Metropolitan Museum of Art, 1927.

Davies, Norman and Nina de G. *The Tombs of Menkheperrasonb*. London: Egypt Exploration Society, 1933.

Davy, Humphry. *The Collected Works*. Nine volumes. Edited by John Davy. London: Smith, Elder and Co., 1839–40.

Dawkins, Peter. *Building Paradise*. Warwick: The Francis Bacon Research Trust, 2001.

Dawkins, Richard. *The God Delusion*. London: Transworld Publishers, 2007.

Day, Peter. *The Philosopher's Tree: A Selection of Michael Faraday's Writings*. Bristol: Institute of Physics, 1999.

Dee, John. *Monas Hieroglyphica*. Translated by C. H. Josten in 'A Translation of John Dee's *Monas Hieroglyphica* (Antwerp, 1564), with an Introduction and Annotations.' *Ambix,* 12 (1964).

Delaporte, L. *Catalogue des Cylindres Orientaux,* II. Paris: Libraire Hachette, 1923.

Descartes, René. *The Philosophical Works of Descartes*. Edited by Elizabeth S. Haldane and G. R. T. Ross. Cambridge: Cambridge University Press, 1980.

—— *Selections*. Edited by Ralph M. Eaton. London: Charles Scribner's Sons, 1927.

—— *Philosophical Writings*. Edited by Elizabeth Anscombe and Peter Thomas Geach. Sunbury-on-Thames: Thomas Nelson and Sons, 1970.

—— *Philosophical Letters*. Translated and edited by Anthony Kenny. Oxford: Clarendon Press, 1970.

Deschanel, A. P. *Elementary Treatise on Natural Philosophy, Part 3: Electricity and Magnetism*. New York: D. Appleton and Co., 1876.

Dijksterhuis, E. J. *The Mechanization of the World Picture*. Oxford: Oxford University Press, 1961.

Dijksterhuis, E. J. and Forbes, R. J. *A History of Science and Technology*. 2 vols. Harmondsworth: Penguin, 1963.

Dillon, John. *The Middle Platonists*. London: Duckworth, 1996.

Dionysius the Areopagite. Pseudo-Dionysius, *The Complete Works*. Translated by Colm Luibheid and Paul Rorem. New York: Paulist Press, 1987.

Doppelmayr, J. G. *Neu-entdeckte Phaenomena von bewunderswüdigen Wirkungen der Natur* Nurenburg, 1774.

Dorn, Gerhard. *De Tenebris Contra Naturam* ('Of Darkness Against Nature'). In L. Zetzner (ed.), *Theatrum Chemicum*. Ursellis,1659, volume 1. Translated by Paul Ferguson at <http://www. independent.academia.edu/PaulFerguson2>

—— *De Spagirico Artificio Johannes Trithemii Sententia* ('On the Propositions of Johannes Trithemius concerning the Spagyric Art').). In L. Zetzner (ed.), *Theatrum Chemicum*. Ursellis,1659, volume 1. Translated by Paul Ferguson at <http://www. independent.academia.edu/PaulFerguson2>

—— *De Duello Animi cum Corpore* ('Of the Conflict of the Soul and the Body'). In L. Zetzner (ed.), *Theatrum Chemicum*. Ursellis,1659, volume 1. Translated by Paul Ferguson at <http://www. independent.academia.edu/PaulFerguson2>

—— *The Monarchy of the Ternary in Union Versus the Monomachia of the Dyad in Confusion*. Translated from the French by Daniel Willens in *Alexandria*, 2 (1993).

Durant, William. *The Age of Napoleon*. New York: Simon and Schuster, 1953.

Durgin, William A. *Electricity: Its History and Development*. Chicago: A. C. McClurg, 1912.

Eliade, Mircea. *The Forge and the Crucible*. Chicago: University of Chicago Press, 1978.

Ellard, Peter. *The Sacred Cosmos*. Scranton: University of Scranton Press, 2007.

Emberson, Paul. *From Gondhishapur to Silicon* Valley. Tobermory: Etheric Dimensions Press, 2009.

Engels, Friedrich. *The Condition of the Working Class in England*. New York: Cosimo, 2008.

Erman, Adolf. *Life in Ancient Egypt* (1894). New York: Dover, 1971.

Essinger, James. *Jacquard's Web*. Oxford: Oxford University Press, 2004.

Evans, Joan (ed.). *The Flowering of the Middle Ages*. London: Thames and Hudson, 1966.

Fahie, J. J. *A History of Electric Telegraphy to the Year 1837*. London: E. and F. N. Spon, 1884.

Fara, Patricia. *An Entertainment for Angels: Electricity in the Enlightenment*. Duxford: Icon Books, 2002.

Faraday, Michael. 'Experimental Researches in Electricity.' *Philosophical Transactions of the Royal Society*, 122 (1832).

—— *Experimental Researches in Electricity*, Volume 1. London: Taylor and Francis, 1839.

—— 'A Speculation touching Electric Conduction and the Nature of Matter.' *Philosophical Magazine*, Series 3, Vol. 24, Issue 157 (1844).

—— *Experimental Researches in Electricity*, Volume 2. London: Richard and John Edward Taylor, 1844.

—— 'Thoughts on Ray Vibrations.' *Philosophical Magazine* Series 3, Vol. 28, Issue 188 (1846).

—— *Experimental Researches in Electricity*, Volume 3. London: Taylor and Francis, 1855.

Farmer, H. G. *The Organ of the Ancients from Eastern Sources*. London: W. Reeves, 1931.

Farré-Olivé, Eduard. 'A Medieval Catalan Clepsydra and Carillon'. *Antiquarian Horology*, 18.4 (1989).

Faulkner, R. O. *The Ancient Egyptian Book of the Dead*. London: British Museum Publications, 1985.

Fideler, David (ed.). *The Pythagorean Sourcebook*. Grand Rapids, Michigan: Phanes Press, 1988.

Figuier, Louis. *Les Merveilles de la Science*. Six volumes. Paris: Furne, 1867–1891.

Finocchiaro, Maurice A. *The Galileo Affair*. Berkeley: University of California Press, 1989.

Firstenberg, Arthur. *The Invisible Rainbow: A History of Electricity and Life*. Santa Fe: AGB Press, 2017.

Fitzgerald, Allan D. *Augustine Through the Ages*. Michigan, Grand Rapids: Wm. R. Eerdmans Publishing Co., 1999.

Flaceliere, Robert. *Greek Oracles*. London: Elek Books, 1965.

Forbes, R. J. and Dijksterhuis, E. J. *A History of Science and Technology*. 2 vols. Harmondsworth: Penguin, 1963.

Forbes, R. J. 'Power', in Singer, et al., *A History of Technology*, vol. 2.

Forshaw, Peter J. 'The Early Alchemical Reception of John Dee's *Monas Hieroglyphica'*. *Ambix,* 52, 3 (2005).

Frankfort, H. and H. A. 'Myth and Reality' in Henri Frankfort et al., *Before Philosophy*. Harmondsworth: Penguin, 1949.

Frankfort, Henri. *Kingship and the Gods*. Chicago: University of Chicago Press, 1978.

Franz, Marie Louise von. *Dreams*. Boston: Shambhala, 1998.

Freeth, T. et al. 'Decoding the ancient Greek astronomical calculator known as the Antikythera Mechanism'. *Nature*, vol. 444, Issue 7119 (2006).

Frege, Gottlob. 'What is a Function?' (1904). In Peter Geach and Max Black, *Translations from the Philosophical Writings of Gottlob Frege*. Oxford: Basil Blackwell, 1966.

Galilei, Galileo. *Discoveries and Opinions*. Translated by Stillman Drake. New York: Anchor Books, 1957.

—— *Dialogue Concerning the Two Chief World Systems*. Berkeley: University of California Press, 1962.

Gallon, Jean-Gaffin. *Machines et Inventions Approuvées par l'Académie Royale des Sciences*, vol. 4. Paris, 1735.

Galvani, Luigi. *Commentary on the Effect of Electricity on Muscular Motion (De Viribus Electricitatis in Motu Musculari Commentarius)* (1791). Translated by Robert Montraville Green. Cambridge MA: Elizabeth Licht, 1953.

Gange, Hedley. 'A Comprehensive Approach to Electricity'. *Science Forum*, 1 (1979).

Gardiner, Alan. *Egyptian Grammar*. Oxford: Oxford University Press, 1927.

Gardner, Martin. *Logic Machines and Diagrams*. Brighton: Harvester Press, 1982.

Geach, P. and Black, M. *Translations from the Philosophical Writings of Gottlob Frege*. Oxford: Basil Blackwell, 1966.

Gentili, Bruno. *Poetry and its Public in Ancient Greece*. Baltimore: The Johns Hopkins University Press, 1988.

Gibson, Margaret. 'A Picture of *Sapientia* from S. Sulpice, Bourges'. In *Transactions of the Cambridge Bibliographical Society*, vol. 6, no. 7 (1973).

Gilbert, Allan H. 'Milton and Galileo'. *Studies in Philology*, 19, 2 (1922).

Gilbert, William. *On the Magnet (De Magnete)*. Translated by S. P. Thompson. London: Chiswick Press, 1900.

Gilson, Etienne. *History of Christian Philosophy in the Middle Ages*. London: Sheed and Ward, 1980.

Gimpel, Jean. *The Medieval Machine: The Industrial Revolution of the Middle Ages*. London: Futura, 1979.

Gladstone, J. H. *Michael Faraday*. New York: Harper and Brothers, 1872.

Glaser, Anton. *History of Binary and Other Nondecimal Numeration*. Los Angeles: Tomash Publishers, 1981.

Goff, Jacques Le. *Time, Work and Culture in the Middle Ages*. Chicago: University of Chicago Press, 1980.

Goldfarb, Warren. 'Frege's Conception of Logic'. In Michael Beaney and Erich H. Reck (eds.), *Frege's Philosophy of Logic*. London: Routledge, 2005.

Goodrick-Clarke, Nicholas. 'The Esoteric Uses of Electricity: Theologies of Electricity from Swabian Pietism to Ariosophy.' *Aries*, 4, 1 (2004).

Grant, Edward. 'Medieval Explanations and Interpretations of the Dictum that "Nature Abhors a Vacuum" '. *Traditio*, 29 (1973).

—— *Physical Science in the Middle Ages*. Cambridge: Cambridge University Press, 1977.

Graves, Robert. *The Greek Myths*. Two volumes. Harmondsworth: Penguin, 1955.

Green, Christopher D. 'Was Babbage's Analytical Engine Intended to be a Mechanical Model of the Mind?' in *History of Psychology*, 8, 1 (2005).

Greenaway, F. *John Dalton and the Atom*. London: Heinemann, 1966.

Greenfield, Liah. *Nationalism: Five Roads to Modernity*. Cambridge, MA: Harvard University Press, 1992.

Gribbin, John. *Science: A History, 1543–2001*. London: Penguin, 2003.

Grosseteste, Robert. *On Light (De Luce)*. Translated by Clare C. Reidl. Milwaukee: Marquette University Press, 1942.

Guénon, René. *The Reign of Quantity and the Signs of the Times*. London: Luzac, 1953.

Gurney, O. R. 'The Babylonians and Hittites', in Carmen Blacker and Michael Loewe (eds.), *Divination and Oracles*. London: George Allen and Unwin, 1981.

Guthrie, W. K. C. *The Greeks and Their Gods*. Boston: Beacon Press, 1955.

—— *History of Greek Philosophy*, vol.1. Cambridge: Cambridge University Press, 1962.

Hamblin, C. L. 'An Improved *Pons Asinorum*?' *Journal of the History of Philosophy*, 14 (1976).

Handorf, D. E. von. 'The Baghdad Battery: Myth or Reality?' in *Plating and Surface Finishing*, 89 (2002).

Harari, Yuval Noah. *Homo Deus*. London: Penguin Random House, 2016.

Harris, W. V. *Ancient Literacy*. Cambridge, Mas: Harvard University Press, 1989.

Harrison, Jane. *Myths of the Odyssey in Art and Literature*. London: Rivingtons, 1882.

—— *Themis: A Study of the Social Origins of Greek Religion*. London: Merlin Press, 1977.

Harrison, Peter. *The Bible, Protestantism and the Rise of Natural Science*. Cambridge: Cambridge University Press, 2001.

—— 'The Bible and the Emergence of Modern Science'. *Science and Christian Belief*, 18.2 (October 2006).

Harthan, John. *Books of Hours*. London: Thames and Hudson, 1977.

Haskins, C. H. *The Renaissance of the Twelfth Century*. New York: Meridian Books, 1957.

Haslam, John. *Illustrations of Madness* (1810). Edited with an introduction by Roy Porter. London: Routledge, 1988.

Heidegger, Martin. *What is Called Thinking?* New York: Harper and Row, 1968.

—— *Poetry, Language and Thought*. New York: HarperCollins, 1971.

—— *The End of Philosophy*. New York: Harper and Row, 1973.

—— *Basic Writings*. Edited by D. F. Krill. New York: HarperCollins, 1993.

—— 'Traditional Language and Technological Language' (1962). Translated by Wanda Torres Gregory in *Journal of Philosophical Research*, 23 (1998).

Heidel, Alexander. *The Gilgamesh Epic and Old Testament Parallels*. Chicago: University of Chicago Press, 1949.

Heijenoort, Jean van. *From Frege to Gödel: A Source Book in Mathematical Logic, 1879–1931*. Cambridge, MA: Harvard University Press, 1990.

Heilbron, J. L. *Electricity in the 17th and 18th Centuries: A Study of Early Modern Physics*. Berkeley: University of California Press, 1979.

Heliodorus, *Aethiopica*. Athens: Athenian Society, 1897.

Henry, D. P. *Medieval Logic and Metaphysics*. London: Hutchinson, 1972.

Henry, Joseph. 'On a Reciprocating Motion Produced by Magnetic Attraction and Repulsion.' *American Journal of Science and Arts*, 20 (July, 1831).

Herbert, Robert L. (ed.). *The Art Criticism of John Ruskin*. New York: Plenum, 1964.

Herington, John. *Poetry into Drama: Early Tragedy and the Greek Poetic Tradition*. Berkeley: University of California Press, 1985.

Herodotus, *Histories*. Translated by Aubrey de Sélincourt. Harmondsworth: Penguin, 1972.

Hertz, Heinrich. *Electric Waves: Researches on the Propagation of Electric Action with Finite Velocity Through Space*. Translated by D. E. Jones. London: MacMillan, 1893.

—— *Miscellaneous Papers*. Translated by D. E. Jones and G. A. Schott. London: MacMillan, 1896.

Hesiod. *Works and Days*. Translated by Glenn W. Most. Cambridge MA and London: Harvard University Press, 2006.

—— *Theogony*. Translated by Glenn W. Most. Cambridge MA and London: Harvard University Press, 2006.

Hobbes, Thomas. *Leviathan* (1651). London: J.M. Dent and Sons, 1973.

Hodges, Henry. *Technology in the Ancient World*. Harmondsworth: Penguin, 1971.

Holmes, Richard. *The Age of Wonder: How the Romantic Generation Discovered the Beauty and Terror of Science*. London: HarperCollins, 2008.

Homer. *Odyssey*. Translation by A. T. Murray. London: Harvard University Press, 1995.

Hornung, Erik. *Idea into Image: Essays on Ancient Egyptian Thought*. New York: Timken Publishers, 1992.

Hyman, Anthony. *Charles Babbage: Pioneer of the Computer*. Oxford: Oxford University Press, 1982.

Irvine, Martin. *The Making of Textual Culture: 'Grammatica' and Literary Theory, 350–1100*. Cambridge: Cambridge University Press, 1994.

Isidore of Seville. *The Etymologies of Isidore of Seville*. Translated by Stephen A. Barney et al. Cambridge: Cambridge University Press, 2006.

Jacobsen, Thorkild. *The Treasures of Darkness*. New Haven and London: Yale University Press, 1976.

Jaeger, Werner. *Paideia*, vol. 1. Oxford: Oxford University Press, 1965.

James, Frank A. L. (ed.). *The Correspondence of Michael Faraday, Volume 3: 1841–1848.* London: The Institution of Engineering and Technology, 1996.

—— *Michael Faraday: A Very Short Introduction.* Oxford: Oxford University Press, 2010.

Jastrow, Morris. *Religious Belief in Babylonia and Assyria.* New York: Benjamin Blom, 1911.

Jeauneau, Édouard. *Rethinking the School of Chartres.* Toronto: University of Toronto Press, 2009.

Jeyes, Ullah. 'The Act of Extispicy in Ancient Mesopotamia: An Outline' in Bendt Alster (ed.), *Assyriological Miscellanies*, vol. 1. Copenhagen: University of Copenhagen Institute of Assyriology, 1980.

—— *Old Babylonian Extispicy: Omen Texts in the British Museum.* Istanbul: Nederlands Historich-Archaeologisch Institut, 1989.

John of Salisbury. *Metalogicon.* Translated by Daniel D. McGarry. Philadelphia: Paul Dry Books, 2009.

Josten, C. H. 'A Translation of John Dee's *Monas Hieroglyphica* (Antwerp, 1564), with an Introduction and Annotations.' *Ambix* 12 (1964).

Jung, C. G. *Symbols of Transformation* (CW 5). Translated by R. F. C. Hull. London: Routledge and Kegan Paul, 1956.

—— *The Structure and Dynamics of the Psyche* (CW 8). Translated by R. F. C. Hull. London: Routledge and Kegan Paul, 1970.

—— *The Archetypes and the Collective Unconscious* (CW 9.i). Translated by R. F. C. Hull. London: Routledge, 1980.

—— *Aion* (CW 9.ii). Translated by R. F. C. Hull. Princeton: Princeton University Press, 1959.

—— *Civilization in Transition* (CW 10). Translated by R. F. C. Hull. London: Routledge, 1964.

—— *Psychology and Religion: West and East* (CW 11). Translated by R. F. C. Hull. London: Routledge and Kegan Paul, 1958.

—— *Alchemical Studies* (CW 13). Translated by R. F. C. Hull. Princeton: Princeton University Press, 1967.

Kahn, Charles H. *Pythagoras and the Pythagoreans: A Brief History.* Indianapolis: Hackett Publishing Company, 2001.

Katzenellenbogen, Adolf. 'The Representation of the Seven Liberal Arts'. In Marshall Clagett et al, (eds). *Twelfth Century Europe and the Foundations of Modern Society* Madison: University of Wisconsin Press, 1961.

Kay, James Phillips. *The Moral and Physical Condition of the Working Classes Employed in the Cotton Manufacture in Manchester.* London: James Ridgway, 1832.

Kemeny, John. 'Man Viewed as a Machine'. *Scientific American,* 192 (April, 1955).

Kenny, Anthony. *Frege: An Introduction to the Founder of Modern Analytic Philosophy.* Oxford: Blackwell, 2000.

—— *A New History of Western Philosophy.* Oxford: Oxford University Press, 2010.

Kerényi, Carl. *The Gods of the Greeks.* Harmondsworth: Penguin Books, 1958.

—— *Eleusis: Archetypal Image of Mother and Daughter.* Princeton: Princeton University Press, 1967.

King, Peter. *Western Monasticism.* Kalamazoo, Michigan: Cistercian Publications, 1999.

Kirk, G. S. and Raven, J. E. *The Presocratic Philosophers.* Cambridge: Cambridge University Press, 1957.

Klein, Ernest. *A Comprehensive Etymological Dictionary of the English Language.* Bingley: Emerald, 2008.

Kneale, William and Martha. *The Development of Logic*. Oxford: Oxford University Press, 1962.

Koetsier, T. 'On the prehistory of Programmable Machines: musical automata, looms, calculators'. *Mechanism and Machine Theory*, 36 (2001).

Koster, Jan. 'From Carillon to IBM: The Musical Roots of Information Technology' published online, 2003, at <http://www. let.rug.nl/koster>

Kramer, Samuel Noah. *The Sumerians: Their History, Culture and Character*. Chicago: University of Chicago Press, 1963.

Krüger, Kristina. *Monasteries and Monastic Orders*. Potsdam: Ullmann, 2008.

Kurzweil, Ray. *The Age of Spiritual Machines*. London: Penguin, 2000.

—— *The Singularity is Near*. London: Duckworth, 2005.

Lai, Yueng-Ting. 'Leibniz and Chinese Thought'. In A. P. Coudert, et al. (eds.), *Leibniz, Mysticism and Religion*. Dordrecht: Kluwer Academic Publishers, 1998.

Laufer, Berthold. *The Prehistory of Aviation*. Chicago: Field Museum of Natural History, 1928.

Lawrence, C. H. *Medieval Monasticism*. London: Longman, 1989.

Lehner, Mark. *The Complete Pyramids*. London: Thames and Hudson, 1997.

Leclercq, Jean. *The Love of Learning and the Desire for God*. London: SPCK, 1978.

Leff, Gordon. *The Dissolution of the Medieval Outlook*. New York: Harper and Row, 1976.

Lehrs, Ernst. *Man or Matter*. London: Rudolf Steiner Press, 1985.

Leibniz, Gottfried Wilhelm. *Philosophical Papers and Letters* Edited by Leroy E. Loemker. Dordrecht: D. Reidel, 1969.

—— *Philosophical Writings*. Edited by G. H. R. Parkinson. London and Toronto: J. M. Dent and Sons, 1973.

—— *Logical Papers*. Edited by G. H. R. Parkinson. Oxford: Clarendon Press, 1966.

—— *Philosophical Essays*. Edited by R. Ariew and D. Garber. Indianapolis: Hacket Publishing Co., 1989.

—— *Selections*. Edited by Philip P. Weiner. New York: C. Scribner's Sons, 1951.

Lewis, M. J. T. 'Gearing in the Ancient World'. *Endeavour*, 17.3 (1993).

Lichtheim, Miriam. *Ancient Egyptian Literature*, vol. 1. Berkeley: University of California Press, 1975.

Liddell, H. G. and Scott, R. *Greek-English Lexicon*. Revised and augmented throughout by H. S. Jones. Oxford: Oxford University Press, 1968.

Lindberg, David C. *The Beginnings of Western Science*. Chicago: University of Chicago Press, 2007.

Lovejoy, Arthur O. *The Great Chain of Being*. Cambridge MA: Harvard University Press, 1978.

Lovelace, Ada. 'Notes Upon the Memoir' appended to L. F. Menabrea, *Sketch of the Analytical Engine Invented by Charles Babbage*. In Philip Morrison and Emily Morrison (eds.), *Charles Babbage and His Calculating Engines: Selected Writings by Charles Babbage and Others*. New York: Dover, 1961.

Lucas, Adam Robert. 'Industrial Milling in the Ancient and Medieval Worlds' in *Technology and Culture*, 46 (January, 2005).

Luscombe, David. 'Peter Abelard and the Poets'. In John Marenbon (ed.), *Poetry and Philosophy in the Middle Ages: A Festschrift for Peter Dronke*. Leiden: Brill, 2001.

Marchant and Charles. *Cassell's Latin Dictionary*. London: Cassell, 1915.

Marciszewski, W. and Murawski, R. *Mechanization of Reasoning in a Historical Perspective*. Amsterdam: Editions Rodopi, 1995.

Marenbon, John (ed.). *Poetry and Philosophy in the Middle Ages: A Festschrift for Peter Dronke*. Leiden: Brill, 2001.

Maritain, Jacques. *The Dream of Descartes*. London: Nicholson and Watson, 1946.

Mathias, Peter. *The First Industrial Nation*. London: Routledge, 1983.

Maxwell, James Clerk. *A Treatise on Electricity and Magnetism,* Volume 1. Oxford: Oxford University Press, 1873.

—— *The Scientific Papers*. Two volumes. Edited by W. D. Niven. Cambridge: Cambridge University Press, 1890.

McEvoy, James. *Robert Grosseteste*. Oxford: Oxford University Press, 2000.

Menabrea, L. F. *Sketch of the Analytical Engine Invented by Charles Babbage* (1842). In Philip Morrison and Emily Morrison (eds.), *Charles Babbage and His Calculating Engines: Selected Writings by Charles Babbage and Others*. New York: Dover, 1961.

Mercer, Samuel A. B. *The Religion of Egypt*. London: Luzac, 1949.

Merchant, Carolyn. *The Death of Nature*. London: Wildwood House, 1980.

Mettrie, Julien Offray de la. *Man a Machine (L'Homme Machine)*. Translated by Richard A. Watson and Maya Rybalka. Cambridge, Indianapolis: Hackett, 1994.

Meyer, Herbert W. *A History of Electricity and Magnetism*. Norwalk, Connecticut: Burndy Library, Massachusetts Institute of Technology, 1971.

Middleton, W. E. Knowles. *The Experimenters: A Study of the Accademia del Cimento*, Baltimore and London: Johns Hopkins Press, 1971.

Milton, John. *Paradise Lost*. Illustrated by Gustave Doré. London: Cassell, Petter and Galpin, 1866.

—— *The Complete Poetical Works of John Milton*. Edited by H. C. Beeching, London: Henry Frowde, 1911.

Moody, Ernest A. 'Okham, Buridan and Nicholas Autrecourt'. In James F. Ross (ed.), *Inquiries into Medieval Philosophy*. Westport, Connecticut: Greenwood, 1971.

Morrison, Philip and Emily Morrison (eds.), *Charles Babbage and His Calculating Engines: Selected Writings by Charles Babbage and Others*. New York: Dover, 1961.

Mumford, Lewis. *Technics and Civilization*. Chicago and London: University of Chicago Press, 2010.

Mure, G. R. G. *Retreat from Truth*. Oxford: Blackwell, 1958.

Muri, Alison. *The Enlightenment Cyborg: A History of Communications and Control in the Human Machine, 1660–1830*. Toronto: University of Toronto Press, 2007.

Murray, Oswyn. *Early Greece*. London: Fontana, 1980.

Murray, Penelope (ed.). *Plato on Poetry*. Cambridge: Cambridge University Press, 1996.

Nasr, S. H. 'Scientia Sacra'. In Martin Lings and Clinton Minnaar (eds.), *The Underlying Religion: An Introduction to the Perennial Philosophy*. Bloomington, Indiana: World Wisdom, 2007.

Naydler, Jeremy. 'The Regeneration of Realism and the Recovery of a Science of Qualities'. *The International Philosophical Quarterly*, XXIII, 2 (June, 1983).

—— *Temple of the Cosmos: The Ancient Egyptian Experience of the Sacred*. Rochester, Vt.: Inner Traditions International, 1996.

—— *The Future of the Ancient World: Essays on the History of Consciousness*. Rochester Vt: Inner Traditions, 2009.

—— *Shamanic Wisdom in the Pyramid Texts: The Mystical tradition of Ancient Egypt* (Rochester, Vt: Inner Traditions, 2005.

Neckam, Alexander. *De Naturis Rerum et De Laudibus Divinae Sapientiae*. Edited by Thomas Wright. London: Longman, Roberts and Green, 1863.

Nicholson, William. 'Account of the new Electrical Apparatus of Sig. Alex. Volta, and Experiments performed with the same' in *A Journal of Natural Philosophy, Chemistry and the Arts*, vol.4 (July, 1800).

Nidditch, P. H. *The Development of Mathematical Logic*. London: Routledge and Kegan Paul, 1962.

Nilsson, Martin. *Greek Folk Religion*. New York: Harper, 1961.

Nolte, Rudolf August. *Leibniz Mathematischer Beweis der Ershaffung und Ordnung der Welt in einem Medallion an den Herrn Rudolf August*. Leipzig: J. C. Langenheim, 1734.

Ohly, Friedrich. *Sensus Spiritualis: Studies in Medieval Significs and the Philology of Culture*. Chicago and London: University of Chicago Press, 2005.

Oppenheim, Leo. *The Interpretation of Dreams in the Ancient Near East*. Transactions of the American Philosophical Society, vol.46. Philadelphia: The American Philosophical Society, 1956.

——— *Ancient Mesopotamia: Portrait of a Dead Civilization*. Chicago: University of Chicago Press, 1977.

Oppenheimer, J. R. *Science and the Common Understanding*. Oxford: Oxford University Press, 1954.

Origen. *The Song of Songs: Commentary and Homilies*. Translated by R. P. Lawson. Mahway, N.J.: Paulist Press, 1957.

Otto, Rudolf. *The Idea of the Holy*. Harmondsworth: Penguin, 1959.

Pascal, Blaise. *Thoughts, Letters and Minor Works*. Selected by Charles W. Eliot. New York: P. F. Collier and Son, 1910.

——— *Pensées*. Translated by A. J. Krailsheimer. Harmondsworth: Penguin, 1966.

Passmore, John. *A Hundred Years of Philosophy*. Harmondsworth: Penguin Books, 1968.

Peregrinus, Petrus. *The Letter of Petrus Peregrinus on the Magnet (1269) (Epistola de Magnete)*. Translated by Brother Arnold. New York: McGraw Publishing Co., 1904.

Plato. *Ion*. In Penelope Murray (ed.), *Plato on Poetry*. Cambridge: Cambridge University Press, 1996.

——— *Euthydemus*. In *The Dialogues of Plato*, vol. 2. Translated by Benjamin Jowett. London: Sphere Books, 1970.

——— *Republic*. Translated by Paul Shorey. London: William Heinemann, 1942.

——— *Timaeus*. Translated by R. G. Bury. Cambridge MA and London: Harvard University Press, 2005.

Plotinus, *Enneads*. Seven volumes. Translated by A. H. Armstrong. London: William Heinemann, 1966–1988.

Plutarch, *De Isis et Osiride*. Translated by J. Gwyn Griffiths. Swansea: University of Wales Press, 1970.

——— *The Rise and Fall of Athens*. Translated by Ian Scott-Kilvert. Harmondsworth: Penguin, 1960.

——— *Life of Marcellus*. Translated by Bernadotte Perrin in Plutarch, *Lives*, volume 5. Harvard University Press, 1917.

Price, Derek J. de Solla. 'On the Origin of Clockwork, Perpetual Motion Devices and the Compass.' In *Contributions from the Museum of History and Technology*. Washington DC: Smithsonian Institution, 1959.

Pseudo-Plutarch. *Sentiments Concerning Nature With Which Philosophers Were Delighted*. In *The Complete Works of Plutarch: Essays and Miscellanies*. New York: Crowell, 1909.

Ragozin, Z. A. *Chaldea: from the earliest times to the rise of Assyria*. London: T. Fisher Unwin, 1886.

Raizman-Kedar, Yael. 'Plotinus's Conception of Unity and Multiplicity as the Root to the Medieval Distinction between *Lux* and *Lumen*' in *Studies in History and Philosophy of Science,* 37.3 (2006).

Reynolds, Terry S. 'The Medieval Roots of the Industrial Revolution'. *Scientific American,* 251 (July, 1984).

Rorem, Paul, *Hugh of St Victor.* Oxford: Oxford University Press, 2009.

Rose, H. J. *A Handbook of Greek Mythology.* London: Methuen, 1964.

Ross, D. *Aristotle.* London: Methuen, 1971.

Rossum, Gerhard Dohrn-van. *History of the Hour: Clocks and Modern Temporal Orders.* Chicago and London: University of Chicago Press, 1996.

Roux, George. *Ancient Iraq.* Harmondsworth: Penguin, 1980.

Ruppert, Kathleen. 'Gilded Age Art and Literature'. In Kevin Hillstrom and Laurie Collier Hillstrom (eds.), *The Industrial Revolution in America: Communications.* Santa Barbara: ABC-CLIO, 2007.

Russell, Jeffrey Burton. *The Devil: Perceptions of Evil from Antiquity to Primitive Christianity.* New York: Cornell University Press, 1977.

Schiffer, Michael Brian. *Draw the Lightning Down: Benjamin Franklin and Electrical Technology in the Age of Enlightenment.* Berkeley: University of California Press, 2003.

Schmidt, Robert W. *The Domain of Logic According to St Thomas Aquinas.* The Hague: Martinus Nijhoff, 1966.

Schroeder, F. M. *Form and Transformation: A Study in the Philosophy of Plotinus.* Montreal/Kingston, Ontario: McGill-Queen's University Press, 1992.

Schwaller de Lubicz, R. A. *The Temple of Man.* Two volumes. Rochester, Vt.: Inner Traditions International, 1998.

Shannon, Claude E. 'A Symbolic Analysis of Relay and Switching Circuits'. *Transactions of the American Institute of Electrical Engineers,* 57 (1938).

Shelley, Mary. *Frankenstein.* Harmondsworth: Penguin Books, 1985.

Sherman, William H. 'How to Make Anything Signify Anything.' *Cabinet Magazine,* 40 (2010/11).

Silverstein, Theodore. 'The Fabulous Cosmology of Bernardus Silvestris', in *Modern Philology,* 46 (1948–9).

Sinclair, W. A. *The Traditional Formal Logic.* London: Methuen, 1945.

Singer, Charles, et al. (eds.). *A History of Technology,* 8 vols. Oxford: Oxford University Press, 1954–1984.

Singer, Katie. *An Electronic Silent Spring.* Great Barrington, MA: Portal Books, 2014.

Sophocles, *The Theban Plays.* Translated by E. F. Watling. Harmondsworth: Penguin, 1947.

Stahl, William Harris. *Martianus Capella and the Seven Liberal Arts.* Two volumes. New York: Columbia University Press, 1971 and 1977.

Starr, Ivan. *The Rituals of the Diviner.* Malibu: Undena Publications, 1983.

Steiner, Rudolf. *Anthroposophical Leading Thoughts.* Forest Row: Rudolf Steiner Press, 1973.

—— *The Problem of Faust.* Lecture 1, 30 September, 1916, Dornach. Unpublished typescript.

—— *Faust and the Mothers.* A lecture given on 2 November, 1917, Dornach. Unpublished typescript.

—— *The Mission of the Archangel Michael.* New York: Steiner Books, 1961.

—— *The Etherization of the Blood.* London: Rudolf Steiner Press, 1971.

—— *The Riddles of Philosophy*. New York: The Anthroposophic Press, 1973.

—— *The Incarnation of Ahriman*. Forest Row: Rudolf Steiner Press, 2006.

—— *The Fall of the Spirits of Darkness*. Forest Row: Rudolf Steiner Press, 1993.

Strandh, Sigvard. *A History of the Machine*. London: Hutchinson, 1984.

Strouhal, Eugen. *Life in Ancient Egypt*. London: Opus Publishing Ltd., 1992.

Stuckrad, Kocku von. *Western Esotericism: A Brief History of Secret Knowledge*. Durham: Acumen Publishing, 2013.

Sturgeon, W. 'Improved Electro Magnetic Apparatus' in *Transactions of the Royal Society of Arts, Manufactures, and Commerce*, 43 (1824).

Sullivan, Arthur. *Logicism and the Philosophy of Language: Selections from Frege and Russell*. Toronto: Broadview Press, 2003.

Suto, Taki. *Boethius on Mind, Grammar and Logic*. Leiden: Brill, 2012.

Swade, Doron. *The Cogwheel Brain*. London: Little, Brown and Company, 2000.

Swetz, Frank J. 'Leibniz, the Yijing, and the Religious Conversion of the Chinese' in *Mathematics Magazine*, 76.4 (2003).

Talbott, Stephen. *The Future Does Not Compute*. Sebastopol, CA: O'Reilly and Associates, 1995.

—— *Devices of the Soul*. Sebastopol, CA: O'Reilly Media, 2007.

Tausk, Viktor. 'On the Origin of the 'Influencing Machine' in Schizophrenia' (1919). Translated in *Psychoanalytic Quarterly* (1933), 2.

Taylor, F. Sherwood. *Science Past and Present*. London: Heinemann, 1945.

—— *The World of Science*. London: Heinemann, 1950.

Theophrastus. *On Stones*. Greek text, translation and commentary by John F. C. Richards and Earle R. Caley. Columbus, Ohio: Ohio State University, 1956.

Thomas, Rosalind. *Literacy and Orality in Ancient Greece*. Cambridge: Cambridge University Press, 1992.

Thorndike, Lynn. *The Sphere of Sacrobosco and Its Commentators*. Chicago: University of Chicago Press, 1949.

Thurston, Robert H. *A History of the Growth of the Steam-Engine*. New York: D. Appleton and Co., 1886.

Trentman, John A. 'Logic'. In *Contemporary Philosophy. A New Survey*, Vol. 6: *Philosophy and Science in the Middle Ages*, Part 2. Dordrecht: Kluwer Academic Publishers, 1990.

Turing, Alan. 'Can Digital Computers Think', in B. Jack Copeland (ed.), *The Essential Turing*. Oxford: Clarendon Press, 2004.

Uphill, Eric. 'The Egyptian Sed-Festival Rites'. *Journal of Near Eastern Studies*, 24 (1965).

Vaughan, Thomas. 'Anima Magica Abscondita'. In Arthur Edward Waite (ed.), *The Works of Thomas Vaughan*. London: Theosophical Publishing House, 1919.

Velde, H. te. *Seth, God of Confusion*. Lieden: E. J. Brill, 1967.

Vernant, J-P. *Myth and Thought Among the Greeks*. London: Routledge and Kegan Paul, 1983.

Volta, Alessandro. 'On the Electricity Excited by the Mere Contact of Conducting Substances of Different Kinds' in *Philosophical Transactions of the Royal Society*, vol.90, 1800.

Wadler, Philip. 'Proofs are Programs: 19th Century Logic and 21st Century Computing' (November, 2000). Accessed 31 January, 2017 from <http://www.homepages.inf.ed.ac.uk/wadler/papers/frege/frege.pdf>.

Waite, Arthur Edward (ed.). *The Works of Thomas Vaughan.* London: Theosophical Publishing House, 1919.

Walker, Adam. *A System of Familiar Philosophy.* Two volumes. London, 1802.

Walters, Kerry S. *Benjamin Franklin and His Gods.* Urbana: University of Illinois Press, 1999.

Warren, Michelle R. and Glimp, David (eds.). *Arts of Calculation: Quantifying Thought in Early Modern Europe.* New York: Palgrave Macmillan, 2004.

Wasserman, James (ed.). *The Egyptian Book of the Dead.* Translated by R. O. Faulkner with an introduction and commentary by Ogden Goelet. San Francisco: Chronicle Books, 1994.

Whidden, David L. *Christ the Light: The Theology of Light and Illumination in Thomas Aquinas.* Minneapolis, MN: Fortress Press, 2014.

Wikander, Örjan. *Handbook of Ancient Water Technology.* Leiden: Brill, 2000.

Wilkinson, Richard. *Symbol and Magic in Egyptian Art.* London: Thames and Hudson, 1994.

Wolf, A. *A History of Science, Technology, and Philosophy in the 16th and 17th Centuries.* London: George Allen and Unwin, 1935.

—— *A History of Science, Technology and Philosophy in the 18th Century* (1938). Two volumes. London: George Allen and Unwin, 1962.

Wolf, Robert, et al., *Paradise Lost: The Poem and its Illustrators.* Grasmere: The Wordsworth Trust, 2004.

Wolkstein, Diane and Kramer, Samuel Noah. *Inanna: Queen of Heaven and Earth.* London: Hutchinson, 1984.

Woolley, C. L. *The Royal Cemetery: Ur Excavations*, vol. 2. London: British Museum, 1934.

Yates, Frances A. *The Occult Philosophy.* Routledge and Kegan Paul, 1983.

Youmans, William J. (ed.). *Popular Science Monthly*, vol. 39. New York: Appletone Co., 1891.

Zarnecki, George. 'The Monastic World'. In Joan Evans (ed.), *The Flowering of the Middle Ages.* London: Thames and Hudson, 1966.

Zeller, Eduard. *Outlines of the History of Greek Philosophy.* New York: Dover, 1980.

ILLUSTRATION SOURCES

While every effort has been made to obtain permission to use images not in the public domain, this has not been possible in every case. We apologize for any omissions and will rectify them in future editions.

2.1 Two leopard-lions. Vignette to Chapter 17, Book of the Dead. Papyrus of Ani. Budge, *The Book of the Dead*, p. 94.

2.2 An early Mesopotamian *shaduf*. Akkadian cylinder seal. Late third ml., BC. Singer et al., *A History of Technology*, vol. 1, p. 524, fig. 346. Drawing by D. E. Woodall, after Delaporte, *Catalogue des Cylindres Orientaux*, II, pl.72, A.156.

2.3 Drawing water in pots from a lily pond. Tomb of Rekhmire. New Kingdom, circa 1450 BC. Restored drawing by Nina de G. Davies. Singer et al., op. cit., p. 522, fig. 343.

2.4 Hermaphrodite figure of Hapy. Temple of Horus, Edfu. Erman, *Life in Ancient Egypt*, p. 424.

2.5 The *shaduf* depicted in an eighteenth dynasty tomb painting. Tomb of Neferhotep, Chief Scribe of Amun (*Th 49*). Eighteenth Dynasty, circa 1325 BC. After Davies, *Two Ramesside Tombs*, pl. XXVIII.

2.6 Four-wheeled war chariots. Royal Standard of Ur. Circa 2700 BC. Drawing by D. E. Woodall, after Woolley, *The Royal Cemetery*, pl.XCII. Singer et al., op. cit., p. 718, fig. 517.

2.7 New Kingdom wheelwrights. From a Theban tomb. Circa 1475 BC. After Davies, *The Tombs of Menkheperrasonb*, pl. XII. Singer et al., op. cit., p. 213, fig. 134.

2.8 Inanna, clutching lightning bolts. Mesopotamian cylinder seal, circa 2500 BC. Drawing by Tessa Rickards. Black and Green, *Gods, Demons and Symbols*, p. 52.

2.9 The storm-god Ninurta or Marduk grasps thunderbolts. Temple of Ninurta. Nimrud, Iraq. Ragozin, *Chaldea*, p. 291.

2.10 Seth. Brodrick and Morton, *A Concise Dictionary*, p. 161.

2.11 Seth animal determinative above sky hieroglyph. Clark, *Myth and Symbol*, p. 115.

2.12 Cutting implement hieroglyph. Gardiner, *Egyptian Grammar*, pp. 542.

2.13 Evolution of the *was* sceptre. Schwaller de Lubicz, *The Temple of Man*, vol. 2, p. 1001.

3.1 Greek pottery painting depicting a music class. fifth c. BC. From a pot in the British Museum, E171. Beazley, *Attic Red-Figure Vase-Painters*, p. 579.

3.2 Roman mosaic of a Cyclops. fourth c. AD. Villa Romana del Casale, Sicily. Author's drawing.

3.3 The blinding of Polyphemus. Fragment of seventh c. BC vase. Archaeological Museum of Argos. Author's drawing.

5.1 The watermill described by Vitruvius. Author's drawing.

5.2 The operation of the cam turns rotary motion into linear motion. Reynolds, 'The Medieval Roots', p. 111.

5.3 Schematic diagram of the Antikythera mechanism. Wikimedia Commons, 2009.

6.1 Zeus hurls a thunderbolt at Typhoeus. Black-figure vase. Circa 550 BC. After A. Rumpf, *Chalikidische Vasen* (Berlin, 1927). Kerenyi, *The Gods of the Greeks*, p. 23.

7.1 Grammar personified as a beautiful young lady. Tenth century Carolingian manuscript of Martianus Capella, *The Marriage of Philology and Mercury*. BnF, ms Latin 7900, A fol.127v. Wikimedia Commons.

7.2 Grammar and Logic, either side of Philosophy. Detail of late twelfth century French manuscript. Bibliotheque Nationale, Paris, ms.lat.3110, fol.60r. Author's drawing.

7.3 Dialectic holds a snake. From twelfth century French manuscript of Boethius' Commentary on Porphyry's *Isagoge*. Darmstadt, Landesbibliothek, ms 2282. Author's drawing.

7.4 Dialectic, with a snake in one hand and a bundle of five discs. Twelfth century manuscript of Priscian's *Partitiones* and *Institutiones Grammaticae,* Benedictine monastery of Saint Sulplice in Bourges, now in Cambridge University Library, ms Cg.2.32. Author's drawing.

7.5 Sculpture of Eve holding a dragon. Rheims Cathedral, Circa 1275. Author's drawing.

7.6 Dialectic, with her serpent-dragon. From the West Porch of the Royal Portal, Chartres Cathedral. Circa 1150. Author's drawing.

8.1 The Square of Opposition. Ninth century manuscript of Boethius, *In Librum Aristotelis de Interpretatione (Periermenias Aristotelis).* Library of the Abbey of Fleury, Orléans, ms.277, f.37v.

8.2 The Square of Opposition. Sinclair, *The Traditional Formal Logic*, p. 98.

8.3 Floating watermills. Early fourteenth century manuscript. After H. Martin, *Legende de Saint Denis* (Paris: Champion, 1908), pl. 44. Singer, *A History of Technology*, 2, p. 608.

8.4 A watermill rotates a camshaft. Agricola, *De Re Metallica*, p. 284.

8.5 The mechanism of the self-playing organ and flute. After Farmer, *The Organ of the Ancients*, p. 101.

8.6 The melody of the carillon bells. Adapted from the Museum Guide of the Utrecht Museum, Holland.

9.1 A section of Giovanni de' Dondi's weight-driven astronomical clock. Fifteenth century manuscript. Bodleian Library, Oxford: MS Laud. misc. 620, folio 10. Wikimedia Commons.

9.2 The two parts of the verge escapement. After Rossum, *History of the Hour*, p. 49, fig. 7.

9.3 The verge escapement with the crown wheel. After Rossum, op. cit., p. 49, fig. 7.

9.4 An early tower warden clock with an alarm bell mechanism. Drawing by G. Oestmann, after C. Sandon, *Les Horloges et les Maîtres Horlogogers á Bescançon* (1905*),* reproduced in Rossum, op. cit., p. 104, fig. 25. Adapted by author.

9.5 Levity and Gravity: two complementary principles. Author's drawing.

11.1 The medieval cosmos. Sixteenth century diagram illustrating Peter Appian, *Cosmographia* (1539). Taylor, *Science Past and Present*, p. 21, fig. 4.

11.2 The upper and lower borders of nature. Author's drawing.

11.3 Gilbert's spherical lodestone or terrella, *De Magnete,* Book 1, Chapter 3. Benjamin, *A History of Electricity*, p. 277.

12.1 Von Guericke's improved air-pump mounted on a tripod. Wolf, *A History of Science, Technology, and Philosophy in the 16th and 17th Centuries*, p. 106, illustr. 66.

12.2 Von Guericke's demonstration of the pressure of the power of the vacuum. Louis Figuier, *Les Merveilles de la Science,* vol. 1, fig. 23.

12.3 Von Guericke's sulphur globe and electric machine. Otto von Guericke, *Experimenta Nova* (1672), IV, ch. 15. Welcome Collection.

12.4 Francis Hauksbee's electrical machine. Francis Hauksbee, *Physico-Mechanical Experiments* (1709). Welcome Collection.

20.1 General Plan of the Analytical Engine. Design drawing: Plan 25, 6 August 1840. Science Museum, London.

21.1 Galvani's experiment on the effect of atmospheric electricity on the amputated limbs of frogs. Galvani, *De Viribus Electricitatis*, 1791. Wikimedia Commons.

21.2 An illustration from Galvani's *De Viribus Electricitatis*, 1791, showing a range of experiments carried out on the amputated legs of frogs. Wikimedia Commons.

21.3 The application of electricity to the corpses of convicted criminals. Figuier, *Les Merveilles de la Science*, 1, p. 653, fig. 333.

21.4 Volta's first 'pile', 1800. After Volta, 'On the Electricity Excited by the Mere Contact', fig. 2.

21.5 Glass tube used for the observation of electrolysis. From Benjamin, *The Age of Electricity*, p. 183, fig. 87.

22.1 Sömmerring's electric telegraph, 1809. Taylor, *Science Past and Present*, p. 200, fig. 41.

22.2 Oersted's experiment with an electric wire and a compass needle. Durgin, *Electricity*, p. 49.

22.3 William Sturgeon's electromagnet of 1824. Sturgeon, 'Improved Electro Magnetic Apparatus', Plate 3, fig.13.

22.4 Schilling's galvanometer/receiver, circa 1825. Fahie, *A History of Electric Telegraphy*, p. 310, fig. 12.

22.5 Schilling's binary alphabet. Fahie, op. cit., p. 311.

22.6 The binary code used in the Gauss and Weber telegraph signalling system. Fahie, op. cit., p. 325.

22.7 Steinheil's binary code of dots on two lines, one above the other. Fahie, op. cit., p. 342.

22.8 Morse's receiving apparatus, circa 1838. Benjamin, *The Age of Electricity*, p. 229, fig. 101.

22.9 The original Morse Telegraph Alphabet, circa 1890. J. H. Bunnell website: jhbunnell.com/morsecode.shtml.

23.1 The NOT Truth Table.

23.2 The OR Truth Table.

23.3 The AND Truth Table.

23.4 The NOT Logic Operator. Author's drawing.

23.5 The OR Logic Operator. Author's drawing.

23.6 The AND Logic Operator. Author's drawing.

24.1 Faraday's electrical induction experiment, using an iron ring or 'torus'. Author's drawing.

24.2 Faraday's experiment with a bar magnet and a coil of copper wire. Illustration by David Darling.

24.3 Faraday's experiment with a bar magnet and a coil of copper wire. Author's drawing.

24.4 Faraday's experiment with a bar magnet and a coil of copper wire. Author's drawing.

24.5 Faraday's disk dynamo. Author's drawing.

24.6 A simplified diagram of Hertz's apparatus. Author's drawing, after Appleyard, 'Pioneers of Electrical Communication', fig. 8, p. 68.

24.7 Three charts of rapidly expanding electromagnetic waves. Hertz, *Electric Waves*, figs. 27–30, pp. 144–5.

INDEX

A note from the publisher

For more than a quarter of a century, **Temple Lodge Publishing** has made available new thought, ideas and research in the field of spiritual science.

Anthroposophy, as founded by Rudolf Steiner (1861-1925), is commonly known today through its practical applications, principally in education (Steiner-Waldorf schools) and agriculture (biodynamic food and wine). But behind this outer activity stands the core discipline of spiritual science, which continues to be developed and updated. True science can never be static and anthroposophy is living knowledge.

Our list features some of the best contemporary spiritual-scientific work available today, as well as introductory titles. So, visit us online at **www.templelodge.com** and join our emailing list for news on new titles.

If you feel like supporting our work, you can do so by buying our books or making a direct donation (we are a non-profit/ charitable organisation).

office@templelodge.com

TEMPLE LODGE

For the finest books of Science and Spirit